Modern Organ

```
COLLEGIATE SCHOOL ENNISKILLEN
DEPT                        NUMBER
Date issued   Date Ret'd.   Date issued   Date Ret'd.
```

Modern Chemistry Series
Under the supervisory editorship of D. J. Waddington, B Sc, ARCS, DIC, Ph D, Senior Lecturer in Chemistry, University of York, this series is specially designed to meet the demands of the new syllabuses for sixth form, introductory degree and technical college courses. It consists of self-contained major texts in the three principal divisions of the subject, supplemented by short readers and practical books.

Major Texts
Modern Inorganic Chemistry
G. F. Liptrot MA, Ph D

Modern Organic Chemistry
R. O. C. Norman MA, DSc, FRIC, and
D J. Waddington B Sc, ARCS, DIC, Ph D

Modern Physical Chemistry
J. J. Thompson MA, ARIC, and
G. R. Walker BA, B Sc, ARIC

Readers
The Periodic Table
J. S. F. Pode MA, B Sc

A Mechanistic Introduction to Organic Chemistry
Glyn James MA

Investigation of Molecular Structure
B. C. Gilbert MA, D Phil

Practical Books
Organic Chemistry Through Experiment
D. J. Waddington B Sc, ARCS, DIC, Ph D, and
H. S. Finlay B Sc

Inorganic Chemistry Through Experiment
G. F. Liptrot MA, Ph D

Modern Organic Chemistry

R. O. C. Norman, MA, DSc, FRIC
Professor and Head of the Chemistry Department, University of York

D. J. Waddington, BSc, ARCS, DIC, PhD
Formerly Head of the Science Department, Wellington College.
Senior Lecturer in Chemistry, University of York

Mills & Boon Limited
London

First published 1972 in Great Britain by
Mills & Boon Limited, 17–19 Foley Street
London, W1A 1DR

Reprinted 1974
Second edition 1975
Reprinted 1977

© R O C Norman and D J Waddington 1972

ISBN 0 263.06006 3

All rights reserved. No part of this publication
may be reproduced, stored in a retrieval
system, or transmitted in any form or by any
means, electronic, mechanical, photocopying,
recording or otherwise, without the prior
permission of the copyright owners.

Filmset and printed in Great Britain by
BAS Printers Limited, Wallop, Hampshire

Contents

	Page
Preface	vii
Acknowledgements	ix

CHAPTER

1	Introduction to Organic Chemistry	1
2	Preparation and Purification of Organic Compounds	12
3	Determination of the Structure of an Organic Compound	30
4	Bonding in Organic Compounds	47
5	Alkanes	65
6	Alkenes	75
7	Alkynes	89
8	Aromatic Compounds	94
9	Halogen Compounds	111
10	Alcohols and Phenols	137
11	Ethers	159
12	Aldehydes and Ketones	164
13	Carboxylic Acids	182
14	Derivatives of Carboxylic Acids	204
15	Isomerism	228
16	Amines	243
17	Nitro Compounds	265
18	Naturally Occurring Compounds	271
19	Petroleum	290
20	The Petrochemical Industry	301
21	Polymers	311

APPENDIX

I	Questions	329
II	Apparatus and Chemicals	349
III	Suppliers of Apparatus and Chemicals	352
IV	Film Libraries	353
V	Physical Constants	354

INDEX

357

Preface

In writing this textbook for schools and colleges, we have been conscious of the profound changes that have taken place in the position of organic chemistry during the last few years.

For example, the ways in which organic compounds and their reactions play their crucial roles in living organisms are beginning to be understood; DNA, which underlies the transmission of inherited characteristics, is, after all, an organic compound, and when animals need energy they obtain it by the controlled oxidation of another organic compound, glucose. These are but two examples of the ways in which organic chemistry impinges on the work of the biologist.

We have also witnessed the increasingly important part played by the techniques of the physicist in the study of the structures of organic compounds, especially the spectroscopic methods such as infrared and mass spectrometry.

As well as these changes, organic chemistry has been undergoing an internal revolution: we are concerned now with answering not only the question 'What product is formed when A reacts with B?' but also the question '*How* does A react with B?'—a field described as mechanistic organic chemistry.

Finally, the importance of organic compounds in our economy has been rising steadily for many years and will continue to do so. New plastics are being invented, new medicines are being tailor-made for specific requirements, there is a new awareness of the problems of pollution which has meant that the organic chemist is concerned with the discovery of new fuels, detergents and pesticides, and we are perhaps now on the threshold of manufacturing proteins for animal feedstocks which could revolutionise the economies of developing countries.

We have tried to reflect such factors in this book. We have set out the preparations and properties of organic compounds in terms of functional groups and against a background of the principles of bonding, energetics and reaction mechanism. Although we believe that these principles can be all the more easily understood in this way, rather than as isolated topics in a physical chemistry course, the book has been written so that students can leave out some sections if these are outside their immediate scope and can go back later to read the book as a whole.

Chapter 2 describes the modern techniques for the preparation and purification of organic compounds and Chapter 3 is concerned with the methods now employed for studying their structures. In Chapter 18, we have related the chemistry of some of the types of compound which occur naturally to their functions in living systems, while Chapter 21 deals with the man-made macromolecules we use as plastics and fibres. We have shown throughout the book how petroleum, including natural gas, is vital for the chemical industry, and two Chapters, 19 and 20, are specifically devoted to this.

The introduction of new syllabuses at every level is encouraging for all of us who teach and study Chemistry, all the more so since, hand-in-hand with these positive changes in our ideas on how theory should be presented,

PREFACE

comes the desire to illustrate the work experimentally. We have therefore suggested practicals, with very simple apparatus, at the end of the chapters. Most of these practicals should take less than an hour.

The syllabus of A and S examinations, University Scholarships and much of the National Certificate work is covered and we believe our readers will also find the book of great use in the beginning of their University work.

York 1972   R. O. C. N.
                                                                    D. J. W.

In revising this text for its second edition, we have been especially conscious of the changing practices for naming organic compounds. The use of the original, or 'trivial', names is rapidly being superseded by the I.U.P.A.C. nomenclature, and we have followed, with two exceptions, the recommendations based on this nomenclature made by the Association for Science Education in their booklet, 'Chemical nomenclature, symbols and terminology' (1972). The first exception is that we have retained the generic name 'ether'; the alternative—alkoxyalkanes, aryloxyalkanes and aryloxyarenes, as appropriate—is cumbersome. Secondly, we have used the trivial names for the important α-amino-acids since the original names are still widely used in Biology and Biochemistry. Both these exceptions are accepted by I.U.P.A.C. We have also altered, where applicable, the nomenclature used in past examination papers so that it is as consistent as possible with the naming of compounds in this text.

York 1975   R. O. C. N.
                                                                    D. J. W.

Acknowledgements

We are grateful for the generous assistance given to us by many organisations, especially to the Esso Petroleum Co. Ltd., Shell Petroleum Co. Ltd., British Petroleum Co. Ltd., Imperial Chemical Industries Ltd., and the Gas Council for permission to use photographs and for most helpful assistance with technical data.

We acknowledge permission to us to use examination questions by the Colleges of the Universities of Cambridge and Oxford, the Local Examination Syndicate of the University of Cambridge, the Delegates of Local Examinations, University of Oxford, the Oxford and Cambridge Schools Examination Board, the Joint Matriculation Board, the Southern Universities' Joint Board, the Associated Examining Board, the Welsh Joint Education Committee and the School Examinations Department, University of London. Those questions marked London (Nuffield) were used for schools taking part in Nuffield Science Teaching Project Trials, and those marked London (X) are from papers taken by schools who were taking the London A examination following a Nuffield O course.

We thank members of this department, Mr J. Olive for the photographs used in Chapters 1, 2 and 15, Miss Ann Warriss and Miss Morag Brown who tried out experiments we wished to include and for the modifications they suggested and Mr D. J. M. Rowe for many helpful suggestions in revising Chapters 19–21. We thank also Mr H. S. Pickering (Uppingham School) for his advice.

Chapter 1
Introduction to organic chemistry

1.1 Introduction

Over two million compounds are known which contain the element carbon, and about 80,000 new carbon compounds are made each year. It is therefore convenient to study the compounds of carbon separately, and this branch of chemistry is known as **organic chemistry**.

Originally the word organic applied to those substances that were produced by living organisms. Berzelius wrote in 1815 that the essential difference between inorganic and organic compounds was that the formation of organic compounds could only be achieved by the influence of a 'vital force' which was present in nature. No organic material could be synthesised in the laboratory. Sugar, dyes, starch, oils, alcohol, known since the earliest times, could only be made by nature.

A conflicting point of view was put forward by Wöhler in 1828. He found that when an aqueous solution of ammonium cyanate is evaporated to dryness, carbamide (urea) is obtained:

$$NH_4CNO \xrightarrow{\text{heat}} H_2N-\overset{\overset{\displaystyle O}{\|}}{C}-NH_2$$

Ammonium cyanate   Carbamide

Ammonium cyanate is an inorganic compound whereas carbamide is present in the urine of all animals and was therefore particularly well known as an organic compound. It seemed, then, that 'vital force' was unnecessary. However, other scientists argued that, since the ammonia and cyanic acid from which Wöhler had made ammonium cyanate were both of animal origin, the true synthesis of an organic compound had not been achieved. It was not until 1845 that a final refutation of the 'vital force' idea was propounded, when Kolbe prepared the organic compound, ethanoic acid (acetic acid) from its constituent elements, carbon, hydrogen and oxygen.

Today, the distinction between organic and inorganic chemistry is an arbitrary one. Organic chemistry is regarded as the chemistry of compounds of carbon other than its oxides, the metallic carbonates and related compounds.

1.2 Bonding

The bonds which carbon forms are covalent: that is, each bond is formed by the sharing of two electrons, one of which is provided by the carbon atom and one by the other atom. Carbon has four electrons available for sharing, so that it forms four bonds; a fuller description is given later (Chapter 4). It is convenient to represent each pair of electrons which constitutes a bond by a line, —; for example, a bond between carbon and hydrogen is shown as C—H.

The bonds to a carbon atom have particular positions in space in relation to one another. For example, in methane, CH_4, the bonds are directed towards the corners of an (imaginary) regular tetrahedron of which the carbon atom is the centre; the angle between each pair of C—H bonds is 109°28'.

INTRODUCTION TO ORGANIC CHEMISTRY

For simplicity, a two-dimensional structure is usually drawn; for example, methane is written as

$$\begin{array}{c} H \\ | \\ H-C-H \\ | \\ H \end{array}$$

However, two-dimensional representations can be misleading and care must be exercised in their use. For instance, it might appear that there would be two compounds with the formula CH_2Cl_2:

$$\begin{array}{cc} H & Cl \\ | & | \\ Cl-C-Cl \quad & H-C-Cl \\ | & | \\ H & H \end{array}$$

A three-dimensional representation shows that the two planar structures actually represent the same compound, and indeed only one compound of this formula exists.

Plate 1.1. The tetrahedral arrangement of four atoms of hydrogen around a carbon atom in a molecule of methane: (a) ball and spring, (b) space-filling

It is useful to have a simple set of molecular models to consider problems like this; they can provide a quick means of translating the planar representations on paper into the more realistic three-dimensional structures.

The simplest models are called **ball and spring**. The balls are coloured differently to represent different atoms, and have holes drilled in them corresponding to the number of bonds they can form. The balls are joined together by means of stiff springs fitted into the holes which represent the bonds (Plate 1.1). **Space-filling** models are more useful when it is necessary to obtain a more accurate idea of how near together different atoms will be in the compound. In one sort, Stuart models, the atoms (generally constructed in a plastic) are made to scale according to the relative atomic radii of the elements they represent, and they are joined together by clips. The construction of the model for CH_2Cl_2 (Plate 1.2) shows at once that only one compound with this molecular formula exists.

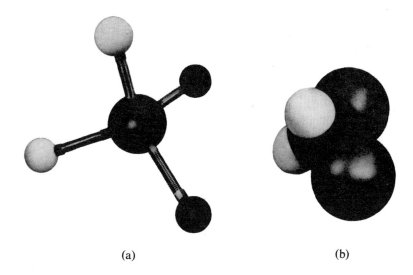

Plate 1.2. Molecular models of dichloromethane, CH_2Cl_2: (a) ball and spring, (b) space filling. The atoms are coloured black (carbon), white (hydrogen), grey (chlorine).

(a)   (b)

A carbon atom can be attached to monovalent atoms other than hydrogen. For example, one chlorine atom can replace one hydrogen atom to give chloromethane (also called methyl chloride):

$$\mathrm{H-\underset{\underset{H}{|}}{\overset{\overset{H}{|}}{C}}-Cl}$$

Again for simplicity, some or all of the bonds are usually omitted in representing these compounds, so that chloromethane is written as CH_3—Cl or CH_3Cl. Replacement of more than one of the hydrogen atoms by chlorine atoms gives the compounds dichloromethane (CH_2Cl_2), trichloromethane ($CHCl_3$) and tetrachloromethane (CCl_4).

Carbon can be bonded to a divalent atom, as in methanol, CH_3OH:

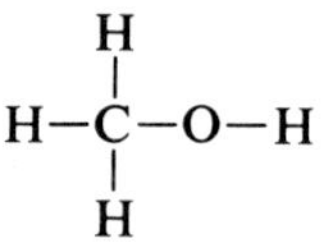

$$\begin{array}{c} H \\ | \\ H-C-O-H \\ | \\ H \end{array}$$

to a trivalent atom, as in methylamine, CH_3NH_2:

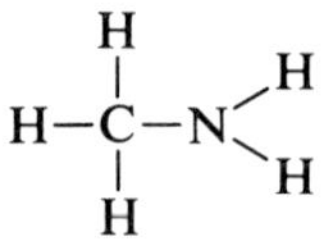

$$\begin{array}{c} H \\ | \\ H-C-N{\diagdown}^H_H \\ | \\ H \end{array}$$

or to another carbon atom, as in ethane, CH_3CH_3:

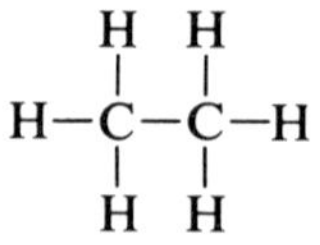

$$\begin{array}{c} H \; H \\ | \; | \\ H-C-C-H \\ | \; | \\ H \; H \end{array}$$

Notice that ethane can be written as above or as C_2H_6, but not as CH_3; although the formula CH_3 describes the *relative* numbers of each kind of atom correctly, it does not give adequate information about the total number of atoms in the molecule. For this purpose, the **molecular formula** must be used, that is, a description of the *actual* number of each kind of atom present.

1.3 The unique nature of carbon

Why is it that carbon forms so many more compounds than all the other elements? The answer can be given in terms of bond energies. It can be shown that it requires about 1,652 kilojoules to break up one mole of methane into its carbon and hydrogen atoms. Since there are four C—H bonds in methane, the bond energy of one C—H bond is one-quarter of 1,652 = 413 kilojoules per mole (abbreviated to 413 kJ per mol). It can also be shown that it requires 2,823 kJ to break up one mole of ethane (CH_3—CH_3) into its constituent atoms. Since this compound contains six C—H bonds, each of which requires 413 kJ per mol for its rupture, the energy of the C—C bond is calculated to be 345 kJ per mol. This is a very high value as compared with those for other elements joined by single bonds (e.g. 163 kJ per mol for N—N and 146 kJ per mol for O—O). Thus, whereas compounds containing O—O and N—N bonds are not very stable, very vigorous conditions—for example, the high temperatures produced in combustion—are necessary to destroy C—C bonds, and this underlies the occurrence of large numbers of stable compounds containing many C—C bonds. For example, in poly(ethene), a plastic (p. 312), many hundreds of carbon atoms are linked together in one molecule. The occurrence of chains of carbon atoms is known as **catenation**.

INTRODUCTION TO ORGANIC CHEMISTRY

As well as forming long chains, carbon atoms can form branched chains, e.g.

$$\begin{array}{c} | \\ -C- \\ | \quad | \quad | \\ -C-C-C- \\ | \quad | \quad | \end{array}$$

There are also compounds in which some of the bonds between carbon atoms are double or triple bonds (1.5) or in which the atoms form rings (1.6). All these possibilities increase still further the number of carbon compounds which can be formed.

1.4 Homologous series and functional groups

The large number of organic compounds fall into a comparatively small number of series, known as **homologous series**. In a particular series, each member has similar methods of preparation and chemical properties to the other members. In the series of alkane hydrocarbons (Chapter 5), the simplest member is methane, CH_4. The next member is ethane, C_2H_6, then propane, C_3H_8, and so on; the general formula of the series is C_nH_{2n+2}. As the series is ascended, a **methylene group**, CH_2, is added to each successive member. As each methylene group is added, the physical properties change slightly. This is demonstrated in detail with the alkanes (Chapter 5).

We have seen already that there is a number of compounds which contain the grouping CH_3. This is called the **methyl group** (sometimes, the **methyl radical**). Other collections of atoms which occur frequently are the amino group, NH_2, and the hydroxyl group, OH. Each of these is able to bond to another group, as in methylamine, CH_3NH_2, and methanol, CH_3OH. They are examples of **functional groups**, which consist of an atom or group of atoms which determine the properties of the homologous series. For example, in the series of alcohols,

CH_3—OH CH_3—CH_2—OH CH_3—CH_2—CH_2—OH
Methanol   Ethanol   Propan-1-ol

CH_3—CH_2—CH_2—CH_2—OH
Butan-1-ol

there is a gradation of physical properties but the chemical properties are very similar.

Several homologous series can be considered to be derived from the alkanes, C_nH_{2n+2}, a hydrogen atom being replaced by a functional group. The group formed from the alkane, C_nH_{2n+1}, is known as the **alkyl group**, and is often represented by the letter R.

The names of alkyl groups are related to the names of the corresponding alkanes:

Alkane C_nH_{2n+2} (RH)	Alkyl group C_nH_{2n+1} (R)
Methane, CH_4	Methyl, CH_3
Ethane, C_2H_6	Ethyl, C_2H_5
Propane, C_3H_8	Propyl, C_3H_7
Butane, C_4H_{10}	Butyl, C_4H_9

1.5 Unsaturated compounds

The compounds we have described so far contain single covalent bonds, and these compounds are described as **saturated**. There are also **unsaturated** compounds in which two atoms share either four or six electrons. For example, the carbon atoms in ethene share four electrons (two originating from each atom); the bond is described as a **double bond**:

$$\underset{H}{\overset{H}{\cdot}}C::C\underset{H}{\overset{H}{\cdot}}$$

As can be seen from Plate 1.3, the carbon and hydrogen atoms are in a plane, with bond angles of 120° (compare the tetrahedral structure of methane). The molecule is described as *planar* and can be represented as

$$\underset{H}{\overset{H}{>}}C=C\underset{H}{\overset{H}{<}}$$

Plate 1.3. A space-filling molecular model of ethene

In ethyne, the carbon atoms share six electrons (three originating from each atom); the bond is described as a **triple bond**:

$$H:C:::C:H$$

The carbon and hydrogen atoms lie in a straight line (Plate 1.4) and the molecule is described as *linear*. A simple representation is

$$H-C\equiv C-H$$

Plate 1.4. A space-filling molecular model of ethyne

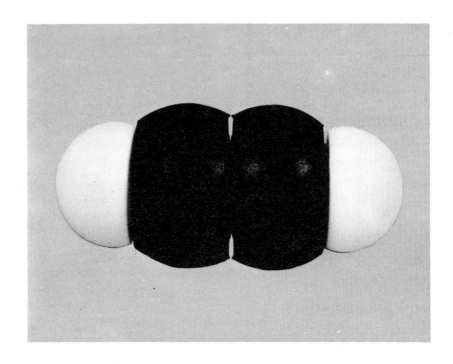

Other common unsaturated groups are C≡N, as in ethanonitrile and C=O, as in ethanal and ethanoic acid.

$$CH_3-C\equiv N \qquad \underset{\underset{O}{\|}}{CH_3-C-H} \qquad \underset{\underset{O}{\|}}{CH_3-C-OH}$$

Ethanonitrile · · · · · · Ethanal · · · · · · Ethanoic acid

The groups C≡N, C=O and CO_2H are called the nitrile, carbonyl and carboxylic acid groups, respectively.

1.6 Aliphatic, alicyclic and aromatic compounds

The compounds we have mentioned so far have had carbon atoms joined in straight chains or branched chains, and these are described as **aliphatic**. There are also compounds in which some of the atoms form a ring, e.g.

Cyclopentane

Rings with from three atoms to very large numbers are known, and these compounds are described as **alicyclic**.

7

INTRODUCTION TO ORGANIC CHEMISTRY

There is a special class of ring compound of which benzene is the parent:

$$
\begin{array}{c}
H \\
| \\
H-C \\
\diagdown C=C \diagup H \\
\diagup   \diagdown \\
H-C C-H \\
\diagdown C \diagup \\
| \\
H
\end{array}
$$

A large number of compounds contain the same carbon ring but have atoms or groups other than hydrogen attached to it. For example, writing benzene as C_6H_6, chlorobenzene is C_6H_5—Cl. Compounds of this nature were originally termed **aromatic** because some of them have pleasant smells (Greek: *aroma*, fragrant smell). The term has been retained because it provides a useful classification; as we shall see, benzene has different properties from the simple unsaturated compound, ethene, and the differences arise because of the special nature of the bonding in benzene (8.2).

1.7 Isomerism

There are two compounds with the molecular formula C_2H_6O:

$$
\begin{array}{cc}
\text{H H} & \text{H H} \\
| | & | | \\
H-C-C-O-H & H-C-O-C-H \\
| | & | | \\
\text{H H} & \text{H H}
\end{array}
$$

Ethanol Dimethyl ether

However, although both compounds have the same number of each kind of atom (two C, six H, one O), they have different physical and chemical properties. They are described as **isomers**, and the phenomenon is described as **isomerism**. Isomerism is said to occur when two or more compounds have the same molecular formula.

The existence of a large number of isomers is illustrated by the alkanes. There is only one compound of molecular formula CH_4 (methane), C_2H_6 (ethane) or C_3H_8 (propane). There are two of molecular formula C_4H_{10}:

$$
CH_3-CH_2-CH_2-CH_3 \qquad
\begin{array}{c}
CH_3 \\
| \\
CH_3-C-H \\
| \\
CH_3
\end{array}
$$

and three of molecular formula C_5H_{12}:

$$
CH_3-CH_2-CH_2-CH_2-CH_3 \qquad
\begin{array}{c}
CH_3 \\
\diagdown \\
CH-CH_2-CH_3 \\
\diagup \\
CH_3
\end{array}
\qquad
\begin{array}{c}
CH_3 \\
| \\
CH_3-C-CH_3 \\
| \\
CH_3
\end{array}
$$

Because carbon atoms can be joined in straight and in branched chains, the number of possible isomers increases very rapidly as the number of carbon atoms increases; for example, there are 5 isomers with molecular formula C_6H_{14} and 18 with molecular formula C_8H_{18}. When other atoms are intro-

INTRODUCTION TO ORGANIC CHEMISTRY

duced, the number of possible compounds with a particular molecular formula increases still further; e.g., there is only one compound with molecular formula C_3H_8 but three with molecular formula C_3H_8O, two being alcohols and one an ether:

$$CH_3-CH_2-CH_2-OH \qquad CH_3-\underset{\underset{OH}{|}}{CH}-CH_3 \qquad CH_3-CH_2-O-CH_3$$

Propan-1-ol　　　　　　Propan-2-ol　　　　　Ethyl methyl ether

1.8 Nomenclature of carbon compounds

As organic chemistry developed, each new compound to be discovered was given its own name, so that a variety of unrelated names quickly grew up. It eventually became necessary to introduce a systematic form of nomenclature in order that the structure of a compound could be readily deduced from its name, and *vice-versa*. The nomenclature at present in use was laid down by the International Union of Pure and Applied Chemistry (I.U.P.A.C.), and the rules for naming some of the simpler compounds are given here.

Methane, as we have seen, is the simplest alkane, and each member of the homologous series of alkanes is given the suffix **-ane**. The first four retain the names originally given to them: methane (CH_4), ethane (C_2H_6), propane (C_3H_8) and butane (C_4H_{10}). After that, the first part of the name is derived from the Greek for the number of carbon atoms in the molecule: pentane (C_5H_{12}), hexane (C_6H_{14}), heptane (C_7H_{16}), octane (C_8H_{18}), and so on. When the chain is branched, the name is taken from that of the longest straight chain of carbon atoms in the molecule; the carbon atoms are numbered from one end of the chain, and the position of the branch and the nature of the group there are indicated by the number of the carbon atom at which branching occurs and the name of the alkyl group which forms the branch. For example,

$$\overset{1}{C}H_3-\overset{2}{C}H_2-\overset{3}{\underset{\underset{CH_3}{|}}{C}H}-\overset{4}{C}H_2-\overset{5}{C}H_3$$

is termed 3-methylpentane. The compound

$$CH_3-\underset{\underset{CH_3}{|}}{CH}-CH_2-CH_2-CH_3$$

might be called 2-methylpentane or 4-methylpentane, depending upon from which end the chain was numbered. The rule is to number from that end which gives the substituent the lower of the two possible numbers, so that the correct name is 2-methylpentane.

The unsaturated compounds with a C=C double bond which are often referred to as olefins are termed alkenes in the I.U.P.A.C. scheme. The number of carbon atoms is described in the same way as for the alkanes. Each member terminates in **-ene**, and the position of the double bond is determined by inserting the lowest possible number before the suffix to describe the carbon atom which forms one end of the double bond relative to its position in the chain; e.g.

$$CH_3-CH_2-CH_2-CH=CH_2$$

is pent-1-ene, not pent-2-ene or 4- or 5-ene.

The unsaturated compounds with a C≡C triple bond, which are often described as acetylenes after the simplest member of the series, acetylene itself (CH≡CH), are termed alkynes in the I.U.P.A.C. scheme, and individual members are described as for the alkenes but with the termination **-yne**.

When an atom other than carbon or hydrogen is present, the name either begins or ends with a description of the type of group which contains this atom, preceded by the number of the carbon atom to which the group is bonded. The remainder of the name is that of the hydrocarbon with the same number of carbon atoms, the final **e** of this name being omitted when a suffix is employed for the substituents. For example, if chlorine is present, **chloro-** is used as a prefix, so that

$$CH_3-\underset{\underset{Cl}{|}}{CH}-CH_3$$

is 2-chloropropane. If a hydroxyl group is present, **-ol** is used as a suffix, so that

$$CH_3-\underset{\underset{OH}{|}}{CH}-CH_2-CH_3$$

is butan-2-ol. Table 1.1 gives the names of some of the homologous series and their functional groups which are met early in this book.

Table 1.1. Nomenclature of organic compounds

HOMOLOGOUS SERIES	FUNCTIONAL GROUP	PREFIX OR SUFFIX	EXAMPLE OF I.U.P.A.C. NOMENCLATURE
Alkanes	(—H)	-ane	$CH_3CHCH_2CH_3$ $\quad\;\;\|$ $\quad\;\;CH_3$ 2-Methylbutane
Alkenes	C=C	-ene	$CH_3CH_2CH=CH_2$ But-1-ene
Alkynes	C≡C	-yne	$CH_3C\equiv CCH_3$ But-2-yne
Alcohols	—OH	-ol	$CH_3CH_2CH_2CH_2OH$ Butan-1-ol
Chloroalkanes (Alkyl chlorides)	—Cl	chloro-	$CH_3CH_2CHCH_3$ $\quad\quad\quad\;\;\|$ $\quad\quad\quad\;\;Cl$ 2-Chlorobutane
Primary amines	—NH₂	amino-	$CH_3CH_2CHCH_2CH_3$ $\quad\quad\quad\;\;\|$ $\quad\quad\quad\;\;NH_2$ 3-Aminopentane
Aldehydes	$-\underset{\underset{O}{\|}}{\overset{}{C}}-H$	-al	CH_3CH_2CHO Propanal
Ketones	$-\underset{\underset{O}{\|}}{\overset{}{C}}-$	-one	$CH_3COCH_2CH_2CH_3$ Pentan-2-one
Carboxylic acids	$-\underset{\underset{O}{\|}}{\overset{}{C}}-OH$	-oic acid	$CH_3CH_2CO_2H$ Propanoic acid
Acid chlorides	$-\underset{\underset{O}{\|}}{\overset{}{C}}-Cl$	-oyl chloride	$CH_3CH_2CH_2COCl$ Butanoyl chloride

The nomenclature for each homologous series is described in detail in the Chapter concerned with that series. Where the older, or 'trivial', names are still often used, they are given, in brackets, after the I.U.P.A.C. name. Two commonly used prefixes in the older nomenclature are **n-**, which indicates an unbranched chain of carbon atoms, and **iso-**, which indicates the presence of the group $(CH_3)_2C$; for example:

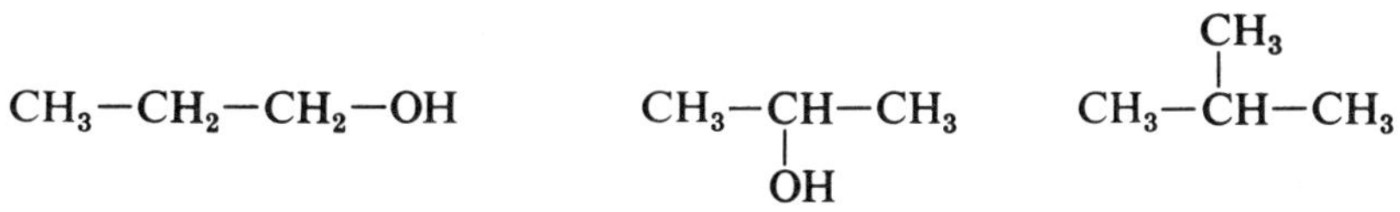

$$CH_3-CH_2-CH_2-OH \qquad CH_3-\underset{\underset{OH}{|}}{CH}-CH_3 \qquad CH_3-\underset{\underset{CH_3}{|}}{\overset{\overset{CH_3}{|}}{CH}}-CH_3$$

Propan-1-ol   Propan-2-ol   2-Methylpropane
(n-Propanol)   (Isopropanol)   (Isobutane)

1.9 Practical work

1. Make models for the following compounds: (a) ethane, (b) chloroethane, (c) ethanol.

2. Make models for compounds with the molecular formulae: (a) C_3H_8, (b) C_3H_6, (c) C_3H_4. Name the compounds.

3. Make models for compounds with the molecular formulae: (a) C_4H_8, (b) C_4H_6. Name the compounds.

1.10 Questions

1. How many isomers would you expect with the following molecular formulae: CH_2Br_2; C_4H_6; $C_4H_{10}O$? How many isomers, each containing the group C=O, would you expect with the molecular formula $C_5H_{10}O$?

2. Write down the structures of the following compounds; 2-methylpentane; 2,2-dimethylpropane; hex-2-ene; 1-bromobutane; 2-methylpropan-2-ol; pentan-3-one.

3. Name the following compounds according to the I.U.P.A.C. rules:

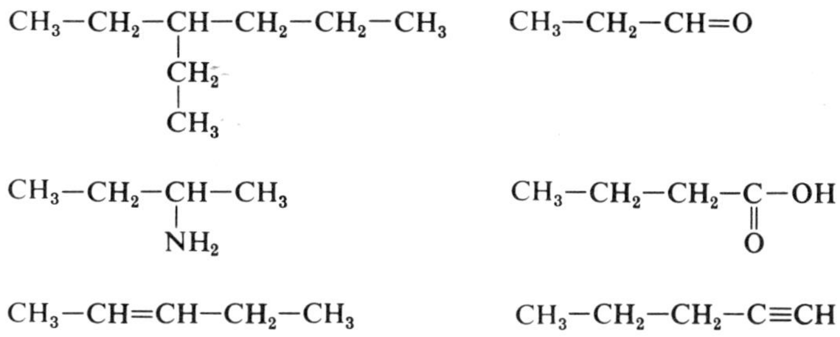

$$CH_3-CH_2-\underset{\underset{\underset{CH_3}{|}}{\underset{CH_2}{|}}}{CH}-CH_2-CH_2-CH_3 \qquad CH_3-CH_2-CH=O$$

$$CH_3-CH_2-\underset{\underset{NH_2}{|}}{CH}-CH_3 \qquad CH_3-CH_2-CH_2-\underset{\underset{O}{\|}}{C}-OH$$

$$CH_3-CH=CH-CH_2-CH_3 \qquad CH_3-CH_2-CH_2-C\equiv CH$$

Chapter 2

Preparation and purification of organic compounds

2.1 Introduction

There are three stages in the preparation of a pure organic compound. First, conditions are found under which the required product is formed in a relatively short time (that is, in a few minutes or at most a few hours, rather than days). Secondly, the product is separated from other materials. Finally, methods are employed to find out if the compound is pure.

The preparations of many organic compounds are described in detail in later chapters. In order to simplify the operations, the number of pieces of apparatus necessary has been kept as small as possible (a list is given in Appendix II). We describe below the more commonly used techniques in the preparation and purification of organic compounds.

2.2 Heating under reflux

Many organic reactions occur very slowly at room temperature. However, the rates of all reactions are increased by raising the temperature, and a reasonably rapid rate can often be achieved by carrying out the reaction at the boiling point of the mixture of reactants, or of the solution of the reactants if they are dissolved in a solvent. The apparatus for this purpose (Fig. 2.1) consists of a **water condenser** attached vertically to the reaction flask; when the mixture in the flask is boiled, its vapour condenses at the cold surface of the condenser and the liquid runs back into the flask. This is known as **heating under reflux**.

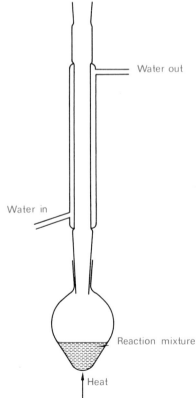

FIG. 2.1. *An apparatus for heating under reflux*

2.3 Distillation

When the product is a liquid or solid with a boiling point below about 250°C and no other such volatile compounds are present, the simplest method of purification is by **distillation**. The apparatus is shown in Figs. 2.2 and 2.3;

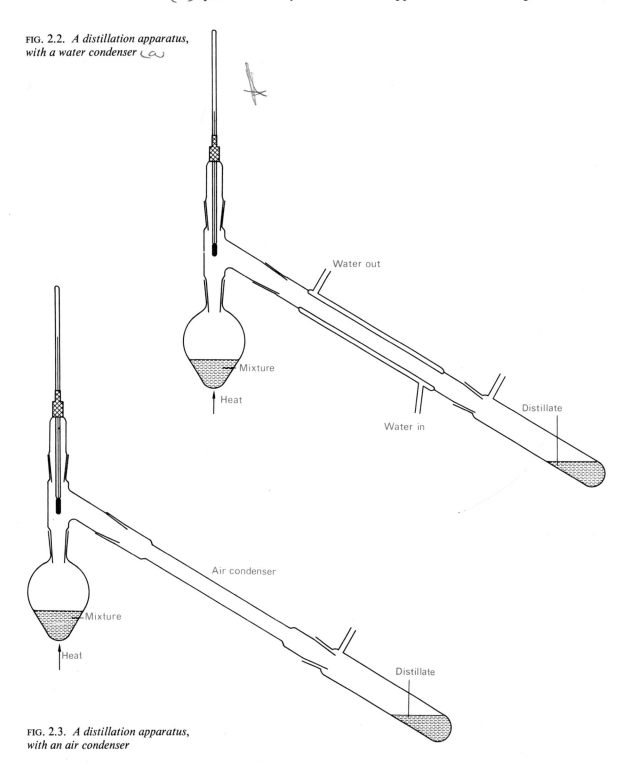

FIG. 2.2. *A distillation apparatus, with a water condenser*

FIG. 2.3. *A distillation apparatus, with an air condenser*

the water condenser is used for compounds boiling up to about 180°C and the air condenser for compounds of higher boiling point. The mixture is heated to boiling and the vapour of the product liquefies in the condenser, the liquid collecting in a receiving flask at the end of the condenser. It is important to place the thermometer so that its bulb is fully immersed in the stream of vapour which is about to enter the condenser; the recorded temperature can then be compared with the quoted boiling point for the required compound. Since the boiling point of a liquid is dependent on the external pressure, the atmospheric pressure should also be measured and recorded when a boiling point is reported; e.g. ethanol, b.p. 78°C/760 mm Hg (the pressure in S.I. units is 101·33 kN m^{-2}). Unless the pressure is stated, you may assume that the reported boiling point was measured at a pressure of 760 mm Hg.

Although some organic liquids decompose before they have a chance to boil at atmospheric pressure, they can still be purified by distillation by reducing the external pressure until their boiling points are below their decomposition temperatures. This is known as **vacuum distillation**. The receiving flask is attached to the condenser and to a thick-walled rubber tube through which air can be withdrawn from the apparatus (Fig. 2.4). By

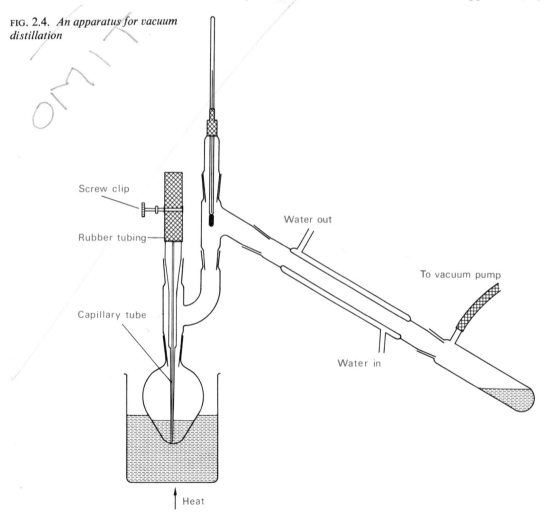

FIG. 2.4. *An apparatus for vacuum distillation*

attaching this tube to a water pump, the pressure can be reduced to a value equal to the vapour pressure of water, which is about 12 mm Hg at room temperature. Much lower pressures can be obtained with a rotary oil pump.

When the required product is accompanied by one or more other compounds of similar boiling point, **fractional distillation** must be carried out. A glass column containing glass or stainless steel coils or beads (Fig. 2.5) is attached to the boiling flask and, at its top outlet, to the condenser. In the apparatus shown, liquids boiling at least 30°C apart (e.g. benzene, b.p. 80°C, and methylbenzene, b.p. 111°C) can be separated successfully but as the difference in boiling points becomes smaller the efficiency of separation decreases and longer columns are necessary. The theoretical basis for fractional distillation is discussed in physical chemistry textbooks.

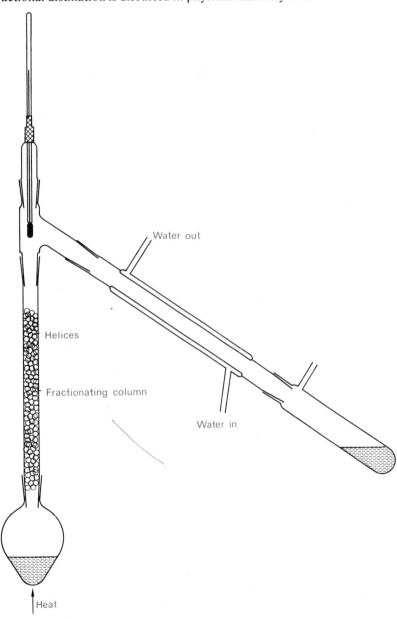

FIG. 2.5. *An apparatus for fractional distillation*

FIG. 2.6. *An apparatus for distillation in steam*

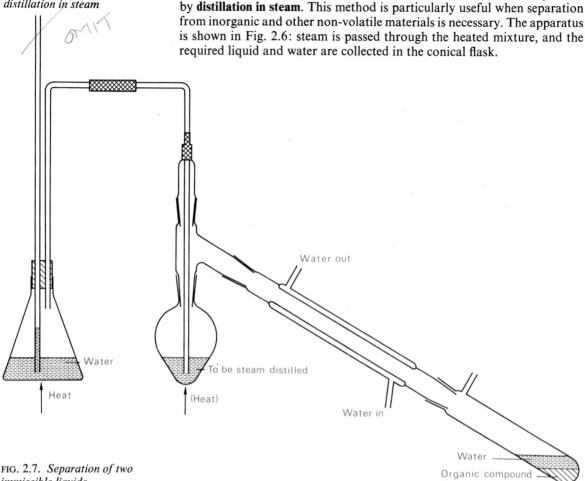

If a liquid of high boiling point is immiscible with water it can be purified by **distillation in steam**. This method is particularly useful when separation from inorganic and other non-volatile materials is necessary. The apparatus is shown in Fig. 2.6: steam is passed through the heated mixture, and the required liquid and water are collected in the conical flask.

FIG. 2.7. *Separation of two immiscible liquids*

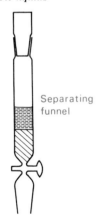

For example, nitrobenzene boils at 210°C. At 98°C, its vapour pressure is about 50 mm, while that of water is 710 mm. The total vapour pressure is equal to the atmospheric pressure, and the mixture boils. Thus, nitrobenzene is distilled at 112°C below its boiling point. It is separated from water by using a separating funnel (Fig. 2.7) from which the denser liquid (in this case nitrobenzene) is run off first.

2.4 Extraction

If the material to be purified is soluble in one solvent, while the impurities are not, the mixture can be partitioned between two immiscible solvents.

For example, sometimes an organic compound is obtained as an aqueous solution, which is contaminated with inorganic materials. The aqueous solution is shaken with diethyl ether (often referred to simply as ether) in a separating funnel. Only the organic compound is soluble in ether, and the residue remains in the aqueous layer. The two layers are separated, and the ether solution is shaken in the presence of a solid drying agent such as anhydrous magnesium sulphate to remove the small amount of water which will have dissolved in it. The solution can then be filtered from the drying agent and the organic compound can be separated from the ether by distillation.

Extraction is more efficient if several small quantities of ether are used, rather than one large quantity. This is considered theoretically in many physical chemistry textbooks.

2.5 Recrystallisation

This is the commonest method for purifying a solid. It is based on the fact that the solubility of organic compounds in a particular solvent increases as the temperature is raised. The solvent is heated with that amount of the solvent which gives a nearly saturated solution at the boiling point. The solution must then be filtered very rapidly, and this is done using a fluted filter paper, contained in a glass funnel from which the stem has been removed, through which the filtrate runs into a conical flask; insoluble impurities are thereby removed. The solution is then allowed to cool, and crystals of the solid continue to be deposited until the solution has reached room temperature. The crystals are filtered off in a Buchner funnel (a

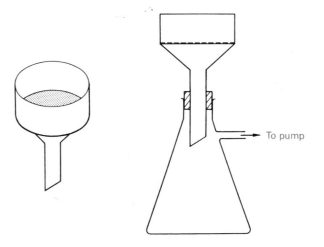

FIG. 2.8. *A Buchner funnel and flask*

porcelain funnel with a large number of holes in its base on which a filter paper rests), which is attached to the filter flask (a Buchner flask) whose side-tube is connected to a water-pump. When filtration is complete, the crystals are washed with a small quantity of the pure solvent to remove any impurities which might have deposited on the surface. The crystals are then dried on a watch-glass to remove the solvent; drying may be at room temperature if the solvent is a particularly volatile one, but it is usually quicker to place the watch-glass in an oven, taking care that the temperature of the oven is below the melting point of the solid.

If one recrystallisation does not yield a pure material, further recrystallisations, preferably with different solvents, should be carried out.

2.6 Chromatography

The chromatographic methods of separation are particularly important for complex mixtures of substances which are not otherwise readily separated. The principal methods are:

(a) Column chromatography
(b) Thin-layer chromatography
(c) Paper chromatography
(d) Gas chromatography

(a) Column chromatography

This is a method for the separation of solids, and can be successful even when the solids have very similar solubilities, when recrystallisation would not be effective.

A glass column about 30 cm long and 2 cm in diameter, which narrows at one end to an outlet tube, is used. A pure liquid such as benzene is run in until the column is about half-full, and then a slurry of a solid adsorbent (such as alumina) in the same solvent is added. The solid sinks to the bottom, forming an evenly packed column immersed in the solvent; a pad of glass wool at the bottom prevents the solid running out, and a disc of filter paper is added at the top of the column to avoid disturbance of the solid when a solution or solvents are added later. The solvent above the column of damp solid is run off, care being taken that the column of solid remains wetted, for otherwise cracks and channels appear and the solution to be added does not run evenly through the column.

A solution of the mixture in the same solvent is then added at the top of the column and the liquid emerging at the bottom is collected; by attaching the receiver to a pump, the rate of passage through the column can be increased (Fig. 2.9). Following addition of the solution, further quantities of the solvent are added, and after a while solutions of the components of the mixture will emerge successively. This process is known as **elution** and the solvents as **elutants**. Each solution is collected in a separate flask, and the solvent is then removed by distillation to leave the pure compound.

If the solutes are coloured, they are seen as they pass down the column. If they are colourless, a physical property, such as fluorescence under ultraviolet light, can be employed to find out how each compound is moving down the column so that a fresh flask can be inserted at the bottom just when the solution of the compound is about to emerge.

The principle of the method is that each compound in the mixture has a particular solubility in the solvent and a particular tendency to be **adsorbed** by the solid in the column; no two compounds behave exactly alike in these respects. Thus, compounds which are readily soluble and not strongly adsorbed move rapidly down the column in the stream of solvent; those which are not so soluble and are more strongly adsorbed are held on the column for longer periods.

In some cases a particular solvent causes one component of a mixture to be eluted rapidly while another component does not travel down the column at a significant speed. It is then helpful to use a second solvent after the first compound has been eluted. Plate 2.1 shows a column packed with alumina in which a solution obtained from spinach is being separated.

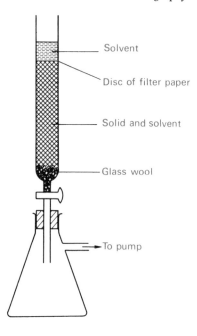

FIG. 2.9. *Column chromatography*

Plate 2.1. Column chromatography. Separation of pigments in a solution of spinach in light petroleum. (a), (b) The column is being developed with benzene. The carotenes (coloured orange) are being eluted. (c) The column is being developed with another solvent, a mixture of butan-1-ol, ethanol and water. The chlorophylls (coloured green) are being eluted

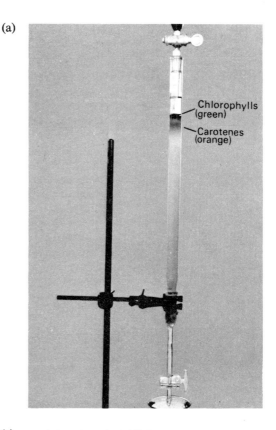
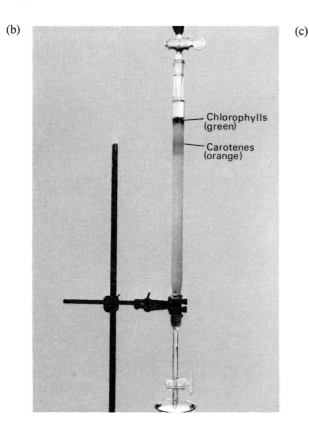
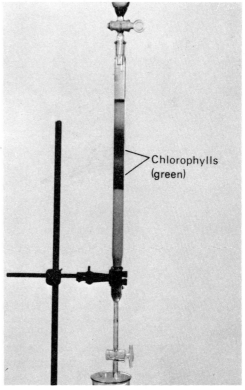

(b) Thin-layer chromatography

This method is similar to column chromatography but on a much smaller scale. It is generally used for the identification of components in a mixture rather than as a method of purification in preparative work.

A solid adsorbent such as silica gel or alumina is spread as a thin, even layer on a glass plate [Plate 2.2(a)]. A solution of the sample to be analysed is placed as a spot on the surface of the plate and near one end. When the solvent has evaporated from the spot, the plate is put in a tall beaker or jar containing the same solvent, so that the spot is a little way above the level of the liquid. The beaker is covered to reduce evaporation. The solvent rises slowly up the plate by capillary action, and after it passes the spot, the components in the mixture begin to move with the solvent at different rates, just as in column chromatography [Plate 2.2(b)].

When the solvent has almost reached the top of the plate, the plate is removed from the beaker and the solvent is allowed to evaporate. If the components of the mixture are coloured, each will be seen as a coloured area [Plate 2.2(c) and (d)]. If they are colourless, the plate is **developed** by being stood in a beaker containing some crystals of iodine. After a few minutes, a dark spot will appear where each compound is held on the plate.

The distances from the original spot to the solvent front and the position of each compound are measured. It is convenient to calibrate the distance travelled by the solvent from 0–1, and the fraction of this distance travelled by each compound is described as its R_F value. For example, in Fig. 2.10, the mixture was placed on the plate at A, and thus solute X has an R_F value of 0·25 and solute Y has an R_F value of 0·75.

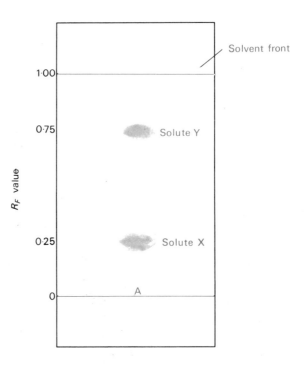

FIG. 2.10. R_F *values*

Different substances have different R_F values for the particular conditions of the experiment, so that comparison of the values from the mixture with those from authentic samples of pure compounds enables the components

Plate 2.2. Thin-layer chromatography. (a) A thin-layer plate. (b) The development of a thin-layer plate. (c) Separation of black ink using butan-1-ol, ethanol and 2M ammonia solution (3:1:1 by volume) as solvent. (d) Separation of pigments in a solution of spinach in light petroleum

FIG. 2.11. *Apparatus for paper chromatography*

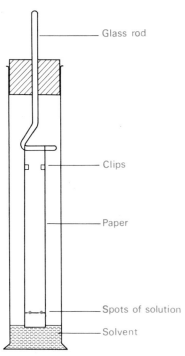

of the mixture to be identified. It is essential that comparison is made for the same solvent and temperature, for the R_F value depends on these as well as on the nature of the compound.

(c) Paper chromatography

This method is like thin-layer chromatography in being suitable for the identification of the components of a very small sample of a mixture. The procedure is similar except that a large rectangular piece of filter paper mounted in a glass tank (Fig. 2.11) is used instead of solid adsorbent on a glass plate.

A pencil line is drawn across the paper near the bottom and a drop of the solution (which contains, for example, a mixture of amino-acids) is placed on the paper at this line by means of a fine glass pipette. Samples of solutions of the compounds believed to be in the mixture are placed alongside so that a direct comparison of R_F values can be made. Solvent is then poured into the tank so that the pencilled line is just above the surface.

When the solvent reaches almost the top of the paper, the paper is removed from the tank and the solvent is allowed to evaporate. If the components are coloured, they are easily seen, but if not, a physical or chemical property must be employed to reveal their positions. In Plate 2.3, a chemical reagent, ninhydrin, was used to form a coloured compound with each of the amino-acids.

The underlying principle in paper chromatography is the **partition** of a solute between two solvents. One solvent is the one which travels up the paper (the eluting solvent), and the other is water, the molecules of which are adsorbed on the cellulose which constitutes the filter paper. Compounds

Plate 2.3. Paper chromatography. Separation of some amino-acids with a mixture of butan-1-ol, ethanoic acid (acetic acid) and water as solvent. The paper has been developed with ninhydrin solution. On the extreme right-hand side, a mixture of amino-acids has been separated, and the R_F values can be compared with those for pure amino-acids developed at the same time

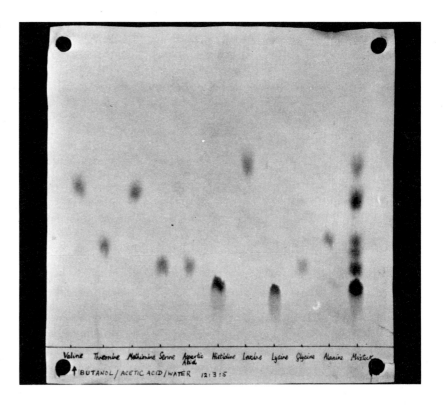

which are relatively more soluble in the eluting solvent travel faster (i.e. have larger R_F values) than the less soluble ones.

In some cases, two or three components in a complex mixture are not separated completely in this way. It is an advantage to elute with one solvent, remove the paper, dry it, turn the paper through 90° and elute with a second solvent.

Thin-layer chromatography is a more rapid method than paper chromatography. However, for compounds which are only slightly soluble in organic solvents but moderately soluble in water, thin-layer chromatography is unsuitable whereas paper chromatography is successful.

(d) Gas chromatography

There are two types of gas chromatography: gas-liquid chromatography and gas-solid chromatography. Each is suitable for separating mixtures of gases, liquids and volatile solids.

In **gas-liquid chromatography**, as in paper chromatography, the compounds in the mixture are partitioned between two solvents. One solvent is fixed in position and is known as the **stationary phase** (cf. water in paper chromatography); it consists of a non-volatile liquid (such as a long-chain alkane) which is coated on an inert solid. The other solvent is a gas which travels through the column containing the stationary phase (cf. the eluting solvent in paper chromatography). It is known as the **carrier gas**, and must be one which does not react with the components of the mixture. Nitrogen is often used.

Gas-solid chromatography differs in employing a solid as the stationary phase. The principle of separation is like that of column chromatography save that the eluting solvent is a gas and not a liquid. The solids generally used are silica gel or alumina.

The gas chromatograph consists of: (i) a supply of carrier gas at constant pressure; (ii) a flow-meter to measure the rate at which the carrier gas passes through the column; (iii) the column, in a thermostat; and (iv) a detector, to determine when each component of the mixture is eluted. At the top of the column is a rubber septum cap through which the solution for analysis is introduced by means of a hypodermic syringe; the cap is self-sealing so that when the syringe is withdrawn the sample and carrier gas cannot escape. Figure 2.12 gives an outline of a gas chromatograph with a column 2 m long and 4 mm in diameter.

The detector makes use of a physical or chemical property of the compounds in the mixture. Two detectors commonly employed are the katharometer and the flame-ionisation detector. The **katharometer** (Fig. 2.12) measures the change in the thermal conductivity of the emerging gas which occurs when an organic compound is eluted in a stream of carrier gas. The **flame-ionisation detector** measures the concentration of ions in a flame. Hydrogen is mixed with the carrier gas as it emerges from the end of the column and is burned at a jet, and the ions formed in the flame are collected on a charged plate above it. While only hydrogen is being burnt, the concentration of these ions remains steady, but when an organic compound is eluted there is a change in the concentration.

These detectors are used in conjunction with a device which enables the changes to be observed. A galvanometer is suitable, but it is helpful to have a permanent record, and a pen-recorder can be used. The result of a typical gas chromatography experiment is shown in Fig. 2.13. There were six compounds in the mixture.

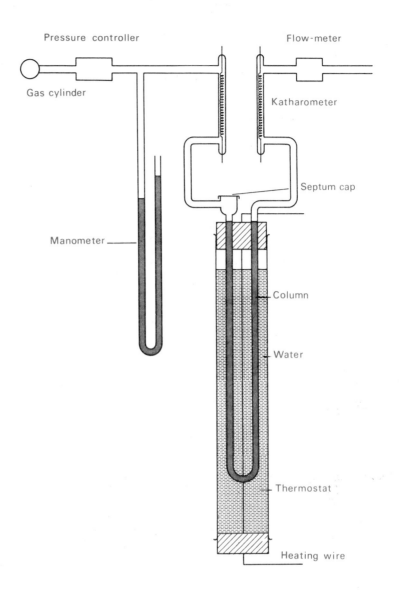

FIG. 2.12. *A gas chromatograph*

The time taken for each solute to pass through the column [for acetone, time (B–A)]—its **retention time**—is characteristic of the compound and the conditions of the experiment (the rate of flow of gas, the nature and concentration of the stationary phase and the temperature). Provided that the conditions are constant, an unknown compound can be identified by comparison of its retention time with those for samples of known compounds. Care must be exercised, because sometimes two compounds have the same retention time on a particular column. It is therefore best to ensure that a compound has the same retention time as an authentic sample on two or three columns packed with different stationary phases. It is still better to measure the mass spectrum of the compound following elution, as described below.

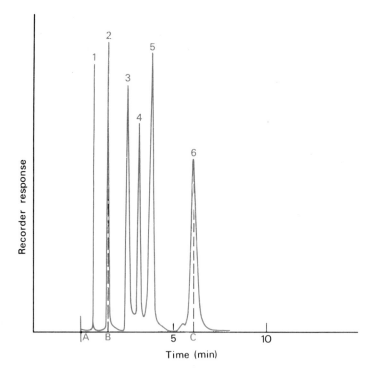

FIG. 2.13. *A gas chromatogram. The separation of a mixture of six substances: 1, Diethyl ether; 2, Propanone; 3, Methanol; 4, Ethanol; 5, Pentan-3-one; 6, Propan-1-ol. The retention time for propanone is $(B - A)$ min, and for propan-1-ol, $(C - A)$ min*

The area under the peak of the chromatogram depends on the amount of the material present, so that the amount in a mixture can be measured by calibrating the chromatogram with injections of measured quantities of the compound and comparison of the resulting areas with that from the mixture.

The gas chromatograph described above is suitable only for analytical purposes, since the quantity of a mixture injected into the column is only about one-hundredth of a gramme. However, it is also possible to use columns of larger diameter which enable several grammes of a sample to be introduced, so that each component can be isolated as well as identified. For this purpose, the emerging gas is divided into two streams, the smaller of which passes into the detector and the larger of which flows through a U-tube surrounded by a freezing mixture. After each peak is registered, a fresh U-tube is inserted to collect the next compound.

A recent development has been to connect the gas chromatograph to a mass spectrometer. As the vapour of each solute is eluted at the end of the column, part of it is channelled into the spectrometer, and by analysis of the mass spectrum the structure of the materials can be deduced (3.7).

2.7 Criteria of purity

Pure solids melt over a very small temperature range (approximately 1–2°C), whereas the presence of even 1 per cent of an impurity can increase the range to several degrees. The melting point therefore provides a very good criterion of purity.

A small sample of the solid is placed in a glass capillary tube which is sealed at its lower end. The tube is attached to a thermometer so that the sample is level with the bulb and the thermometer is suspended in a small beaker of paraffin oil (Fig. 2.14). The oil is heated slowly (not more than 2° per minute near the melting point of the solid) and is stirred at the same time. The temperature at which melting is first observed and that at which it is complete are both noted, and the melting point is then recorded by quoting both temperatures, e.g. ethanamide, m.p. 81–82°C.

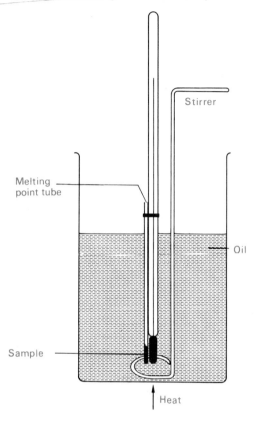

FIG. 2.14. *A simple melting point apparatus*

A melting-point method can also be used to confirm the identity of a compound. Suppose that a compound has been isolated which is thought to be ethanamide and which has the expected melting point for ethanamide of 81–82°C. A small amount is mixed with about the same quantity of a sample from an authentic source of ethanamide. The mixture is ground up and its melting point determined. If the first compound is in fact ethanamide, the melting point of the mixture will still be 81–82°C, but if it is not, then melting will begin at a lower temperature and extend over a wider range of temperature than that for a pure compound. This is known as the **mixed melting point** method.

A second criterion of purity for a solid is that it should give only one spot on a thin-layer or paper chromatogram. Small quantities of impurities can be revealed in this way, but unfortunately it sometimes happens that an impurity has the same R_F value as the required compound and is therefore not revealed. Thus, observation of a single chromatographic spot is a necessary but not a sufficient criterion for purity.

Pure liquids boil within a range of about 1°C. If the temperature recorded during a distillation covers a wider range, the resulting liquid cannot be pure. However, the converse is not necessarily true, because sometimes a distillate which boiled at a constant temperature is found to contain small quantities of other materials. In this respect, the boiling point of a liquid is a less satisfactory criterion of purity than the melting point of a solid, while even fractional distillation of a liquid is not so successful at removing impurities as recrystallisation of a solid.

Two other criteria for the purity of liquids are available. One is the refractive index, which can be measured to five significant figures with a modern refractometer and which is very sensitive to impurities. The second is the combination of the gas chromatogram and the mass spectrum. The chromatogram of a pure compound shows only a single peak, and gas chromatography is so sensitive a method that it is possible to detect less than 1 per cent of an impurity even when only about 1 mg of the compound is available. However, an impurity with the same retention time as that of the compound is not revealed on the chromatogram, and to test for this possibility the eluting compound is fed into the mass spectrometer; an impurity is revealed by the resulting spectrum.

2.8 Summary of the use of chromatographic methods

Between them, the various types of chromatography provide methods for (i) the separation of the components of a mixture on a preparative scale, (ii) the identification of unknown compounds in a mixture, even when only minute quantities (e.g. 1 mg) are available, and (iii) testing the purity of a material.

Chromatographic method	*Separation*	*Identification*	*Purity*
Column	+		
Thin-layer		+	+
Paper		+	+
Gas (small scale)		+	+
Gas (large scale)	+		

2.9 Practical work

Column chromatography

Preparation of a column for the separation of dyes

Place a plug of glass wool in the narrow neck of the bottom of a glass tube (about 30 cm long and 2 cm in diameter; Fig. 2.9). Fill the column half-full of water and pour a slurry of alumina and water into it, tapping the tube. The solid will sink to the bottom, giving a uniform column; there must be no cracks or channels in the solid. Liquid is run out of the column until the solid is just covered. Place a small piece of filter paper on top of the solid to prevent it from being disturbed when the solution is added.

Pipette (carefully, so that the solid is not disturbed) 3 cm^3 of an aqueous solution of two dyes (0·1 per cent w/v of malachite green and methylene blue). Elute with water and, when one dye has passed through, elute with ethanol until the second dye has run off. Make sure that the solid is always covered with solvent, and adjust the water pump so that solvent passes through the column at 5–10 cm^3 per minute.

Preparation of a column for the separation of pigments in spinach

(a) *Preparation of a solution of pigments from spinach*

Crush about 20 g of frozen spinach with about 20 cm^3 of methanol. Decant the methanol and crush the spinach with 50 cm^3 of a mixture (2:1 by volume) of light petroleum (b.p. 60–80°C) and methanol. Filter the mixture, collect the solvents in a flask and transfer them to a separating funnel. Run off the methanol layer, and then pour the petroleum layer into a conical flask. Add about 2 g of anhydrous sodium sulphate and shake for a few minutes to dry the solution.

Decant the solution into a flask and evaporate off the solvent using a hot water bath (cf. Fig. 2.2) until about 5 cm^3 of solution remain.

(b) *Preparation of a column of alumina*

Prepare a column of alumina as described above, using dry benzene instead of water. The empty column should be half-filled with benzene and the slurry made with alumina and benzene.

(c) *Separation of the solution from spinach*

Place 3 cm^3 of the solution from spinach on the column of alumina, and develop the column with benzene until the first band, which are carotenes and coloured orange, passes into the flask. Change the receiver and develop the column with a second solvent containing butan-1-ol, ethanol and water in the ratio of 3:1:1 by volume. This separates the chlorophylls, coloured green, from the xanthophylls, coloured pink-brown.

Thin-layer chromatography

Preparation of thin-layer plates

Make a slurry of silica gel (20 g) in trichloromethane (50 cm^3) in a beaker. Dip two microscope glass slides (15 × 2 cm) back-to-back in the slurry and withdraw them slowly but at a constant rate. Separate the slides and, holding them horizontally, wave them gently while the solvent evaporates. It is important to wipe the back and edges of the slide free of solid.

Scratch a line across the slide about 2 cm from the bottom, leaving spaces where the solution to be analysed is to be placed. Apply the solutions using a fine glass capillary tube; the spot should be no wider than 3–4 mm. Place the slide in a beaker containing solvent; the level of the solvent should be just *below* the level of the line drawn across the slide. When the solvent has risen over three-quarters of the way up the slide, remove the slide and allow the solvent to evaporate.

Separation of pigments from spinach

Prepare a solution of pigments as described on p. 26. Develop the slide with trichloromethane. Note the number of spots developed, measure their R_F values and describe their colours. Place the slide in a beaker in which there are two or three crystals of iodine. Cover the beaker with a watch-glass and note whether any additional spots are developed.

Separation of products from the nitration of phenol

Place 3 spots on a thin-layer plate: (a) a solution of 2-nitrophenol in ethanol, made up by dissolving a few crystals of the solid in 5 drops of ethanol; (b) a solution of 4-nitrophenol in ethanol; (c) a solution of the reaction products from the nitration of phenol (p. 269) (or an artificial mixture of 2- and 4-nitrophenol) in ethanol.

Develop the plate using trichloromethane.

Separation of dyes in black ink

Apply a small spot of black ink (for example, Quink) to a thin-layer plate and develop it using a mixed solvent, butan-1-ol, ethanol and 2M ammonia solution in the ratio of 3:1:1 by volume.

Paper chromatography

Apparatus for paper chromatography (ascending solvent front)

A gas jar can be used as a chromatograph tank (Fig. 2.11). The paper is held in position by a piece of glass rod and clips.

Cut strips of Whatman No. 1 filter paper about 4 cm wide and 30 cm long. Draw a pencil line across the paper near the bottom and apply the solutions with a fine capillary tube (a melting-point tube is useful). To prevent the spot spreading, it should be dried quickly (with a hair dryer, for example).

Pour about 40 cm^3 of solvent in the cylinder and place the end of the paper just below the level of the solvent.

Separation of amino-acids

Make up 100 cm^3 of solvent by shaking together 40 cm^3 of butan-1-ol and 50 cm^3 of water in a separating funnel for about 10 minutes and then adding 10 cm^3 of ethanoic acid. Shake the mixture again. On standing, the mixture will separate into two layers. The top layer should be used as the solvent.

Make up solutions of (a) 0·1 g of glycine in 10 cm^3 of water, (b) 0·1 g of proline in 10 cm^3 of water, (c) a mixture of 0·1 g of glycine and 0·1 g of proline in 10 cm^3 of water.

Place 3 spots, (a), (b), and (c) on the paper, set up the chromatography apparatus and allow the solvent front to move at least three-quarters of the way up the paper. Dry the paper (preferably in an oven at 100°C) and spray it with a solution of ninhydrin (0·2 g of ninhydrin in 99 cm^3 of butan-1-ol and 1 cm^3 of ethanoic acid). If no spray is available (a cheap scent spray is useful), draw the paper through a shallow bath of ninhydrin solution. Dry the paper for about 2 minutes at 100°C.

Glycine and proline react with ninhydrin to form a blue and a yellow compound, respectively. The R_F values of the spots formed in the control experiments (a) and (b) may be compared with those formed by the mixture.

Gas chromatography

Several types of apparatus which can be constructed in a school laboratory are described in *Organic Chemistry Through Experiment* (p. 173–184).

2.10 Further reading

Organic Chemistry Through Experiment. D. J. Waddington and H. S. Finlay (4th ed. 1977). Mills and Boon Ltd.
Chromatography and Electrophoresis on Paper. J. G. Feinberg and I. Smith (2nd ed. 1965). Shandon Scientific Company. Paperback, Longman 1972.
Thin-layer chromatography. D. A. Stephens, *School Science Review*, Vol. 48, p. 376 (1967).
Simple chromatographic demonstrations. M. Taylor, *School Science Review*, Vol. 45, p. 75 (1963).
Experimental gas chromatography. D. E. P. Hughes, *School Science Review*, Vol. 47, p. 125 (1967).
Demonstration chromatograph with a flame ionisation detector. G. D. Brabson *Journal of Chemical Education*, Vol. 49, p. 71 (1972).
A demonstration experiment in gas chromatography. G. R. Fitch and D. J. S. Sharp, *Education in Chemistry*, Vol. 7, p. 242 (1970).

2.11 Films

Chromatography. I.C.I. Films Library, Imperial Chemical Industries Limited.*
Distillation. I.C.I. Films Library, Imperial Chemical Industries Limited.*

* On free loan.

2.12 Questions

1 Summarise the various methods available for the purification of an organic compound and discuss the physico-chemical principles underlying any **two** of these methods. (W(S))

2 When an organic compound is nitrated under certain conditions, the solid mononitro and a small proportion of solid dinitro derivatives are formed. Describe how the mononitro compound is obtained in the pure state by recrystallisation from ethanol explaining the *reasons* for all the techniques employed.

A melting-point determination is carried out after four successive recrystallisations (a), (b), (c) and (d) with the following results:

    m.p.
(a) 120° ⎫ Melting-point *not sharp*
(b) 125° ⎭
(c) 127° ⎫ Melting point *sharp*
(d) 127° ⎭

Describe, with the aid of a diagram, how the determination is carried out and explain the significance of the results quoted. (S)

Chapter 3
Determination of the structure of an organic compound

3.1 Introduction

The determination of the structure of an organic compound necessitates the following steps:
   1. The purification of the compound. Methods of purifying organic compounds, and of testing their purity, are described in Chapter 2.
   2. Determination of which elements are present: qualitative analysis (3.9).
   3. Determination of the molecular formula. This can be achieved by combination of quantitative analysis to find the proportions of each element in the compound (3.3) and measurement of the formula weight (3.5). In recent years it has become possible to find the molecular formulae of some compounds directly by mass spectrometry (3.6).
   4. Determination of the way in which the atoms are arranged in the molecule. Both chemical methods and physical methods (spectroscopy) are used.

3.2 Qualitative analysis

Practical details for the detection of carbon and hydrogen are given in Section 3.9, together with the Lassaigne test for nitrogen, sulphur and the halogens.

3.3 Quantitative analysis

Carbon, hydrogen and nitrogen

A known weight of the compound is heated to a high temperature in an excess of dry oxygen. The compound burns to form carbon dioxide and water. If nitrogen is present in the organic compound, a mixture of nitrogen oxides (and sometimes nitrogen gas) is also produced; the oxides of nitrogen are subsequently reduced by copper to nitrogen. The weights of carbon dioxide, water and nitrogen are then found and the percentage composition of carbon, hydrogen and nitrogen can be calculated.

FIG. 3.1. *A CHN analyser*

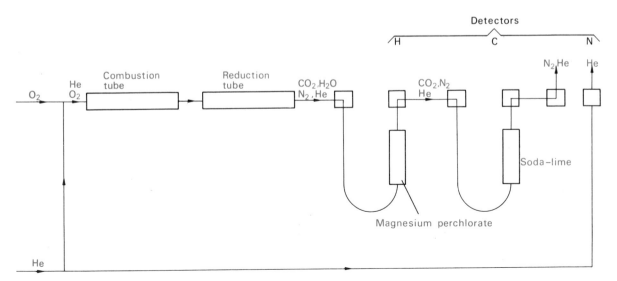

DETERMINATION OF THE STRUCTURE OF AN ORGANIC COMPOUND

FIG. 3.2. *The combustion and reduction tubes used for analysis of carbon, hydrogen and nitrogen. The combustion and reduction tubes are in different furnaces*

Recently, analysers (known as CHN analysers) have been introduced which enable the compound to be analysed automatically. In one of these (Fig. 3.1), the sample (about 1–3 mg) is weighed in a platinum boat placed in a stainless steel **combustion tube** (Fig. 3.2). The tube is heated in a furnace at about 900°C in a stream of dry oxygen and a carrier gas (helium). The compound burns, and the products pass along the tube. They first pass over three materials which are present to remove elements which would otherwise interfere with the analysis for C, H and N: silver removes halogens as

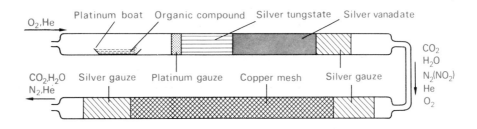

involatile silver halides, and silver tungstate and silver vanadate remove elements such as phosphorus and sulphur with which they combine to form involatile inorganic salts. The gases then pass through a second tube, the **reduction tube**, which is filled with copper and heated to 650°C. This causes the oxides of nitrogen to be reduced to nitrogen and removes oxygen by formation of copper(II) oxide. The gas, which now contains only carbon dioxide, water, nitrogen and helium, then passes through three katharometers (p. 22); each of these consists of two thermal conductivity cells

FIG. 3.3. *Analysis of a compound containing C, H, N. The percentage composition of the elements is found from the heights of the peaks, subtracting values for the blanks (values obtained when the boat is empty). The height for carbon dioxide is* $(IJ - CD)$

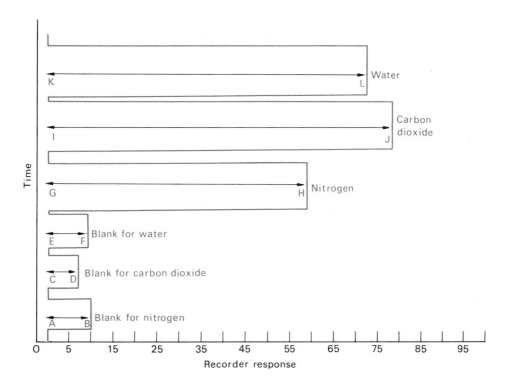

which contain heated platinum wires, the two wires forming two arms of a balanced Wheatstone bridge circuit. The gases pass through the first cell of the first katharometer, then through a tube of magnesium perchlorate to absorb the water vapour, and then through the second cell of the same katharometer. The conductivity of the gases in the two cells is therefore different, so that the Wheatstone bridge circuit becomes out of balance and a current flows; the current, which is proportional to the weight of water absorbed, is recorded automatically (Fig. 3.3) so that the weight of water, and hence the percentage composition of **hydrogen** in the compound, can be calculated. The cells of the second katharometer are separated by a tube of soda-lime which absorbs the carbon dioxide; thus, the percentage of **carbon** can be calculated from the Wheatstone bridge current in the same way as that of hydrogen. The remaining mixture of nitrogen and helium passes into one cell of the third katharometer while a second stream of helium, at the same flow-rate as in the mixture of gases, passes through the second cell. The current in the Wheatstone bridge circuit is proportional to the concentration of nitrogen, so that the percentage composition of **nitrogen** can be calculated.

Nitrogen by Kjeldhal's method

Another method for determining the percentage composition of nitrogen in an organic compound is to boil a weighed sample with concentrated sulphuric acid and anhydrous potassium sulphate (to raise the boiling point of the acid). A catalyst (for example, mercury to which a little selenium is added) is sometimes used. The resulting solution of ammonium sulphate is diluted, excess of sodium hydroxide is added and the liberated ammonia is distilled into excess of standard acid. The acid is then titrated against standard sodium hydroxide to determine the amount of acid originally neutralised by ammonia.

Halogens by Carius's method

A weighed quantity of the compound is heated strongly with concentrated nitric acid and solid silver nitrate in a sealed tube for some hours. The tube is cooled and then broken under water. The silver halide is filtered, washed, dried and weighed.

3.4 Calculation of the empirical formula

Once the percentage composition of each element is known, the ratio of the numbers of atoms of each element present in the compound can be calculated. This is the **empirical formula**.

The method is to divide the percentage composition of each element by its atomic weight and to factorise the resulting numbers so as to obtain simple whole numbers. For example, a compound X, a white solid, was found by analysis to contain 23·30 per cent carbon, 4·85 per cent hydrogen and 40·78 per cent nitrogen. It was known to contain no other elements except oxygen, so that the composition of oxygen was

$$100 - (23 \cdot 30 + 4 \cdot 85 + 40 \cdot 78) = 31 \cdot 07 \text{ per cent.}$$

DETERMINATION OF THE STRUCTURE OF AN ORGANIC COMPOUND

Then:

Element	% Composition	Atomic weight	Atomic ratio	Simplest atomic ratio
Carbon	23·30	12	1·94	2
Hydrogen	4·85	1	4·85	5
Nitrogen	40·78	14	2·91	3
Oxygen	31·07	16	1·94	2

The empirical formula of X is $C_2H_5N_3O_2$. To determine the molecular formula, the formula weight must be found.

3.5 Determination of a formula weight

The determination of formula weights of compounds is described in detail in texts on Physical Chemistry. Those of **gases** are generally determined by the limiting density method, using a gas density balance, while those for **volatile liquids and solids** are found by an adaptation of the traditional Victor Meyer's method in which the volume of vapour from a known weight of compound is determined. The formula weight of an **involatile liquid or solid** is often found from the depression of the freezing point of a solvent.

In some cases, the formula weight can be found quickly and with a very high degree of precision by mass spectrometry. The vapour of the substance is bombarded, at a low pressure (10^{-5}–10^{-6} mm Hg), by a stream of electrons (*ca.* 70 kV energy). This can cause the loss of one electron from a molecule of the substance, giving the **molecular ion** (sometimes called the **parent ion**):

$$M + e \rightarrow M^+ + 2e$$

This ion is accelerated in an electric field and is then deflected and focused in a magnetic field (Fig. 3.4), the amount of deflection depending on the mass of the ion and hence giving the formula weight directly. With modern spectrometers, the formula weight can be measured to the fourth decimal place.

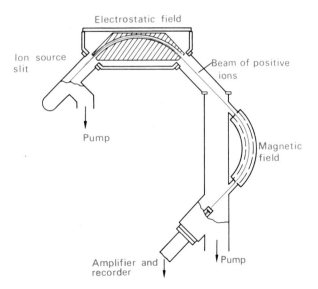

FIG. 3.4. *A mass spectrometer*

3.6 Determination of the molecular formula

The yield of the molecular ion is generally small, while some compounds are unable to form a stable molecular ion, the ion decomposing very rapidly into smaller particles (p. 35). The formula weight cannot be determined by this method for such compounds, although important information about the structure of the compound can still be obtained from identification of the fragments (3.7).

The molecular formula of a compound—the **actual** number of each kind of atom in the molecule—can be determined from the empirical formula and the formula weight.

For example, the empirical formula of the compound X in Section 3.4 was found to be $C_2H_5N_3O_2$. The formula weight (determined, for example, by freezing-point depression) is 103. In this case, therefore, the empirical formula is also the molecular formula. On the other hand, if the formula weight had been found to be 206, the molecular formula would have been $C_4H_{10}N_6O_4$.

If a molecular ion is formed in the mass spectrometer, the molecular formula can be found without determination of the empirical formula or any other study of the compound. Thus, the molecular ion of the compound X was measured as 103·0382. Its molecular formula must therefore be $C_2H_5N_3O_2$ since no other possible combination of elements corresponds to this mass. (Table 3.1 lists compounds of formula weights between 103·017046 and 103·049406, based on $C = 12·000000$; see Appendix V.)

Table 3.1. Organic compounds of formula weight ca. 103

MASS	NUMBER OF ATOMS IN THE MOLECULE			
	C	H	N	O
103·017046	5	1	3	–
103·018388	7	3	–	1
103·025597	1	3	4	2
103·026940	3	5	1	3
103·029622	6	3	2	–
103·038173	**2**	**5**	**3**	**2**
103·039515	4	7	–	3
103·042197	7	5	1	–
103·049406	1	5	5	1

3.7 Determination of the structural formula

The structural formula shows us which atoms are joined to which in the molecule. Strictly speaking, a structural formula requires a three-dimensional representation, but planar, two-dimensional representations are usually more convenient (1.2).

In elucidating the structural formula, it is usual to consider the possible structures based on the molecular formula. For example, the compound X has the molecular formula $C_2H_5N_3O_2$; a possible structural formula is:

$$\begin{array}{c} O H O \\ \| | \| \\ H_2N-C-N-C-NH_2 \end{array}$$

DETERMINATION OF THE STRUCTURE OF AN ORGANIC COMPOUND

In order to confirm this structure, we examine both the physical and the chemical properties of the compound. By studying the chemical reactions, the presence or absence of the functional groups can be determined. For example, for X, we would test for the presence of the groups $-NH_2$, $-NH-$, and $C=O$. There is also a special chemical test for the $-NH-$ group which is adjacent to $C=O$ (p. 221).

However, in more complex molecules, physical methods for the determination of structure are always used as well, in particular three spectroscopic techniques: mass spectroscopy, infrared spectroscopy and nuclear magnetic resonance (NMR) spectroscopy.

Mass spectroscopy

When molecules, M, are bombarded by electrons in the mass spectrometer, molecular ions, M^+, are usually formed (p. 33). A proportion of these fragment into smaller species, one of which carries a positive charge and the other of which is a neutral radical:

$$M^+ \to X^+ + Y.$$

These fragments can in turn break down into smaller ones. Each of the positively charged fragments is recorded by the spectrometer, so that its mass can be determined. The resulting array of detected ions is described as the **fragmentation pattern**.

The abundances of the different positively charged fragments vary widely. For convenience in interpreting the spectrum, the relative abundances of the fragments are plotted against their masses, the most abundant ion being given an arbitrary height of 100 units (the **base peak**); the plot is known as a **stick diagram**. The molecular ion is rarely the most abundant one; indeed, with some compounds, the molecular ion is not detected and the formula weight cannot then be determined (3.5).

The fragmentation pattern of a compound depends on its structure. The presence of particular groupings is associated with specific fragmentation patterns, and so the determination of the fragmentation pattern for a compound of unknown structure can often enable the structure to be deduced.

FIG. 3.5. *The mass spectrum of hexadecane (formula weight 226)*

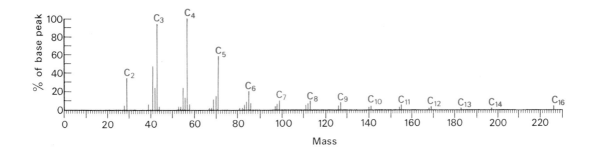

For example, the stick diagram for hexadecane is shown in Fig. 3.5. The molecular ion is at 226, and the principal fragments are at 29, 43, 57, 71, 85, 99, 113, 127. There is a difference of 14 units between these fragments, characteristic of the successive loss of CH_2 groups, for example:

$$CH_3-CH_2-CH_2-\overset{+}{C}H_2 \rightarrow CH_3-CH_2-\overset{+}{C}H_2 + :CH_2$$
$$(57) \qquad\qquad\qquad\qquad (43)$$

$$CH_3-CH_2-\overset{+}{C}H_2 \rightarrow CH_3-\overset{+}{C}H_2 + :CH_2$$
$$(43) \qquad\qquad\qquad (29)$$

If there had been branching in the carbon chain, fragmentation would have occurred preferentially at the branch [as a more stable, secondary carbonium ion is formed (p. 63)]. For example, 2-methylheptane has a molecular ion at 114 and a significant fragment is at 99 (Fig. 3.6).

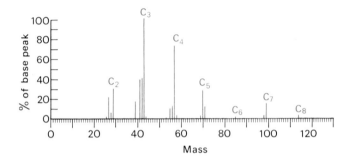

FIG. 3.6. *The mass spectrum of 2-methylheptane (formula weight 114)*

$$\left[CH_3-(CH_2)_4-\overset{\overset{\displaystyle CH_3}{|}}{C}H-CH_3\right]^+ \rightarrow CH_3-(CH_2)_4-\overset{+}{C}H-CH_3 + \cdot CH_3$$
$$(114) \qquad\qquad\qquad\qquad\qquad (99)$$

In contrast, for the isomer, 2-ethylhexane, there is a major fragment at 85:

$$\left[CH_3-(CH_2)_3-\overset{\overset{\displaystyle CH_2CH_3}{|}}{C}H-CH_3\right]^+ \rightarrow CH_3-(CH_2)_3-\overset{+}{C}H-CH_3 + \cdot C_2H_5$$
$$(114) \qquad\qquad\qquad\qquad\qquad (85)$$

Infrared spectroscopy

The bonds in organic compounds undergo various types of vibration. For example, a C—H bond can stretch:

or bend:

DETERMINATION OF THE STRUCTURE OF AN ORGANIC COMPOUND

For a particular bond, only a particular set of vibrational frequencies is possible. Suppose a bond is vibrating at frequency v_1, and its next available higher frequency is v_2; then, if electromagnetic radiation with frequency $(v_2 - v_1)$ is incident on the compound containing this bond, some of the radiation is absorbed and some of the bonds then vibrate at the higher frequency. The appropriate radiation to excite the bonds in organic compounds in this way corresponds to the infrared region of the spectrum, that is, to frequencies in the range $7 \cdot 5 \times 10^6$ to 2×10^7 s^{-1}. (The frequency is usually quoted in reciprocal cm (cm^{-1}); a frequency of v s$^{-1} = 1/\lambda$ cm^{-1} where λ is the wavelength of the radiation and is related to v by the expression $\lambda v = c$, the velocity of light (3×10^{10} cm s^{-1}). For example, a frequency of 6×10^{13} s^{-1} corresponds to 2000 cm^{-1} or a wavelength of 5×10^{-4} cm.)

The excitation frequency for a particular grouping is, to a first approximation, independent of the other groups in the compound. Therefore, determination of the frequencies in the infrared region which are absorbed by a compound gives information about the types of groups which are present.

The apparatus for this purpose—an **infrared spectrophotometer**—consists of a cell (usually made of rock salt as glass absorbs infrared radiation) which contains the organic compound, a source of infrared radiation such as a Nernst glower (a rod made of metal oxides which is heated at 1500°C), a means of separating the infrared beam into its constituent frequencies (a prism or diffraction grating) and a detector for infrared light (a sensitive thermocouple). The infrared light is passed through the separator and a particular emergent frequency then passes through the cell; behind the cell is the detector. By rotating the prism or grating, all the infrared frequencies are passed in turn through the cell and, by coupling the detector to a pen recorder, a record of the behaviour of the compound towards each frequency—the **infrared spectrum**—is obtained.

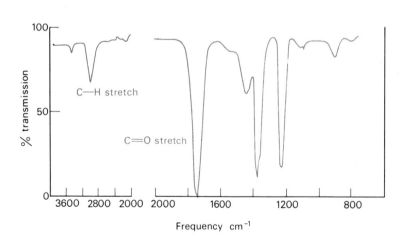

FIG. 3.7. *The infrared spectrum of propanone*

The characteristic frequencies for various types of organic group are in Table 3.2. Thus, a compound which is found to absorb in the region 1700–1750 cm^{-1} can be deduced to contain the carbonyl group, C=O (for example, propanone, the infrared spectrum of which is in Fig. 3.7).

Table 3.2. Characteristic infrared frequencies for organic bonds

BOND	COMPOUND	FREQUENCY (CM^{-1})
C—H	Alkane	2850–2960
C—H	Alkene	3010–3100
C—H	Alkyne	3300
C=C	Alkene	1620–1680
C≡C	Alkyne	2100–2260
C—O	Alcohol, ether	1000–1300
C=O	Aldehyde, ketone, carboxylic acid	1700–1750
O—H	Alcohol, phenol	3590–3650
N—H	Amine	3300–3500

Nuclear magnetic resonance spectroscopy

The atoms of some of the elements have a nuclear structure which causes them to behave like small magnets. In organic compounds, the commonest of these elements is hydrogen; the hydrogen nucleus—the proton—when placed in a magnetic field, aligns itself so that its magnetic moment (sometimes referred to as its 'spin') is either in the same direction as the applied field or in the opposite direction.

The energies of these two possible states differ. The protons whose spins are aligned *with* the field have a lower energy than those aligned *against* the field; the energy difference, ΔE, is proportional to the strength of the field, H:

$$\Delta E = \frac{\gamma h H}{2\pi}$$

where γ is a constant for a particular nucleus and h is Planck's constant.

If electromagnetic radiation is incident on the protons, then, providing that its frequency ν is such that its energy, $h\nu$, is exactly equal to the energy difference, ΔE, for the proton's two orientations, some of the radiation will be absorbed and some of the protons in the orientation of lower energy will 'flip' over to the orientation of higher energy; the total energy of the system remains constant. Thus, the condition for absorption of radiation (the **'resonance' condition**) is

$$\nu = \frac{\gamma H}{2\pi}$$

This phenomenon can be studied with a **nuclear magnetic resonance spectrometer**, shown schematically in Fig. 3.8. A sample of the compound containing hydrogen atoms is placed in a cell between the pole pieces of a magnet and is irradiated; a detector records whether a particular frequency of the radiation is absorbed or not. For the proton, the size of γ is such that, with a magnetic field of about 1300 T, the frequency absorbed is 60 MHz, which is within the range of radio waves; it is usual to keep this frequency constant and to vary the magnetic field (e.g. by varying the current in an electromagnet) while searching for absorption of the radiation.

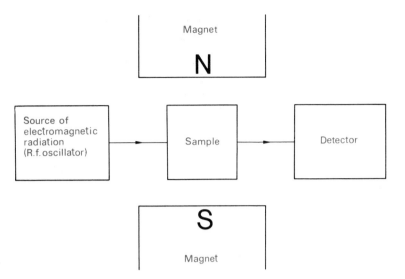

FIG. 3.8. *An NMR spectrometer*

So far, we have considered only the resonance condition for a proton which experiences the applied field. What makes NMR spectroscopy so useful in organic chemistry is the fact that, when a hydrogen atom is bonded to another atom or group, the magnetic field to which the proton is subjected is not the same as the field which is applied. This is because the applied field induces a circulatory motion of the electrons in the bond around the nucleus, and this motion in turn produces a small magnetic field which opposes the applied field. Thus, the magnetic field experienced by the proton is slightly *less* than the applied field. Consequently, for the resonance condition to be fulfilled, the applied field must be increased slightly.

The magnetic field induced at the proton by the bonding electrons decreases as the electrons are moved away from the proton. Now, the average position of the pair of electrons in a bond relative to each nucleus depends on the nature of the atoms concerned (4.8); for example, the electrons are relatively further from the proton in the O—H bond than in the C—H bond. Therefore, the applied magnetic field at which the resonance condition is met is smaller for a proton in an O—H bond than for one in a C—H bond. This is shown in the NMR spectrum of methanol (Fig. 3.9); the area under

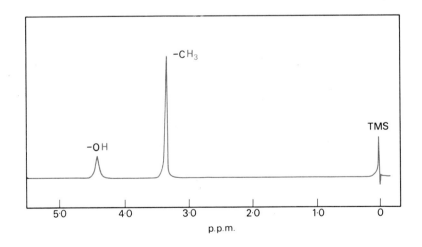

FIG. 3.9. *The NMR spectrum of methanol. Methanol has been dissolved in tetrachloromethane and TMS added as an internal reference standard*

the peak corresponding to resonance for the C—H protons is three-times that for the O—H proton since there are three times as many protons in the former environment.

Since the NMR characteristics of protons in a wide variety of environments are known, the measurement of the NMR spectrum of a compound of unknown structure reveals the bonds present between hydrogen and other groups and also, from their relative peak areas, the relative numbers of these bonds. It is convenient to measure the NMR spectrum relative to a standard; tetramethylsilane (TMS), $(CH_3)_4Si$, is usually chosen since, having only one type of proton, it gives only one absorption peak and so causes minimum interference with the peaks from the unknown compound. Suppose that, for a given frequency, the protons in tetramethylsilane come into resonance at an applied field H_1 and the proton in another bond comes into resonance at an applied field H_2. Then the **chemical shift**, δ, for the latter proton is defined as

$$\delta = \frac{H_1 - H_2}{H_1}$$

To obtain simple numbers for δ, the value is usually multiplied by 10^6 and then expressed as parts per million (p.p.m.). The chemical shifts are independent of the frequency used for the NMR measurements and therefore provide a common scale. Values for protons in environments which occur commonly in organic compounds are in Table 3.3.

Table 3.3. Chemical shifts for protons

TYPE OF PROTON	CHEMICAL SHIFT (P.P.M.)
$R-CH_3$	0·9
$R-CH_2-R$	1·3
$R-CH(R)-R$	2·0
$R-C(=O)-CH_3$	2·3
$R-O-CH_3$	3·8
$R_2C=CH_2$	5·0
$R-O-H$	ca. 5
$R-C(=O)-H$	9·7

DETERMINATION OF THE STRUCTURE OF AN ORGANIC COMPOUND

Still further information can be obtained from the NMR spectrum by increasing the resolution of the spectrometer. Figures 3.10 and 3.11 show the spectrum of ethanol. The former spectrum has three peaks, corresponding to the OH, CH_2 and CH_3 protons (relative areas, 1:2:3). In the latter spectrum, measured with higher resolution, the CH_2 and CH_3 peaks are split into four lines and three lines, respectively.

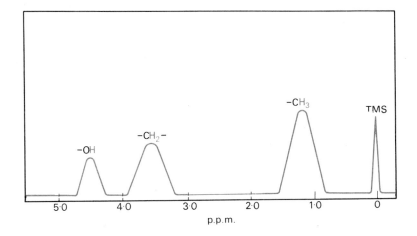

FIG. 3.10. *The NMR spectrum of ethanol at low resolution. Ethanol has been dissolved in tetrachloromethane and TMS added as an internal reference standard*

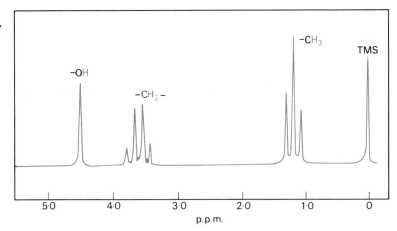

FIG. 3.11. *The NMR spectrum of ethanol at high resolution showing spin-spin coupling. Ethanol has been dissolved in tetrachloromethane and TMS added as an internal reference standard*

The splitting is the result of **spin-spin coupling** and has the following basis. Consider the conditions for resonance of the protons in the methyl group in a compound of the type

$$CH_3-\overset{|}{\underset{|}{CH}}$$

The magnetic moment of the single proton on the adjacent carbon atom can be aligned either with or against the applied magnetic field. For the former orientation, the protons in the methyl group experience a magnetic field slightly *greater* than the applied field, and for the latter orientation they experience a magnetic field correspondingly slightly *less*. Consequently, there are *two* values of the applied field at which the resonance condition for the methyl protons is met; one is slightly less than would be the case in the absence of the single adjacent proton, and the other is correspondingly slightly greater. Thus, the methyl group appears as a **doublet.**

Consider now the condition for resonance of the single proton. The magnetic moments of the protons in the methyl group can all be aligned with the applied field; two can be aligned with, and one against, the applied field; one can be aligned with and two against, the applied field; and all three can be aligned against the applied field. There are therefore four possible magnetic environments for the methyl protons, and so there are four values of the applied field at which resonance occurs; the CH group appears as a **quartet**. However, the resulting four peaks in the spectrum do not have equal areas. If we designate as → or ← a proton whose magnetic moment is aligned respectively with or against the applied field, then we see that there are three times as many ways in which two are aligned with the field and one against it (→→←, →←→, ←→→) or the converse (←←→, ←→←, →←←) as there are ways in which all three are aligned either with (→→→) or against (←←←) the field. Therefore the four peaks have relative areas 1:3:3:1. Likewise, it can be shown that two protons interact to give a 1:2:1 **triplet** pattern. Thus, in the spectrum of ethanol, the quartet corresponding to the CH_2 group results from coupling with the three methyl protons, and the triplet corresponding to the CH_3 group results from coupling with the two CH_2 protons (Fig. 3.12).

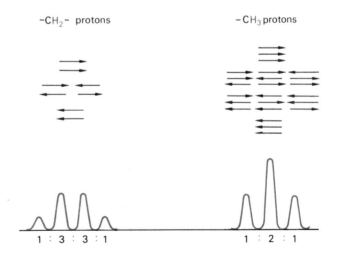

FIG. 3.12. *Possible orientations of nuclear spins of the protons in an ethyl, CH_3—CH_2, group and the expected spin-spin coupling*

In summary, while the chemical shift of a proton gives information about the nature of the group containing that proton, the spin-spin coupling pattern gives information about other protons in adjacent groups.

3.8 Summary

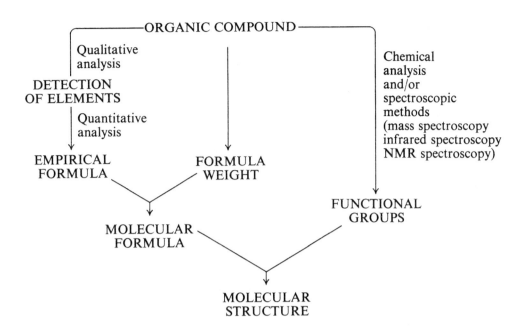

3.9 Practical work: Qualitative analysis of an organic compound

Carbon and hydrogen

Heat a mixture of the compound and an excess of dried copper(II) oxide in a pyrex test-tube. Pass the gases evolved into a solution of lime-water (Fig. 3.13). If the solution turns milky, **carbon** is present in the compound.

If a liquid forms on the side of the tube, test it with anhydrous copper(II) sulphate. If it turns blue, the liquid is probably water and the compound contains **hydrogen**.

FIG. 3.13. *Testing for carbon and hydrogen in an unknown compound*

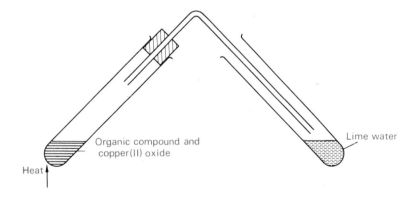

Lassaigne sodium test for nitrogen, halogens and sulphur

Take great care and wear safety glasses.

Place about 0·2 g of the compound in an ignition tube together with a small pellet of sodium. Hold the tube in an almost horizontal position and

DETERMINATION OF THE STRUCTURE OF AN ORGANIC COMPOUND

warm it gently so that the metal becomes molten. Then hold the tube vertically and heat it strongly. Plunge the tube into a small beaker containing about 3 cm³ of water. Boil the mixture for a few minutes, filter it and divide the filtrate into three parts.

Some of the elements in the organic compounds have now been converted into inorganic salts of sodium. If nitrogen is present in the compound, it will have formed cyanide ions, the halogens will have formed halide ions and sulphur will have formed sulphide ions.

Sulphur

To one part of the filtrate, add a drop of an aqueous solution of disodium pentacyanonitrosylferrate(III) ('sodium nitroprusside'). A purple colour indicates the presence of **sulphur**.

Nitrogen

To the second part of the filtrate, add an equal volume of a fresh solution of iron(II) sulphate. The mixture now contains a green precipitate of iron(II) hydroxide. Boil the mixture for a few minutes, then add 2 or 3 drops of iron(III) chloride solution and acidify the mixture with dilute hydrochloric acid. Centrifuge (or filter) the mixture. A residue of Prussian blue indicates the presence of **nitrogen** in the compound:

$$Fe^{2+} \xrightarrow{6CN^-} Fe(CN)_6^{4-} \xrightarrow{Fe^{3+}} Fe[Fe(CN)_6]^-$$

Halogens

(a) *If nitrogen is present.* Cyanide ions interfere with the test for halide ions and must therefore be removed. If it has already been shown that nitrogen is present, acidify the third part of the filtrate with dilute nitric acid and evaporate it to about a quarter of its original volume. Add a few drops of silver nitrate solution.

A white precipitate which is soluble in an excess of dilute ammonia solution indicates the presence of **chlorine** in the compound:

$$Ag^+ \xrightarrow{Cl^-} AgCl\downarrow \xrightarrow{2NH_3} Ag(NH_3)_2^+ + Cl^-$$

A pale yellow precipitate which is soluble in an excess of concentrated ammonia solution indicates the presence of **bromine**.

A yellow precipitate which is insoluble in an excess of concentrated ammonia solution indicates the presence of **iodine**.

(b) *If nitrogen is not present.* Acidify the third portion of the filtrate from the sodium fusion with dilute nitric acid, add a few drops of silver nitrate solution and proceed as in (a).

3.10 Further reading

The Physics of Chemical Structure. R. J. Taylor (1968). Unilever Educational Booklet. Advanced Series No. 2.

3.11 Films

Analysis by Mass. A.E.I.-G.E.C. Film. Guild Sound and Vision Ltd.
Molecular spectroscopy (CHEM-study series). Guild Sound and Vision Ltd.

3.12 Questions

1 A compound **P** contains 85·7 per cent carbon and 14·3 per cent hydrogen. After reaction with trioxygen and then with water, two of the compounds formed, **Q** and **R**, were distilled and purified. Both compounds absorbed infrared radiation at about 1700 cm^{-1}. Mass spectra were obtained for **Q** and **R**:

$$\mathbf{Q}\begin{cases}\text{Mass} & 29 \quad 44 \quad 43 \quad 42 \\ \text{Abundance (\%)} & 100 \quad 89 \quad 50 \quad 15\end{cases}$$

$$\mathbf{R}\begin{cases}\text{Mass} & 43 \quad 58 \quad 15 \\ \text{Abundance (\%)} & 100 \quad 33 \quad 30\end{cases}$$

What are **P**, **Q** and **R**? Describe carefully how you elucidated the structures of **Q** and **R** from the evidence given.

2 Two hydrocarbons, **X** and **Y**, contain 83·3 per cent carbon and 16·7 per cent hydrogen. They have the following mass spectra:

$$\mathbf{X}\begin{cases}\text{Mass} & 43 \quad 42 \quad 41 \quad 57 \quad 29 \quad 72 \\ \text{Abundance (\%)} & 100 \quad 86 \quad 67 \quad 54 \quad 46 \quad 6\end{cases}$$

$$\mathbf{Y}\begin{cases}\text{Mass} & 43 \quad 42 \quad 41 \quad 27 \quad 29 \quad 57 \quad 72 \\ \text{Abundance (\%)} & 100 \quad 58 \quad 41 \quad 35 \quad 24 \quad 13 \quad 9\end{cases}$$

Write down the structural formulae for **X** and **Y**, giving reasons for your choice.

3 Compound **A** contains 22·2 per cent carbon, 4·6 per cent hydrogen and 73·2 per cent bromine. The mass spectrum for compound **A** was:

$$\begin{cases}\text{Mass} & 108 \quad 110 \quad 29 \quad 79 \quad 81 \\ \text{Abundance (\%)} & 100 \quad 97 \quad 51 \quad 4 \quad 4\end{cases}$$

The following NMR spectrum (with TMS as standard) was obtained for **A**:

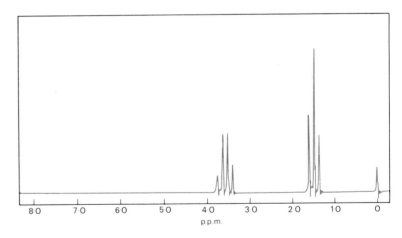

How do you account for the physical evidence given?

4 The chemical shift, δ, for the CH$_2$ protons for the following compounds are:

CH$_3$—CH$_2$—Cl	3·57 p.p.m.
CH$_3$—CH$_2$—Br	3·43 p.p.m.
CH$_3$—CH$_2$—I	3·20 p.p.m.

How do you account for these results?

DETERMINATION OF THE STRUCTURE OF AN ORGANIC COMPOUND

5 A compound, **F**, contains 38·2 per cent carbon, 4·9 per cent hydrogen and 56·9 per cent chlorine. On reduction with hydrogen, it forms **G** which has the following mass spectrum:

Mass	64	28	29	27	66	26	49	51
Abundance (%)	100	90	84	75	32	29	25	8

The NMR spectrum of **G** (with TMS as standard) is shown below. Give the structural formulae for **F** and **G** and account for the physical data given.

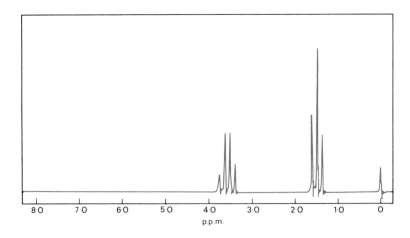

Chapter 4

Bonding in organic compounds

4.1 Introduction

Chemistry is concerned with the making and breaking of bonds between atoms, and the bonds are associated with the electrons that surround the nucleus. Present ideas of the nature of chemical bonding are based on both experiments and mathematical theory. We are only able to describe the conclusions in a non-mathematical way, but this is still a valuable exercise because many experimental observations can be rationalised if we understand the principles of the theory.

4.2 Atomic orbitals

Our understanding of the behaviour of the electrons in atoms and molecules has evolved from quantum theory. Four especially important principles are involved.

First, an electron can only possess particular energies; for example, it might have energy *a*, *b*, *c*, etc., but could not have an energy intermediate between *a* and *b* or between *b* and *c*. That is, it has various **quanta** of energy.

Secondly, the behaviour of electrons can be described by the same equations as describe a wave motion. This is not to say that electrons *are* waves, but only that they behave in the same way as waves. An electron of mass *m* and velocity *v* has a wave-length, λ, defined by de Broglie's equation: $\lambda = h/mv$, where *h* is a constant (Planck's constant).

Thirdly, it is not possible to describe simultaneously both the precise position and the momentum of an electron (Heisenberg's *Uncertainty Principle*); if the momentum is determined with a high degree of precision, then the position is known only approximately, and *vice-versa*.

Consider an electron in an atom. It can possess one of various specific energies, each of which, because of de Broglie's relationship, is associated with a particular wave-length. Thus, the energy of an electron can be defined by a series of equations which describe wave motion—**wave functions**—and it is customary to refer to these wave functions as **orbitals**. Now, because the energy of the electron is defined precisely, it follows from the Uncertainty Principle that its position cannot be known with certainty. It is possible only to say that there is a particular probability of finding the electron at a given point, or to describe a volume of space in which there is, say, 99 per cent probability that the electron will be found.

The fourth principle is that a particular orbital can be associated with a maximum of two electrons (*Pauli's Principle*); they are described as having opposite spins, and are sometimes represented as ↑ and ↓. When two electrons are present, they are described as **paired**; when only one is present it is described as **unpaired**. Now, since systems tend to adopt states in which their potential energy is minimised, it follows from Pauli's principle that the electrons in an atom are associated with the orbitals of lowest energy, each orbital being associated with not more than two electrons.

The hydrogen atom has one electron. It is associated with the orbital of lowest energy, which is spherically symmetrical about the nucleus. A contour diagram is shown in Fig. 4.1; if lines are drawn from the nucleus, the electron is as likely to be found at a particular distance along one line as at the same distance along any other. However, the probability that the

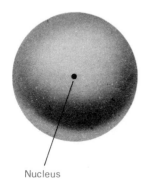

FIG. 4.1. *The 1s atomic orbital*

Nucleus

electron will be found at a particular distance from the nucleus varies with the distance, rising to a maximum at a distance of 0·5 Å (1 Å = 10^{-10} m) and then decreasing again.

It must be noted that the diagram in Fig. 4.1 represents an orbital (wave function) and not the volume in which the electron is most likely to be found. However, the probability of finding the electron at a given point is related to the wave function (it is actually proportional to the square of the wave function), so that the diagram gives an indication of the likely 'distribution' of the electron. It must also be emphasised that, strictly, it is incorrect to describe an electron as 'occupying an orbital'; it is simply a useful short-hand notation to describe it in this way, remembering that an orbital, as a wave function, is not like the orbit which describes the motion of a planet.

The orbital shown in Fig. 4.1 is described as the 1s orbital; the electron in the hydrogen atom which occupies it is unpaired. In the next element, helium, there are two electrons in the 1s orbital, with their spins paired. The next element, lithium, has three electrons. Two are in the 1s orbital and the third is in the next lowest orbital in the energy scale. This is the 2s orbital, and it is also spherically symmetrical about the nucleus (as are all orbitals of s type). However, it is larger than the 1s orbital, and the distance from the nucleus at which the electron has the maximum probability of being found is greater for the 2s than for the 1s orbital. The next element, beryllium, has two electrons in the 1s and two electrons in the 2s orbital, each of which is therefore complete.

Of the five electrons in the next element, boron, four are in the 1s and 2s orbitals and the fifth is in an orbital of different symmetry, namely, a 2p orbital. There are three 2p orbitals which are mutually perpendicular and are described as $2p_x$, $2p_y$ and $2p_z$ orbitals (Fig. 4.2); their shapes are identical and are illustrated in Fig. 4.3. The electron is as likely to be found on one side of the nucleus as on the opposite side, but there is zero probability of its being found at the nucleus (known as the **node**).

The three 2p orbitals have the same energy, so that the choice between the $2p_x$, $2p_y$ and $2p_z$ orbitals for the fifth electron in boron is an arbitrary one. However, with the next element, carbon, a different choice is available: the sixth electron could either go into the same 2p orbital as the fifth or into one of the other two 2p orbitals. The latter is found, and moreover the spins of the two 2p electrons are the same (*Hund's rule*) (Fig. 4.4). With the next element, nitrogen, the third 2p orbital is occupied. In successive ele-

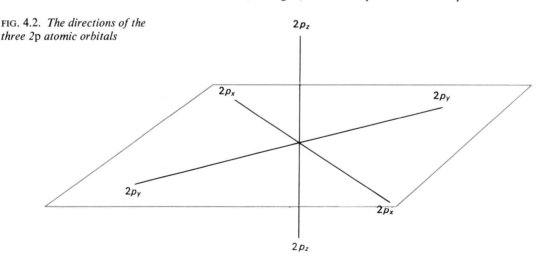

FIG. 4.2. *The directions of the three 2p atomic orbitals*

FIG. 4.3. *The 2p atomic orbitals*

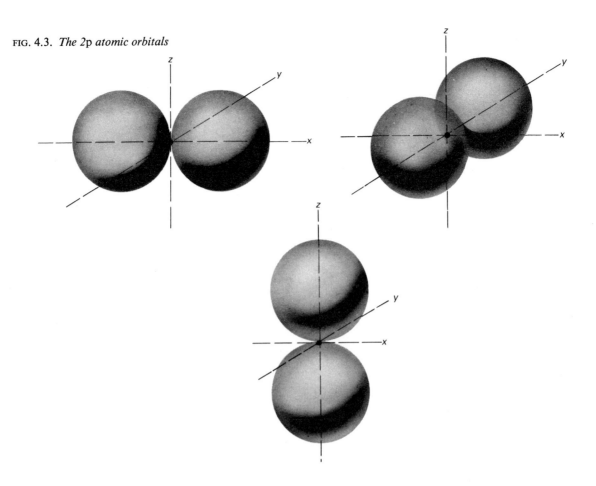

ments, the three 2p orbitals are completed by the introduction of one more electron into each, then the 3s orbital is occupied, after which three 3p orbitals and the 3d orbitals (of different symmetry from the s and p orbitals) are available.

The electronic configuration of an atom is conveniently described as in Table 4.1; the superscript ^{2} indicates that there are two electrons in the orbital concerned, and where there is no superscript the number is one.

Table 4.1. Electronic configurations

ELEMENT	ATOMIC NUMBER Z	ELECTRONIC CONFIGURATION
Hydrogen	1	$1s$
Helium	2	$1s^2$
Lithium	3	$1s^2, 2s$
Beryllium	4	$1s^2, 2s^2$
Boron	5	$1s^2, 2s^2, 2p_x$
Carbon	6	$1s^2, 2s^2, 2p_x, 2p_y$
Nitrogen	7	$1s^2, 2s^2, 2p_x, 2p_y, 2p_z$
Oxygen	8	$1s^2, 2s^2, 2p_x^2, 2p_y, 2p_z$
Fluorine	9	$1s^2, 2s^2, 2p_x^2, 2p_y^2, 2p_z$
Neon	10	$1s^2, 2s^2, 2p_x^2, 2p_y^2, 2p_z^2$

4.3 Chemical bonding

In general, atoms have a tendency to adopt an electronic configuration in which each orbital has its full complement of two electrons. Two atoms will therefore form a bond by the transference of an electron from one to the other or by the sharing of two electrons, one from each atom, so that singly occupied orbitals become filled.

An **electrovalent (ionic) bond** is formed when one atom donates one or more electrons to an atom of a different element, forming charged particles known as ions. This happens, for example, with lithium and fluorine: the

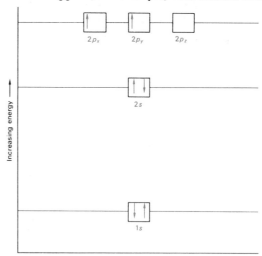

FIG. 4.4. *The electronic configuration of carbon*

lithium atom donates its one $2s$ electron to the fluorine atom, each ion then having an electronic configuration of filled orbitals. The ions, being oppositely charged, are held together by electrostatic forces. These forces act equally in all directions, so that the electrovalent bond is **non-directional**; for example, in the crystal of lithium fluoride, each lithium ion (Li^+) is equidistant from six fluoride ions (F^-), and *vice-versa* (Fig. 4.5).

A **covalent bond** is formed by the sharing of two electrons, one being contributed by each atom.

Suppose that two hydrogen atoms approach each other. The $1s$ atomic orbitals of each atom can overlap, with the result that two molecular orbitals are formed; these are similar to atomic orbitals except that they are associated with two nuclei instead of one. One of the molecular orbitals is of lower energy than the atomic orbitals and is described as a **bonding molecular orbital**; the other, of higher energy, is an **antibonding molecular orbital**. The two $1s$ electrons from the hydrogen atoms occupy the lower energy, bonding orbital, with their spins paired, and the antibonding orbital remains empty. Consequently, the energy of the system is lower than that of the separate atoms, and the molecule is more stable than the two atoms. The formation of the bonding molecular orbital is shown in Fig. 4.6.

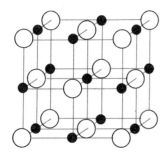

FIG. 4.5. *The lithium fluoride lattice*

FIG. 4.6. *The formation of the s-s molecular orbital from two s atomic orbitals*

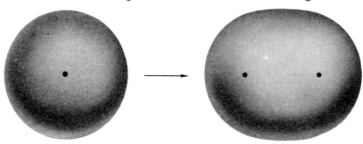

s atomic orbital   *s* atomic orbital   molecular orbital

Note that the two hydrogen nuclei remain separated in the molecule. This is because, as the nuclei approach each other, on the one hand, the degree of overlap between the atomic orbitals increases and so therefore does the effectiveness of the bonding, but on the other hand the repulsive force between the nuclei themselves increases. Consequently, there is an optimum distance of separation at which the total energy reaches a minimum; this is the most stable situation. Therefore, each covalent bond is characterised by a particular **bond length** and has a particular **bond energy**. Bond lengths can be measured by X-ray crystallography and by microwave spectroscopy; they are mostly 1–2 Å and some typical values are in Table 4.2. Bond energies can be measured by calorimetry and spectroscopic methods; typical values are in Table 4.3.

Table 4.2. Some bond lengths

	Å
C—H	1·09
C—C	1·54
C=C	1·34
C≡C	1·19
C—O	1·43
C=O	1·22
C—N	1·47
C=N	1·30
C≡N	1·16
O—H	0·96

Table 4.3. Some bond energies

	kJ/mol	kcal/mol		kJ/mol	kcal/mol
C—C	345	83	C—F	450	108
C=C	610	146	C—Cl	339	81
C≡C	835	200	C—Br	284	68
C—H	413	99	C—I	213	51
C—N	304	73	N—H	391	93
C≡N	889	213	O—H	490	117
C—O	358	86	N—N	163	39
C=O	749	179	O—O	146	35

Consider next the approach of two helium atoms to each other. As with hydrogen atoms, the 1s orbitals of each overlap to form two molecular orbitals, one bonding and one antibonding. In this case, however, there are four 1s electrons—two from each atom—to occupy the molecular orbitals; consequently, each of these contains a pair of electrons, and the effectiveness of the pair in the bonding orbital in holding the atoms together is nullified by the effect of the pair in the antibonding orbital, so that no bond is formed between two helium atoms.

From this example, it can be appreciated that, for the overlap of two atomic orbitals to result in the formation of a bond, each should contain only one electron; then the bonding molecular orbital will be filled and the antibonding one empty. Therefore we can expect the number of bonds which an element forms to be equal to the number of unpaired electrons in its atomic structure. For example, the hydrogen atom has one unpaired electron and forms one bond, as in H_2; the nitrogen atom has three unpaired electrons and forms three bonds, as in NH_3. However, a problem is posed by the carbon atom, which has two unpaired electrons and yet forms four bonds. The explanation is that one of the 2s electrons in carbon is transferred to the 2p orbital, thereby yielding four unpaired electrons so that four bonds can be formed:

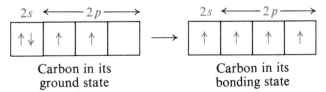

Although the transfer of the electron from the 2s to the 2p orbital (known as **promotion**) requires energy, it is more than compensated by the release

BONDING IN ORGANIC COMPOUNDS

of energy which accompanies the formation of four bonds compared with two. This can be represented as in Fig. 4.7. Although the energy change in forming 2 C—H bonds from a carbon atom in the ground state is not known, it is not likely to be more than half the value for the formation of 4 C—H bonds from a carbon atom in its bonding state.

FIG. 4.7. *An energy diagram showing that carbon prefers to form four bonds*

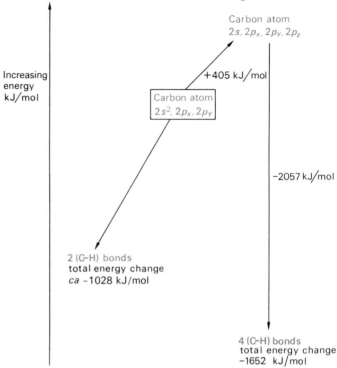

4.4 Saturated carbon compounds

In a simple molecule like methane it would appear at first sight that there would be two types of bonds: one type would be formed by overlap of the singly occupied $2s$ orbital of carbon with the singly occupied $1s$ orbital of a hydrogen atom, and the other would be formed by overlap of each of the three singly occupied $2p$ orbitals of carbon with the $1s$ orbital of each of three hydrogen atoms. However, methane is known to be a symmetrical molecule, containing four C—H bonds of equal length and at equal angles to each other; if the carbon atom were placed at the centre of a regular tetrahedron, the four hydrogen atoms would be at the four corners. This can be understood by considering the four unfilled carbon orbitals to be

FIG. 4.8. *The* sp^3 *hybridised atomic orbital.* (a) *Cross-section,* (b) *Shape*

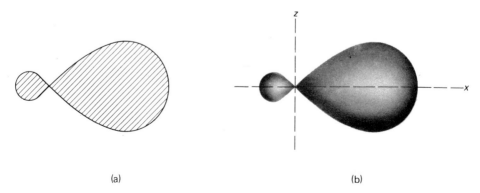

(a)   (b)

52

BONDING IN ORGANIC COMPOUNDS

'mixed' so that each has $1/4$ s character and $3/4$ p character; the process of mixing is described as **hybridisation** and the resulting orbitals as sp^3 orbitals. These orbitals have a different shape compared with both s and p orbitals. Like p orbitals, they are directional, but unlike p orbitals, one lobe is larger than the other (Fig. 4.8). It is the larger lobe which overlaps with another atomic orbital, such as the $1s$ orbital of a hydrogen atom, to form molecular orbitals. For example, the formation of one C—H bond in methane can be represented as in Fig. 4.9. The process of hybridisation occurs because

FIG. 4.9. *The formation of a C—H bond from an s atomic orbital (hydrogen) and an sp^3 hybridised atomic orbital (carbon). Cross-section of the atomic and molecular orbitals*

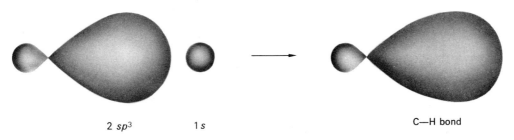

2 sp^3 1s C—H bond

the four directional orbitals which result allow greater overlap with the atomic orbitals of other atoms than if carbon formed its bonds with one $2s$ and three $2p$ atomic orbitals; in this way, the total bonding is increased and therefore the potential energy of the system is decreased.

4.5 Unsaturated carbon compounds

There are two other ways in which the atomic orbitals of carbon can be hybridised. First, the one singly occupied $2s$ orbital and two of the three singly occupied $2p$ orbitals can be hybridised to give three sp^2 orbitals, leaving the remaining $2p$ orbital intact. The three sp^2 orbitals are arranged symmetrically in a plane, making angles of 120° with each other; the unaltered $2p$ orbital is perpendicular to this plane (Fig. 4.10).

FIG. 4.10. *A carbon atom with three sp^2 hybridised atomic orbitals and one p_z orbital. The three sp^2 orbitals, represented by lines from the carbon nucleus, are in a plane at angles of 120° to one another*

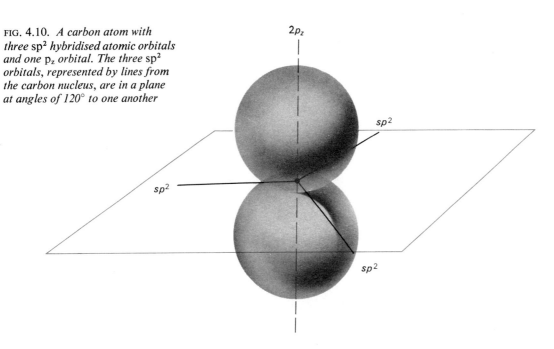

53

BONDING IN ORGANIC COMPOUNDS

This is the state of hybridisation adopted by the carbon atoms in ethene, C_2H_4. Each carbon atom forms one bond with the other carbon atom and two to hydrogen atoms by means of its three sp^2 orbitals. This leaves a singly occupied p orbital on each carbon atom (Fig. 4.11), and these two p orbitals can overlap laterally with each other to form a bonding molecular orbital between the carbon atoms (Fig. 4.12).

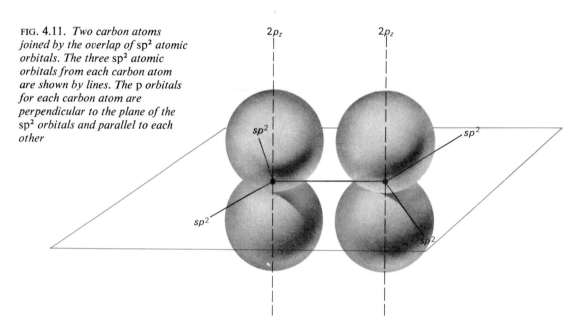

FIG. 4.11. *Two carbon atoms joined by the overlap of* sp^2 *atomic orbitals. The three* sp^2 *atomic orbitals from each carbon atom are shown by lines. The p orbitals for each carbon atom are perpendicular to the plane of the* sp^2 *orbitals and parallel to each other*

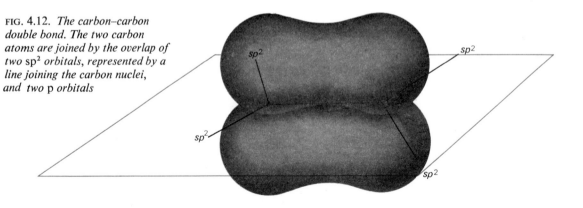

FIG. 4.12. *The carbon–carbon double bond. The two carbon atoms are joined by the overlap of two* sp^2 *orbitals, represented by a line joining the carbon nuclei, and two p orbitals*

Two properties of the carbon–carbon double bond in ethene and its derivatives can be understood immediately in the light of this description. First, the bond is shorter than a carbon–carbon single bond (Table 4.2). This is because relatively more bonding is gained by the overlap of *two* sets of orbitals as the nuclei approach each other in forming C=C than in the overlap of one set of orbitals in formation of C—C, while the repulsive forces between the nuclei are the same in each case. Secondly, the carbon–carbon double bond is resistant to rotation, because this reduces the extent of overlap of the *p* orbitals; if the bond was twisted by 90°, the *p*-orbital overlap would be reduced to zero and the bond energy would be that of a single bond, so that about 265 kJ per mol would be required. It is because of this that **geometrical isomerism** (15.4) occurs; for example, the compounds

$$\begin{array}{cc} CH_3 \quad CH_3 \\ \backslash / \\ C=C \\ / \backslash \\ H H \end{array} \qquad \begin{array}{cc} CH_3 \quad H \\ \backslash / \\ C=C \\ / \backslash \\ H CH_3 \end{array}$$

do not interconvert because the process of interconversion would necessitate destruction of the bond formed by *p*-orbital overlap.

The final type of hybridisation for carbon involves the mixing of one 2*s* orbital with one 2*p* orbital; two *sp*-hybridised orbitals are formed, the larger lobes of which point in diametrically opposite directions from the nucleus. The remaining unpaired electrons are in the original 2*p* orbitals, which are perpendicular to each other and to the direction of the *sp* orbitals.

Ethyne, C_2H_2, is the simplest compound formed from *sp*-hybridised orbitals. Each carbon atom bonds with the other carbon atom and with one hydrogen atom by use of its *sp* orbitals, the two bonds being at an angle of 180°, and forms two further bonds with the other carbon atom by *p*-orbital overlap (Fig. 4.13). Just as the C=C bond is shorter than C—C, so C≡C is shorter than C=C (Table 4.2).

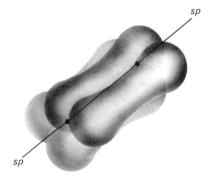

FIG. 4.13. *The carbon–carbon triple bond. The two carbon atoms are joined by the overlap of two sp orbitals, represented by a line joining the carbon nuclei, and four p orbitals, two from each carbon atom*

Bonds between two atoms which are symmetrical about the axis joining the nuclei of the atoms are described as **sigma bonds (σ-bonds)**; examples are the bonds formed by *s* atomic orbitals (e.g. in H_2) and by sp^3, sp^2 and *sp* hybridised orbitals (e.g. in methane, ethene and ethyne), summarised in Fig. 4.14. Bonds formed by lateral *p* orbital overlap are not symmetrical about the axis joining the nuclei; they are described as **pi-bonds (π-bonds)**.

FIG. 4.14. *The three types of hybridised orbitals for the carbon atom. (a) Tetrahedral* sp³ *hybrid orbital; (b) Coplanar* sp² *hybrid orbital; (c) Collinear* sp *hybrid orbital*

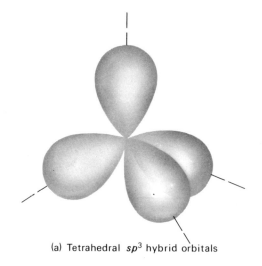

(a) Tetrahedral *sp³* hybrid orbitals

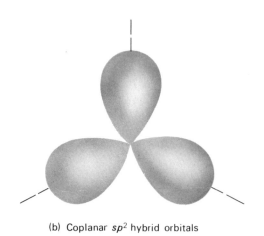

(b) Coplanar *sp²* hybrid orbitals

(c) Collinear *sp* hybrid orbitals

4.6 Delocalised bonds

So far, all the molecular orbitals we have described have been constituted from the overlap of two atomic orbitals and are centred around two nuclei. The bonds to which these molecular orbitals correspond are known as **localised** bonds.

There are also **delocalised** bonds, in which pairs of electrons are associated with bonding molecular orbitals which extend over three or more atoms. They occur less commonly than localised bonds, but they confer special properties on the compounds containing them.

Benzene, C_6H_6, contains delocalised bonds. The six carbon atoms are arranged in the form of a regular hexagon and each forms three localised bonds with sp^2-hybridised orbitals, two to other carbon atoms and one to a hydrogen atom. This leaves a singly occupied *p* orbital on each carbon atom, and each *p* orbital overlaps with the *p* orbital on either side of it (Fig. 4.15). The overlap of these six atomic *p* orbitals gives rise to six molecular π orbitals, of which three are of bonding and three are of antibonding type. The six electrons from the six atomic *p* orbitals then occupy the three bonding molecular π orbitals as three pairs; thus, there are three delocalised π bonds. The shape of one of these three molecular orbitals is shown in Fig. 4.15; the overall effect of filling each with a pair of electrons is to give the same distribution of electron density between each pair of carbon atoms.

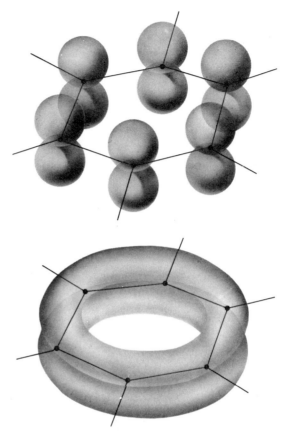

FIG. 4.15. *Structure of benzene. Each carbon atom forms three σ bonds, two to other carbon atoms and one to a hydrogen atom (represented by lines). In addition the six atomic p orbitals shown in the upper diagram overlap laterally to give delocalised π molecular orbitals, one of which is shown in the lower diagram*

Delocalised π orbitals are much larger than the localised π orbitals in, say, ethene. The electrons in them can be found in a greater volume than for localised orbitals; in a sense, they have a greater freedom of movement, and it is a general principle that, the greater the freedom of movement of an electron, the lower is its energy. Hence, a special property of compounds containing delocalised bonds is that they are more stable than similar compounds which only contain localised bonds. The experimental evidence that benzene is stabilised in this way, and fuller details of its structure, are discussed later (8.2).

There is an alternative method of representing delocalised bonds. Thus, if we were to represent benzene as it would imply that three localised π-bonds are present. Instead, a widely used convention is to draw *two* structures:

The double-headed arrow ↔ is taken to mean that the actual structure lies between the two representations; in other words, each C—C bond is neither a simple single nor a simple double bond but is of intermediate type. This method is sometimes described as **mesomerism** ('in-betweenness'), and benzene is described as a **resonance hybrid** of the two structures. It is important to realise that benzene does not oscillate between these structures; it exists in only one form, in which the six C—C bonds are of identical type.

In this text, delocalisation is sometimes described in terms of molecular orbitals, but in other cases it is more convenient to employ the resonance-hybrid method. In equations, benzene is represented for simplicity as:

4.7 The inductive effect

The two electrons in a bond between two atoms are attracted by the two nuclei. In a diatomic molecule in which the two atoms are the same, such as H_2 or Cl_2, attraction by one nucleus is as strong as attraction by the other and so it is as likely that the electrons will be found a particular distance from one nucleus as from the other; the bond is symmetrical in the sense that the centres of gravity of the negative and positive charges coincide. However, in a bond between unlike atoms, the nucleus of one atom exerts a stronger attractive force than that of the other; the electrons are more likely to be found nearer the former nucleus and the centres of gravity of the negative and positive charges do not coincide. For example, in hydrogen chloride the centre of gravity of the negative charges is nearer to chlorine than the centre of gravity of the positive charges, an effect which is represented as H→Cl where → represents the tendency of the electrons to lie nearer chlorine.

The magnitude of the opposite charges multiplied by the separation of their centres of gravity is defined as the **dipole moment** of the compound. This can be measured by finding how the compound affects the capacitance of a condenser; the greater the dipole moment, the more strongly the molecules tend to align themselves between the plates of the condenser (the positive end of the molecule being nearer to the negative plate, and *vice-versa*) and the lower is the capacitance.

The bonds in organic molecules are also polarised in this way by substituents such as the halogens. For example, chloromethane has a dipole moment of which the negative end is the chlorine atom:

BONDING IN ORGANIC COMPOUNDS

It is useful to have a scale for describing the effects of different atoms to attract bonding electrons. For this purpose, hydrogen is chosen as a reference point; an atom or group which attracts the bonding electrons more strongly than hydrogen is described as having an electron-withdrawing **inductive effect**, symbolised as $-I$. For example, since the electrons in the C—Cl bond of CH_3—Cl lie relatively further from carbon than those in the C—H bond in CH_3—H, chlorine is a $-I$ substituent. On the other hand, alkyl groups are electron-releasing ($+I$) compared with hydrogen.

4.8 Organic acids and bases

Some types of organic grouping are acidic and others are basic. The commonest acid group is the carboxylic acid group:

$$-\underset{\underset{O}{\|}}{C}-O-H$$

Carboxylic acids, R—CO_2H, are weak acids compared with the mineral acids such as H_2SO_4; that is, the equilibrium

$$R-\underset{\underset{O}{\|}}{C}-O-H \rightleftharpoons R-\underset{\underset{O}{\|}}{C}-O^- + H^+$$

lies on the left-hand side. For example, the equilibrium constant (described in this case as the acid dissociation constant, K_a) is only $1{\cdot}7 \times 10^{-5}$ mol per dm^3 for ethanoic acid, CH_3CO_2H, i.e.

$$K_a = \frac{[CH_3CO_2^-][H^+]}{[CH_3CO_2H]} = 1{\cdot}7 \times 10^{-5}$$

(the units are usually omitted for brevity). Nevertheless, the compounds are acidic enough to form salts with alkalis, and to turn blue litmus paper red.

Alcohols, ROH, also dissociate:

$$R-O-H \rightleftharpoons R-O^- + H^+$$

However, the dissociation constants are very small and the compounds exhibit few of the properties usually associated with acids; for example, although they form salts and liberate hydrogen when treated with sodium,

$$ROH + Na \rightarrow RO^-Na^+ + \tfrac{1}{2}H_2$$

they do not turn blue litmus paper red.

The reason for the greater acid strength of carboxylic acids than alcohols can be understood by considering the bonding in the corresponding anions, RCO_2^- and RO^-. In the former, the atomic p orbitals on the carbon and two oxygen atoms of the carboxylate group interact to give three delocalised π molecular orbitals; the four p electrons occupy the two lowest energy orbitals of these three, of which one is bonding and one is non-bonding. The shape of the bonding π molecular orbital is shown in Fig. 4.16.

FIG. 4.16. p-*Orbital overlap in a carboxylate ion; the lower diagram shows the delocalised bonding π orbital*

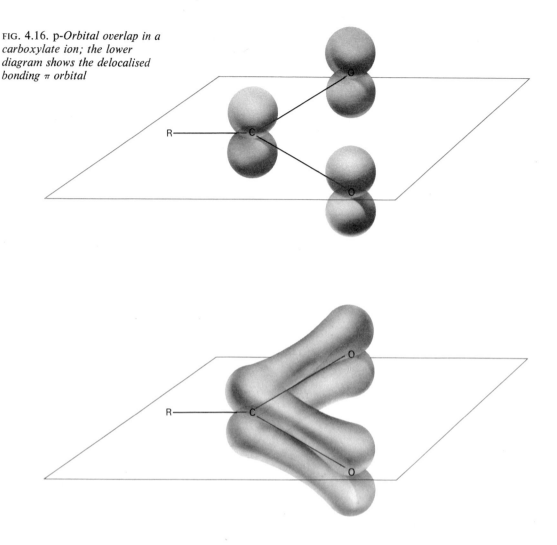

The anion is thus symmetrical, with the negative charge shared equally by the two oxygen atoms. This can perhaps be seen more easily by representing the anion in terms of the theory of mesomerism; two structures can be drawn:

$$R-C\underset{O}{\overset{O^-}{\lessgtr}} \qquad R-C\underset{O^-}{\overset{O}{\lessgtr}}$$

and the actual structure is intermediate between the two, with the equivalent of half a negative charge on each oxygen atom. For the ion RO^-, on the other hand, the charge is localised on one oxygen atom. Now, it is a general principle that the potential of a charged system decreases as the volume associated with the charge becomes larger (in electrostatics, the potential of a charged sphere is inversely proportional to the volume of the sphere). Therefore, the carboxylate ion, RCO_2^-, in which the charge is delocalised, is relatively more stable than the ion RO^- and is formed the more readily.

Another acid grouping is

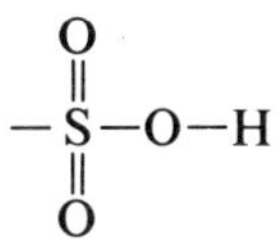

$$-\underset{\underset{O}{\|}}{\overset{\overset{O}{\|}}{S}}-O-H$$

in organic sulphonic acids, R—SO_3H. These acids are much stronger than the carboxylic acids and are as strong as mineral acids like H_2SO_4. This again can be understood in terms of charge delocalisation; in the sulphonate anion, *three* oxygen atoms share the negative charge, as represented by the three structures

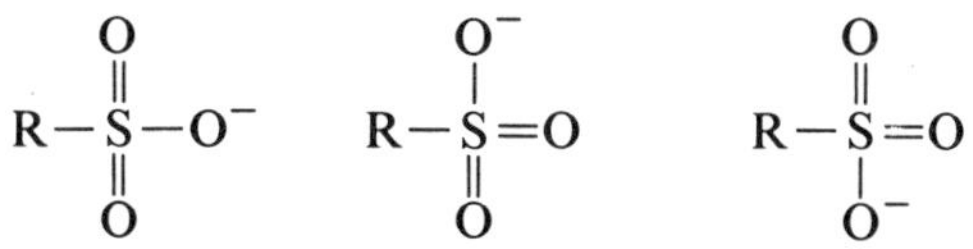

$$R-\underset{\underset{O}{\|}}{\overset{\overset{O}{\|}}{S}}-O^- \qquad R-\underset{\underset{O}{\|}}{\overset{\overset{O^-}{|}}{S}}=O \qquad R-\underset{\underset{O^-}{|}}{\overset{\overset{O}{\|}}{S}}=O$$

In addition to the effect of charge-sharing described above, the strengths of acids are dependent upon the presence of substituents in the rest of the molecule.

For example, chloroethanoic acid ($ClCH_2$—CO_2H) is a stronger acid than ethanoic acid (Table 4.4). This is because chlorine has a $-I$ effect; that is, the chlorine nucleus, by its strong attraction for the electrons in the C—Cl bond, enables the negative charge in the ion Cl—CH_2—CO_2^- to be spread through the molecule more effectively than in the ethanoate ion. Two chlorine substituents are more effective than one, and three are more effective than two.

The fluorine atom, having a stronger attraction for electrons, causes fluoroethanoic acid to be a stronger acid than chloroethanoic acid. Bromoethanoic and iodoethanoic acids are, as expected, weaker acids (Table 4.4).

Table 4.4. **Dissociation constants of some halogeno-substituted carboxylic acids**

ACID	K_a AT 25°C
CH_3—CO_2H	1.7×10^{-5}
$ClCH_2$—CO_2H	1.4×10^{-3}
Cl_2CH—CO_2H	5.1×10^{-2}
Cl_3C—CO_2H	2.2×10^{-1}
FCH_2—CO_2H	2.2×10^{-3}
$ClCH_2$—CO_2H	1.4×10^{-3}
$BrCH_2$—CO_2H	1.3×10^{-3}
ICH_2—CO_2H	6.9×10^{-4}

From such evidence, it can be shown that the following groups exert a $-I$ effect with respect to hydrogen (in descending order of power):

NO_2, CN, F, Cl, Br, I, OH, NH_2, C_6H_5

Groups which are electron-releasing with respect to hydrogen ($+I$ groups) reduce acid dissociation constants (Table 4.5). For example,

Table 4.5. Dissociation constants of some carboxylic acids

ACID	K_a AT 25°C
H—CO_2H	1.7×10^{-4}
CH_3—CO_2H	1.7×10^{-5}
CH_3CH_2—CO_2H	1.3×10^{-5}
C_6H_5—CO_2H	6.3×10^{-5}
$C_6H_5CH_2$—CO_2H	4.9×10^{-5}

ethanoic acid is weaker than methanoic acid (H—CO_2H), and propanoic acid (CH_3—CH_2—CO_2H) is weaker still. Groups that exert a $+I$ effect are, in descending order of power:

$$(CH_3)_3C, \quad (CH_3)_2CH, \quad C_2H_5, \quad CH_3$$

The amino group, —NH_2, is the most commonly found basic group in organic chemistry. It is basic because it can form a bond with a proton by means of the unshared pair of electrons on the nitrogen atom:

$$R-\ddot{N}H_2 + H_2O \rightleftharpoons R-\overset{+}{N}H_3 + OH^-$$

The equilibrium constant for this reaction is given by:

$$K = \frac{[R-\overset{+}{N}H_3][OH^-]}{[R-NH_2][H_2O]}$$

However, usually K is determined for a solution in water, which is in considerable excess over the other components, so that [H_2O] is approximately constant. It is therefore customary to describe the base strength of an amine by K_b, given by:

$$K_b = \frac{[R-\overset{+}{N}H_3][OH^-]}{[R-NH_2]}$$

(the units are again omitted, p. 59).

Amines are weak bases; that is, K_b is small. For example, methylamine (CH_3—NH_2) has K_b (at 25°C) = 4.4×10^{-4}. Thus, they are far weaker bases than inorganic alkalis such as sodium hydroxide, although they are strong enough to form salts with acids.

4.9 Unstable intermediates in organic chemistry

As we have seen, the structures of organic compounds are characterised by the formation of four bonds by each carbon atom. However, there are also species in which the carbon atoms form fewer bonds, and although they are too unstable to exist as compounds which can be isolated, they are nevertheless important in occurring as short-lived intermediates in organic reactions. Two of the more important types with which this book will be concerned are **free radicals**, in which one of the carbon atoms has three bonds and one unpaired electron, and **carbonium ions**, in which one of the carbon atoms has three bonds and possesses a positive charge.

The simplest free radical is methyl, ·CH₃ (the dot signifies an unpaired electron). The radical has a planar structure in which the carbon atom forms three bonds to hydrogen atoms by sp^2-hybridised orbitals and possesses a p orbital with one electron in it (Fig. 4.17).

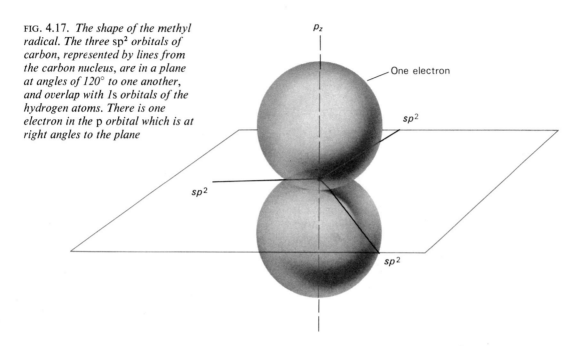

FIG. 4.17. *The shape of the methyl radical. The three* sp^2 *orbitals of carbon, represented by lines from the carbon nucleus, are in a plane at angles of 120° to one another, and overlap with 1s orbitals of the hydrogen atoms. There is one electron in the p orbital which is at right angles to the plane*

Free radicals are particularly important as intermediates in the reactions of alkanes. For example, in the chlorination of methane (5.4), the methyl radical is formed from methane and a chlorine atom:

$$CH_4 + Cl\cdot \rightarrow \cdot CH_3 + HCl$$

and then reacts with a molecule of chlorine:

$$\cdot CH_3 + Cl_2 \rightarrow CH_3Cl + Cl\cdot$$

A carbonium ion has the same structure as a radical except that the p orbital is empty; thus, the carbon atom is associated with a positive charge. Carbonium ions are particularly important as intermediates in the reactions of alkenes, alkyl halides and alcohols. For example, in the addition of hydrogen chloride to ethene, a carbonium ion is formed first:

$$CH_2=CH_2 + HCl \rightarrow CH_3-\overset{+}{C}H_2 + Cl^-$$

and then rapidly reacts with a chloride ion:

$$CH_3-\overset{+}{C}H_2 + Cl^- \rightarrow CH_3-CH_2Cl$$

The substituents in a compound are of importance in determining the ease of formation of a carbonium ion. For example, of the two carbonium ions

$$CH_3-\overset{+}{C}H-CH_3 \qquad CH_3-CH_2-\overset{+}{C}H_2$$

the former is attached to two alkyl substituents which are electron-releasing relative to hydrogen and the latter to only one. Consequently, the concentration of the positive charge is reduced to a greater extent in the former case and the former ion is therefore the more stable. This is important in governing the direction of addition of a proton to propene, CH_3—CH=CH_2 (6.4).

4.10 Film

Chemical Bonding (CHEM-study series). Guild Sound and Vision Ltd.

4.11 Questions

1 Describe the bonding in methane and ethene in terms of orbitals.
    Comment on the following boiling points:
        $CH_3OC_2H_5$, 11°C; C_2H_5SH, 36°C; $(CH_3)_2CO$, 56°C;
        $CH_3CH_2CH_2OH$, 97°C.
(C(T))

2 Discuss or explain the following.
    (i) Ethanol (ethyl alcohol) boils at 78·3°C; dimethyl ether boils at −24·9°C.
    (ii) The pH of a solution containing ethanoic acid and sodium ethanoate is changed only slightly by the addition of a small amount of hydrochloric acid.
    (iii) The effect of a magnetic field on the emissions from radio-active nuclei.
    (iv) The *H—C—H* angle in methane is 109°28′, whilst the *H—C—H* angle in ethene is 120°.
(C(N and T))

3 Explain the following.
    (i) The carbon–carbon bond lengths in benzene are all identical.
    (ii) The carbon–oxygen bond length in dimethyl ether (1·43 Å) is longer than in propanone (1·24 Å).
    (iii) Fluoroethanoic acid is a stronger acid than ethanoic acid.

4 Explain what you understand by the following terms; bond energy, hybridisation, mesomerism, delocalised bonds.

5 What do you understand by the term heat of hydrogenation?
    The heats of hydrogenation of ethene and benzene are −120 and −210 kJ per mol, respectively. Comment.

Chapter 5

Alkanes

General formula

$$C_nH_{2n+2}$$

5.1 Nomenclature

The first four members of the series retain their original names. Alkanes with a straight chain containing five or more carbon atoms are named by combining a prefix derived from the Greek for the length of the chain with the suffix **-ane**.

Table 5.1. Physical properties of some alkanes

NAME	FORMULA	B.P. (°C)	M.P. (°C)	DENSITY g/cm³ (20°C)
Methane	CH_4	−162	−183	gas
Ethane	C_2H_6	−89	−172	gas
Propane	C_3H_8	−42	−188	gas
Butane	C_4H_{10}	−0.5	−135	gas
Pentane	C_5H_{12}	36	−130	0.626
Hexane	C_6H_{14}	69	−95	0.659
Heptane	C_7H_{16}	98	−91	0.684
Octane	C_8H_{18}	126	−57	0.703
Nonane	C_9H_{20}	151	−54	0.718
Decane	$C_{10}H_{22}$	174	−30	0.730
Undecane	$C_{11}H_{24}$	196	−26	0.740
Dodecane	$C_{12}H_{26}$	216	−10	0.749
Triacontane	$C_{30}H_{62}$	343	37	solid

5.2 Physical properties of alkanes

One of the characteristics of a homologous series is that successive members show a gradation of physical properties. This is illustrated for the alkanes in Table 5.1. As the series is ascended, the boiling point and density increase.

The boiling point depends on the attractive forces between the molecules of the liquid; the stronger these are, the more energy is needed to separate the molecules in order to convert the liquid into the vapour, and so the higher is the boiling point. The forces increase with the formula weight of the compound, so that when a **methylene group** (CH_2) is introduced into the alkane chain, the boiling point rises. However, as the molecular weight becomes larger, its percentage increase on introduction of a methylene group becomes smaller, so that the difference in boiling points between consecutive members of the homologous series decreases as the series is ascended, giving the smooth curve in Fig. 5.1 when the boiling point is plotted against the number of carbon atoms in the molecule.

The melting points of the alkanes do not fall on a smooth curve. As Fig. 5.2 shows, two curves can be drawn, one for the alkanes with an even number of carbon atoms and a lower one for those with an odd number. This is because, in the crystalline state, the molecules adopt a highly ordered arrangement in which the carbon chains form a zig-zag pattern. For the even

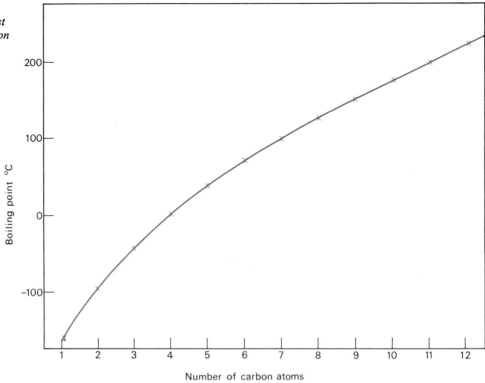

FIG. 5.1. *Plot of boiling point against the number of carbon atoms in straight-chain alkanes*

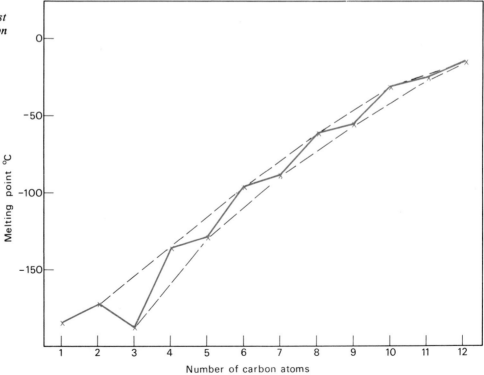

FIG. 5.2. *Plot of melting point against the number of carbon atoms in straight-chain alkanes*

ALKANES

members, different chains pack closer together than for the odd-numbered members, so that the attractive forces are larger for the members of the former group than for members of the latter group of similar size. It follows, in turn, that relatively more energy must be applied to separate the molecules with even numbers of carbon atoms and enable them to adopt the more random arrangement of the liquid state than to separate the molecules with odd numbers of carbon atoms.

Branched-chain alkanes have lower boiling points than their straight-chain isomers, and as branching increases the boiling point decreases still further. The examples in Table 5.2 illustrate the trend.

Table 5.2. Boiling points of the isomeric pentanes

NAME	FORMULA	B.P. (°C)
Pentane	$CH_3CH_2CH_2CH_2CH_3$	36
2-Methylbutane	$CH_3CH(CH_3)CH_2CH_3$	28
2,2-Dimethylpropane	$CH_3-C(CH_3)_2-CH_3$	10

The explanation is that increased branching gives the molecule a more nearly spherical shape and reduces the extent of contact between neighbouring molecules; consequently, the attractive forces are reduced and the boiling point decreases.

There is no regularity in the change of melting point with increasing branching in an alkane. This is because, whereas the linear alkanes all adopt a simple, close-packed structure in a crystal, the type of packing of branched members (and hence the attractive forces between the molecules) is not a simple function of structure.

5.3 Occurrence of alkanes

The principal sources of alkanes are **natural gas** and **petroleum**. Natural gas contains mainly methane, with smaller amounts of the other gaseous alkanes such as ethane, propane and butane. Petroleum contains a wide range of alkanes, from the low molecular weight gases to the high molecular weight solids.

The uses of petroleum, both as a fuel and as a source of chemicals, are of outstanding importance. They are mentioned throughout this book, and are brought together in Chapter 20. The formation of deposits of natural gas and petroleum is discussed in Chapter 19.

5.4 Methane

Structural formula

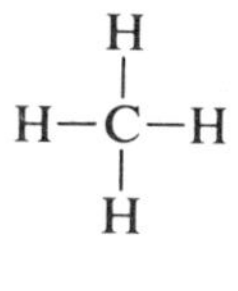

$$\begin{array}{c} H \\ | \\ H-C-H \\ | \\ H \end{array}$$

Plate 5.1. An aerial photograph of the drilling platform 'Staflo' in the North Sea

Occurrence

Natural gas is by far the most important source of methane (19.2). The gas is brought to Britain both by direct pipe-lines from the deposits below the North Sea and by specially constructed tankers which contain gas from the Libyan deposits which has been liquefied.

Chemical properties

1. Like the other alkanes of low formula weight, methane does not react with acids, alkalis or oxidising agents in solution. The lack of reactivity of alkanes towards inorganic reagents led to their being termed originally paraffins (Latin: *parum affinis*, little affinity).

2. Methane burns in air, with a hot, non-luminous flame, to carbon dioxide and water:

$$CH_4 + 2O_2 \rightarrow CO_2 + 2H_2O$$

ALKANES

Under carefully controlled conditions, methane is oxidised to other organic compounds. When a mixture of oxygen is compressed to a high pressure and passed through copper tubes at 200°C, methanol is formed:

$$CH_4 + \tfrac{1}{2}O_2 \rightarrow CH_3OH$$

Methane is now the most important source of ethyne. With a specially designed burner, it can be partially oxidised:

$$CH_4 + \tfrac{1}{2}O_2 \rightarrow CO + 2H_2$$

The heat liberated in this reaction is used to raise the temperature of a further amount of methane, which is heated for a very short time at 1500°C:

$$2CH_4 \rightarrow CH{\equiv}CH + 3H_2$$

Methane is also oxidised when it is mixed with steam and passed over nickel, and this reaction is used as a method for the manufacture of carbon monoxide and hydrogen ('synthesis gas', p. 308):

$$CH_4 + H_2O \rightarrow CO + 3H_2$$

3. When methane and chlorine are mixed together *in the dark*, no reaction occurs. However, if the mixture is either heated or exposed to ultraviolet light (from a mercury lamp), a mixture of products is formed:

$$CH_4 + Cl_2 \rightarrow CH_3Cl + HCl$$
Chloromethane

$$CH_3Cl + Cl_2 \rightarrow CH_2Cl_2 + HCl$$
Dichloromethane

$$CH_2Cl_2 + Cl_2 \rightarrow CHCl_3 + HCl$$
Trichloromethane

$$CHCl_3 + Cl_2 \rightarrow CCl_4 + HCl$$
Tetrachloromethane

With an excess of methane, chloromethane predominates, whereas with an excess of chlorine, tetrachloromethane is the main product.

These are examples of **substitution reactions**: substitution is defined as the replacement of one atom (or group of atoms) in a molecule by another atom (or group). When the reaction is brought about by ultraviolet light, it is described as a **photochemical reaction**.

A detailed examination of the relatively simple reaction between methane and chlorine to form chloromethane and hydrogen chloride:

$$CH_4 + Cl_2 \rightarrow CH_3Cl + HCl \qquad \Delta H = -100 \text{ kJ/mol}$$

illustrates many interesting principles. First, although the reaction is exothermic, energy (in the form of either heat or light) must be supplied to the mixture for reaction to begin. Methane is transparent to ultraviolet light of

Plate 5.2. An undersea pipeline for natural gas. The pipeline (above) being laid in the North Sea (Gas Council) and (below) coming ashore in Norfolk, England (Shell Petroleum Co. Ltd.)

the wavelength used, about 3000 Å, but chlorine is not. Chlorine absorbs the light, the energy of which is equivalent to about 400 kJ/mol; this is considerably greater than the bond strength of the chlorine molecule (242 kJ/mol) which consequently splits into chlorine atoms:

$$Cl\text{—}Cl \rightarrow Cl\cdot + \cdot Cl \tag{1}$$

Each atom retains one electron of the pair which formed the covalent bond. An atom or group of atoms which possesses an unpaired electron is called a **free radical** (p. 62). The energy supplied to the chlorine molecule is not enough to produce ions, Cl^+ and Cl^- (1130 kJ/mol).

Each chlorine atom then reacts with a molecule of methane by abstracting a hydrogen atom to form hydrogen chloride and a methyl radical:

$$Cl\cdot + CH_4 \rightarrow HCl + \cdot CH_3 \tag{2}$$

The methyl radical reacts with a molecule of chlorine to form chloromethane and a chlorine atom:

$$\cdot CH_3 + Cl_2 \rightarrow CH_3Cl + Cl\cdot \tag{3}$$

The chlorine atom produced by reaction (3) can then react with another molecule of methane according to reaction (2); thus reactions (2) and (3) can occur successively once an initial supply of chlorine atoms has been provided.

However, there are other reactions which can intervene to stop the successive occurrence of reactions (2) and (3); these are processes in which two free radicals combine with each other:

$$Cl\cdot + Cl\cdot \rightarrow Cl_2 \tag{4}$$

$$Cl\cdot + \cdot CH_3 \rightarrow CH_3Cl \tag{5}$$

$$\cdot CH_3 + \cdot CH_3 \rightarrow C_2H_6 \tag{6}$$

Now, at any instant, the concentrations of chlorine atoms and methyl radicals are very small compared with the concentration of methane and chlorine. Therefore, the chances of a collision between two atoms, an atom and a radical, or two radicals, are small compared with the chances of a collision between a chlorine atom and a molecule of methane or between a methyl radical and a molecule of chlorine. As a result, many thousands of molecules of chloromethane are formed by reactions (2) and (3) for each molecule of chlorine which is decomposed by reaction (1).

The characteristics described above are typical of a **chain reaction**. There is an **initiating step**—reaction (1); **propagating steps**, which keep the chain reaction in operation—reactions (2) and (3); and **terminating steps**, which bring the chain to an end—reactions (4), (5) and (6).

It is often difficult to test for whether a reaction occurs *via* free radicals. Two methods in this case involve adding another substance. For example, if tetraethyllead vapour is added, the reaction rate is increased considerably. This is because the lead compound decomposes to supply more free radicals:

$$(C_2H_5)_4Pb \rightarrow 4\cdot C_2H_5 + Pb$$

some of which react with chlorine to form chlorine atoms:

$$\cdot C_2H_5 + Cl_2 \rightarrow C_2H_5Cl + Cl\cdot$$

On the other hand, if oxygen is added, the rate is reduced, probably because oxygen reacts with methyl radicals and so prevents them taking any further part in the chain reaction.

In reactions (1), (2) and (3) covalent bonds are broken so that one electron of the pair in each bond becomes associated with each of the atoms or groups. These are examples of **homolysis** or **homolytic fission** (Greek: *lysis*, splitting). A second way in which a bond can undergo fission is for both electrons of the bond to become associated with one of the two atoms or groups:

$$A—B \rightarrow A^+ + B^-$$

Examples of this process, **heterolysis** or **heterolytic fission**, are described in the next chapter.

Uses

The uses of methane are discussed above and in Section 20.3.

5.5 Other alkanes

Ethane, propane and butane are obtained by fractional distillation of 'wet' natural gas, and all the alkanes are obtained during the refining of petroleum (20.3). Ethane and propane are produced on a large scale in the United States and used to make ethene and propene by high temperature reactions in the absence of air:

$$CH_3CH_3 \xrightarrow{700°C} CH_2{=}CH_2 + H_2$$

$$CH_3CH_2CH_3 \xrightarrow[700°C]{Cr_2O_3 + Al_2O_3 \text{ as cat.}} CH_3{-}CH{=}CH_2 + H_2$$

The ethene and propene are used as the starting materials for a wide variety of important products (20.4).

The other properties of ethane and propane are similar to those of methane. They do not react with acids, alkalis or oxidising agents in solution; they burn in air; and they are readily chlorinated. A large number of chloroalkanes can be made; for example, ethane gives chloroethane (C_2H_5Cl), two dichloroethanes ($ClCH_2{-}CH_2Cl$ and $CH_3{-}CHCl_2$), two trichloroethanes ($ClCH_2{-}CHCl_2$ and $CH_3{-}CCl_3$), two tetrachloroethanes ($ClCH_2{-}CCl_3$ and $Cl_2CH{-}CHCl_2$), pentachloroethane ($Cl_2CH{-}CCl_3$) and hexachloroethane ($Cl_3C{-}CCl_3$).

The uses of ethane and higher alkanes are described in Section 20.3.

5.6 Cycloalkanes

Cycloalkanes are classified as **alicyclic** compounds. The simplest is cyclopropane:

```
      H   H
       \ /
        C
   H   / \   H
    \ /   \ /
     C — C
     |   |
     H   H
```

The angles between the atoms in the ring are 60°, whereas sp^3 orbitals are at angles of 109°28′. Consequently, the overlap between the pairs of orbitals is not as complete as in the non-cyclic alkanes, and so the C—C bond strengths are less; the ring is said to be **strained**. This in turn makes cyclopropane more reactive than, for example, propane towards reagents which break C—C bonds. Thus, cyclopropane reacts readily with bromine in the absence of light:

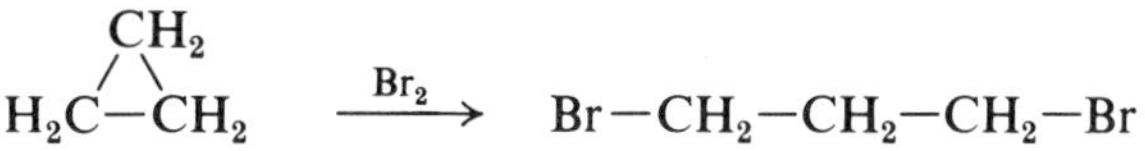

$$\text{cyclopropane} \xrightarrow{Br_2} Br-CH_2-CH_2-CH_2-Br$$
1,3-Dibromopropane

and with sulphuric acid:

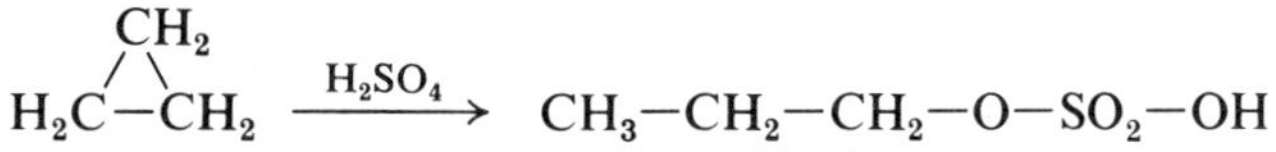

$$\text{cyclopropane} \xrightarrow{H_2SO_4} CH_3-CH_2-CH_2-O-SO_2-OH$$
Propyl hydrogen sulphate

In these respects it resembles an alkene such as propene (6.4).

In cyclobutane, the angles between the carbon atoms are 90°. Thus, the ring is strained, although not so much as in cyclopropane. As would be expected, cyclobutane undergoes the same reactions as cyclopropane but less readily.

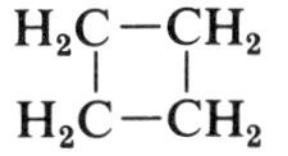

Cyclobutane 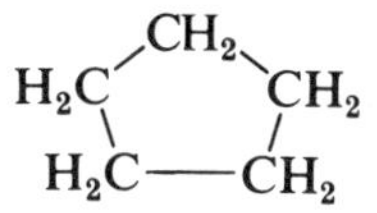   Cyclopentane

Cyclopentane has bond angles of 108°, so close to the tetrahedral value that the ring can be regarded as strainless. As expected, it resembles the non-cyclic alkanes and does not undergo the ring-opening reactions of cyclopropane and cyclobutane.

The structure of cyclohexane is of particular note. Were the carbon atoms in a plane, the bond angles would be 120° and would therefore be strained, though in the opposite sense compared with those in cyclopropane. However, the preferred angle of 109°28′ can be obtained by 'buckling' of the ring, and this is indeed the structure of cyclohexane; a two-dimensional representation is

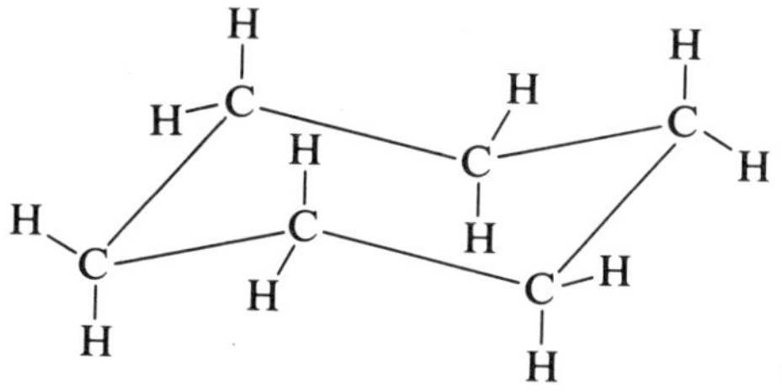

and is sometimes referred to as the **chair** structure. All the cycloalkanes with seven or more members in the ring also have strainless structures, although the exact shapes of the larger ones are not known. They all resemble the non-cyclic alkanes in their properties.

5.7 Practical work

Carry out the following reactions with a liquid alkane (e.g. pentane or hexane) and a liquid cycloalkane (e.g. cyclohexane).

1. Place a few drops of the liquid on a watch-glass or an evaporating basin and apply a lighted splint. Note the colour of the flame.

2. To a few drops of the liquids in separate test-tubes, add:
   (a) 5 drops of an alkaline potassium permanganate solution (made by dissolving about 0·1 g of sodium carbonate in 1 cm^3 of a 1 per cent solution of potassium permanganate), shake the mixture and see whether the liquid is oxidised;
   (b) 5 drops of a solution of bromine in tetrachloromethane, shake the mixture and see whether the bromine is decolorised.

3. Cracking of paraffin oil. See p. 299.

5.8 Questions

1 Outline **two** laboratory methods for the preparation of methane. Give **two** instances of the natural occurrence of methane.

From the properties of methane deduce the characteristic chemical behaviour of the carbon–hydrogen linkage in organic chemistry. How is the behaviour of this linkage modified when it occurs

(a) in the group $-C{\overset{H}{\underset{O}{\diagdown}}}$ (as in ethanal),

(b) in benzene?

To 30 cm^3 of a mixture of methane and carbon monoxide are added 50 cm^3 of oxygen, and the mixture is exploded. After shaking with potassium hydroxide solution, 20 cm^3 of gas are left. Calculate the composition by volume of the original mixture. (All volumes are measured at room temperature and pressure.)
(JMB)

2 Give equations for four methods of preparing ethane, naming the reagents and stating the conditions required.

Indicate briefly how and under what conditions methane reacts with chlorine.

Give the molecular formula for the hydrocarbon of molecular weight 56, and write down structural formulae for the isomers.

3 Give two methods for preparing ethane in the laboratory.

A mixture of 10 cm^3 of a gaseous hydrocarbon and 100 cm^3 of oxygen (excess) was exploded. The volume after explosion was 75 cm^3, and this was reduced to 35 cm^3 on treatment with potassium hydroxide solution. Deduce the molecular formula of the hydrocarbon and give its possible structural formulae. (All measurements were made at the same temperature and atmospheric pressure.)

4 Give equations, and conditions, for the reactions (if any) between methane and (a) chlorine, (b) bromine in the absence of light, (c) sulphuric acid.

How would you expect cyclobutane to react with these reagents?

Chapter 6

Alkenes

6.1 Nomenclature

General formula

$$C_nH_{2n}$$

The compounds are named as for the alkanes, but with the suffix **-ene** instead of -ane and the inclusion before the suffix of a number to describe the position of the double bond in the chain where more than one is possible. For this purpose, the chain is numbered from the end nearer to the double bond and the lower number of the two which describe the positions of the carbon atoms in the double bond is employed. For example:

$$\overset{4}{C}H_3-\overset{3}{C}H_2-\overset{2}{C}H=\overset{1}{C}H_2 \qquad \overset{1}{C}H_3-\overset{2}{C}H=\overset{3}{C}H-\overset{4}{C}H_2-\overset{5}{C}H_3$$
$$\text{But-1-ene} \qquad\qquad \text{Pent-2-ene}$$

$$\overset{1}{C}H_3-\overset{2}{C}H-\overset{3}{C}H=\overset{4}{C}H-\overset{5}{C}H_2-\overset{6}{C}H_3$$
$$\qquad\quad |$$
$$\qquad\; CH_3$$
$$\text{2-Methylhex-3-ene}$$

The two lowest members of the series are sometimes described by their original names: ethene ($CH_2=CH_2$) is known as ethylene, and propene ($CH_3-CH=CH_2$) as propylene. 2-Methylpropene, $(CH_3)_2C=CH_2$, is sometimes referred to as isobutylene or isobutene.

6.2 Physical properties of alkenes

The melting points and boiling points of the alkenes are very close to those of the alkanes with the same number of carbon atoms. Ethene, propene and the butenes are gases at room temperature, and the higher members are liquids (Table 6.1).

Table 6.1. Some alkenes

NAME	FORMULA	B.P. (°C)
Ethene	$CH_2=CH_2$	−102
Propene	$CH_3-CH=CH_2$	−48
But-1-ene	$C_2H_5-CH=CH_2$	−6·5
trans-But-2-ene*	$CH_3-CH=CH-CH_3$	1
cis-But-2-ene*	$CH_3-CH=CH-CH_3$	4
2-Methylpropene	$(CH_3)_2C=CH_2$	−6·5
Pent-1-ene	$C_3H_7-CH=CH_2$	30
Hex-1-ene	$C_4H_9-CH=CH_2$	63
Cyclohexene		83
Phenylethene	$C_6H_5-CH=CH_2$	145

* These are geometrical isomers, which are discussed in Section 15.4.

6.3 Ethene

Structural formula

$$\begin{array}{c}H\\ \end{array}\!\!\!\!\!\!C=C\!\!\!\!\!\!\begin{array}{c}H\\ \end{array}$$
$$\begin{array}{c}H\end{array}\begin{array}{c}H\end{array}$$

Manufacture

1. Ethene is obtained in Western Europe from the naphtha fraction from the distillation of petroleum. This fraction contains straight-chain alkanes with 4–10 carbon atoms, and it is passed with steam through pipes heated at 700–900°C. The ethene formed is purified by fractional distillation (20.4).

2. A method which is of particular importance in the United States is the high-temperature conversion of the ethane derived from 'wet' natural gas (19.2, 20.3).

3. Ethene is a by-product in the cracking of the gas-oil fraction obtained from the distillation of petroleum (19.6).

Chemical properties

(a) Addition reactions

Ethene, and all other alkenes, are characterised by their addition reactions in which the double bond is converted into a single bond and atoms or groups are added to each of the two carbon atoms. The general reaction is:

$$>\!\!C=C\!\!<\; +\; X-Y\; \rightarrow\; -\overset{|}{\underset{X}{C}}-\overset{|}{\underset{Y}{C}}-$$

The following are examples:

1. When ethene is mixed with hydrogen and passed over nickel at 150°C, ethane is formed:

$$CH_2{=}CH_2 + H_2 \xrightarrow{\text{Ni as cat.}} CH_3{-}CH_3$$

The reaction takes place on the surface of the metal, which acts as a catalyst. Finely divided platinum or palladium are more active catalysts, and reaction takes place at room temperature.

2. Ethene reacts with chlorine to form 1,2-dichloroethane:

$$CH_2{=}CH_2 + Cl_2 \rightarrow ClCH_2{-}CH_2Cl$$
1,2-Dichloroethane

Similarly, with bromine, it forms 1,2-dibromoethane:

$$CH_2{=}CH_2 + Br_2 \rightarrow BrCH_2{-}CH_2Br$$
1,2-Dibromoethane

There is considerable evidence that these are *ionic* reactions (compare the *free-radical* reaction by which methane and chlorine give chloromethane, p. 69). Important information about the mechanism is provided by the observation that, if bromine is added to a solution of ethene in which sodium chloride is also present, 1-bromo-2-chloroethane is formed in addition to 1,2-dibromoethane. This points to the mediation of an organic cation which can react with either bromide ion or chloride ion:

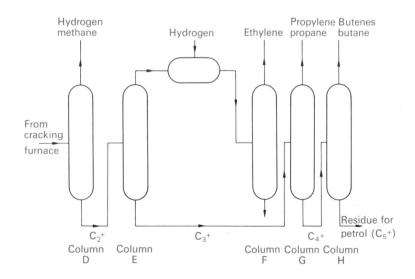

Plate 6.1. Steam cracking of naphtha to form alkenes (principally ethene and propene). A—Furnace to heat naphtha, B—cracking furnaces and C—fractionating column. Other fractionating columns are shown which remove D—methane, E—ethene and ethane, F—ethene from ethane, G—propene and propane, H—butenes and butane (Esso Petroleum Co. Ltd.) The diagram opposite illustrates this

ALKENES

$$CH_2=CH_2 + Br_2 \longrightarrow BrCH_2-\overset{+}{C}H_2 + Br^-$$

$$BrCH_2-\overset{+}{C}H_2 \underset{Cl^-}{\overset{Br^-}{\diagup \diagdown}} \begin{array}{c} BrCH_2-CH_2Br \\ \\ BrCH_2-CH_2Cl \end{array}$$

The cation $BrCH_2-\overset{+}{C}H_2$, and other cations in which a carbon atom bears the positive charge, are given the general name **carbonium ions**. In the first step the two electrons which form the new bond are both provided by the alkene; thus, the reagent (bromine or chlorine) is described as **electrophilic** (i.e. 'electron-seeking').

When the reaction with bromine is carried out in water, the main product is 2-bromoethanol. The first step of the reaction is the same as above, but the carbonium ion is a very reactive species which reacts with a molecule of water almost as readily as with a bromide ion; since there is much more water than bromide ion in the solution, the reaction with water predominates. The reaction occurs by the donation of an unshared pair of electrons on the oxygen atom of water to the electron-deficient carbon atom in the carbonium ion, followed by the loss of a proton:

$$BrCH_2-\overset{+}{C}H_2 + :OH_2 \rightarrow BrCH_2-CH_2-\overset{+}{O}\begin{subarray}{l}H \\ H\end{subarray}$$

$$\rightarrow BrCH_2-CH_2OH + H^+$$
2-Bromoethanol

A similar reaction occurs between ethene and chlorine water, 2-chloroethanol being formed.

3. Ethene reacts with hydrogen chloride to give chloroethane:

$$CH_2=CH_2 + HCl \rightarrow CH_3-CH_2Cl$$

This is also an ionic reaction. The first step is the formation of two ions, the ethyl cation and the chloride anion:

$$CH_2=CH_2 + HCl \rightarrow CH_3-\overset{+}{C}H_2 + Cl^-$$

The two ions then rapidly combine to form the product. In this reaction, hydrogen chloride is the electrophilic reagent.

Hydrogen bromide reacts more readily than hydrogen chloride, giving bromoethane, and hydrogen iodide reacts more readily still to give iodoethane.

4. When ethene is passed through concentrated sulphuric acid, ethyl hydrogen sulphate is formed:

$$CH_2=CH_2 + H_2SO_4 \rightarrow CH_3-CH_2-O-SO_2-OH$$
Ethyl hydrogen sulphate

ALKENES

Once again a carbonium ion is formed first, by reaction with the electrophilic sulphuric acid:

$$CH_2=CH_2 + H_2SO_4 \rightarrow CH_3-\overset{+}{C}H_2 + {}^-O-SO_2-OH$$

This ion then combines with a hydrogen sulphate ion:

$$CH_3-\overset{+}{C}H_2 + {}^-O-SO_2-OH \rightarrow CH_3-CH_2-O-SO_2-OH$$

When the mixture is diluted and warmed, ethanol is produced:

$$CH_3CH_2-O-SO_2-OH + H_2O \rightarrow CH_3CH_2-OH + H_2SO_4$$

This is the basis of a method for manufacturing ethanol from ethene, but it is now being superseded by direct hydration with a solid catalyst (phosphoric acid on silica) at 300°C:

$$CH_2=CH_2 + H_2O \xrightarrow{H_3PO_4/SiO_2 \text{ as cat.}} CH_3CH_2OH$$

5. When ethene is passed through a dilute, alkaline solution of potassium permanganate, ethane-1,2-diol (ethylene glycol) is formed:

$$CH_2=CH_2 + H_2O + \text{'O'} \rightarrow HO-CH_2-CH_2-OH$$
Ethane-1,2-diol

The permanganate is the source of the necessary oxygen atom denoted 'O'. The oxygen atom is never free but is thought to be transferred from the permanganate ion to the alkene *via* a cyclic intermediate which is too unstable towards water to be isolated:

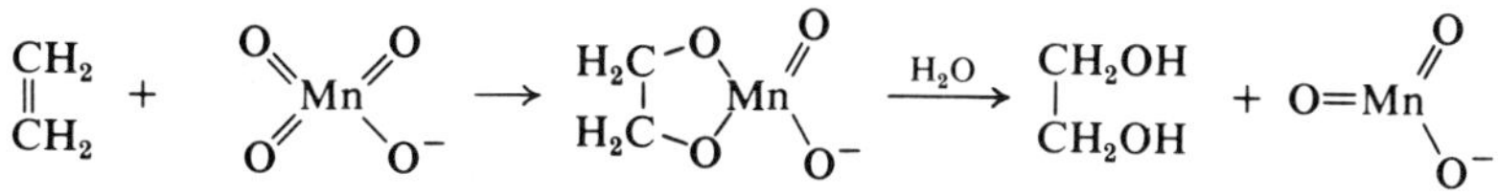

[reaction scheme showing cyclic manganate intermediate converting ethene to ethane-1,2-diol]

(Two MnO_3^- ions disproportionate to one MnO_4^{2-} ion and MnO_2.)

6. When ethene is treated with a **peroxoacid** (a compound containing the group —CO—O—OH), epoxyethane is formed; for example, peroxobenzoic acid can be used:

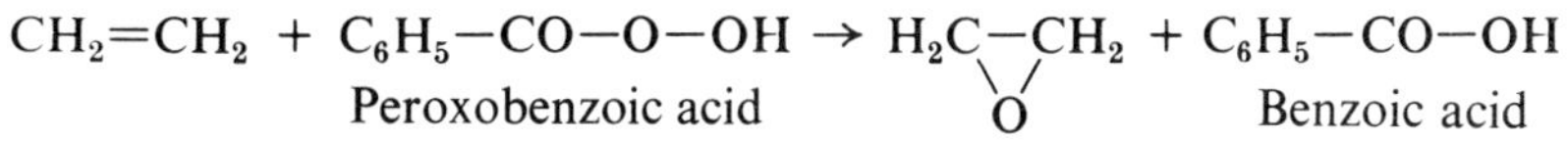

$$CH_2=CH_2 + C_6H_5-CO-O-OH \rightarrow \underset{O}{H_2C\overset{\diagdown\diagup}{-}CH_2} + C_6H_5-CO-OH$$
Peroxobenzoic acid   Epoxyethane   Benzoic acid

Industrially, epoxyethane is made by passing a mixture of ethene and oxygen over a silver catalyst at 250°C:

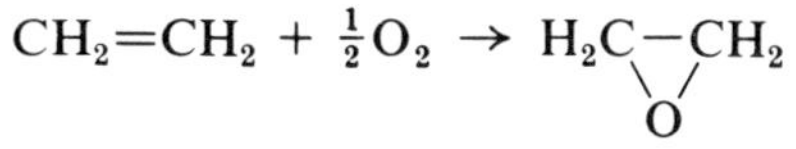

$$CH_2=CH_2 + \tfrac{1}{2}O_2 \rightarrow \underset{O}{H_2C\overset{\diagdown\diagup}{-}CH_2}$$

(b) Other reactions

1. When trioxygen is passed through a solution of ethene in trichloro-

ALKENES

methane, ethene ozonide is formed:

$$CH_2=CH_2 + O_3 \longrightarrow \underset{\text{Ethene ozonide}}{\begin{array}{c}H_2C\overset{O}{\diagup}\diagdown CH_2 \\ O-O\end{array}}$$

It is an unstable, explosive product, which is readily hydrolysed with water to a mixture of methanal and hydrogen peroxide; half the methanal is then oxidised to methanoic acid by the hydrogen peroxide.

$$\begin{array}{c}H_2C\overset{O}{\diagup}\diagdown CH_2 \\ O-O\end{array} + H_2O \longrightarrow 2HCHO + H_2O_2$$
$$\text{Methanal}$$

$$HCHO + H_2O_2 \longrightarrow \underset{\text{Methanoic acid}}{HCO_2H} + H_2O$$

2. When subjected to very high pressures at 120°C in the presence of oxygen as a catalyst, ethene undergoes **polymerisation** to give the polymer, poly(ethene) (also known as polythene) (p. 312). Polymerisation is also catalysed by a mixture of triethylaluminium and titanium(IV) chloride, and this gives a type of poly(ethene) with more useful properties (p. 312).

Plate 6.2. Hortonspheres used to store liquefied gases (for example, ethene, propane and butanes, chloroethene) (Esso Petroleum Co. Ltd.)

Uses

Ethene is one of the most important raw materials for the chemical industry, particularly in making plastics (p. 303). One of these, poly(ethene), is made directly from ethene. A second, poly(chloroethene) (PVC), is made from ethene *via* chloroethene, into which ethene is converted by two processes. In the first, 1,2-dichloroethane is formed by addition of chlorine to ethene in the liquid or vapour phase and is then heated at 600°C over pumice:

$$CH_2=CH_2 + Cl_2 \rightarrow ClCH_2-CH_2Cl$$

$$ClCH_2-CH_2Cl \xrightarrow{600°C} CH_2=CHCl + HCl$$
$$\text{Chloroethene}$$

In the second, ethene, hydrogen chloride and oxygen react at 250°C in the presence of copper(II) chloride:

$$CH_2=CH_2 + HCl + \tfrac{1}{2}O_2 \xrightarrow[250°C]{CuCl_2 \text{ as cat.}} CH_2=CHCl + H_2O$$

By operating both processes, the hydrogen chloride released in the first can be utilised in the second.

Ethene is also made into ethanol (used as a solvent and as a starting material for other products), epoxyethane (used in the manufacture of detergents and of ethane-1,2-diol), ethanal (used in the manufacture of ethanoic acid and ethanoic anhydride) and higher straight-chain alkenes, used to make detergents.

The uses of ethene are further discussed in Section 20.4.

6.4 Propene

Structural formula

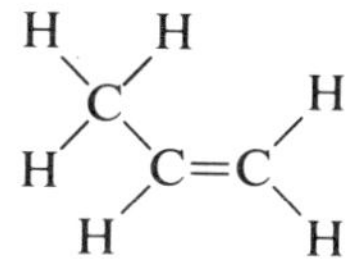

Manufacture

Propene is obtained from propane (p. 72) and from the naphtha fraction obtained by the distillation of petroleum in the same way as ethene.

Chemical properties

The chemistry of propene is similar to that of ethene. It burns in air, can be reduced with hydrogen over metal catalysts, is polymerised to poly(propene) (p. 312) and undergoes addition reactions with the halogen acids, the halogens, sulphuric acid, potassium permanganate and peroxoacids. However, there is a feature in its reactions with acidic reagents which does not apply to ethene; thus, reaction with an acid of general formula HX could give either of two products:

$$CH_3-CH=CH_2 + HX \rightarrow CH_3-CH_2-CH_2X \text{ or } CH_3-CHX-CH_3$$

In practice, the second of these products is the major one. With more highly substituted alkenes, the major products are as follows:

$$\begin{array}{c} R \\ R \end{array}\!\!\!>\!\!C=CH_2 + HX \xrightarrow{HX} R-\underset{\underset{X}{|}}{\overset{\overset{R}{|}}{C}}-CH_3$$

$$\begin{array}{c} R \\ R \end{array}\!\!\!>\!\!C=CHR + HX \xrightarrow{HX} R-\underset{\underset{X}{|}}{\overset{\overset{R}{|}}{C}}-CH_2R$$

These observations were first made by Markownikoff and can be summarised as an empirical rule known as **Markownikoff's rule**: when an acid HX adds to the double bond of an alkene, the hydrogen atom becomes attached to the carbon atom which has the larger number of hydrogen atoms.

The principle which underlies this empirical rule is as follows. The proportion of each of the two products from an alkene is determined by the relative rates of the addition of the proton to each carbon atom, for once this step has occurred the uptake of an anion X^- follows immediately. Now, of the two possible carbonium ions which can be formed during the first step of the reaction of propene,

$$CH_3-CH_2-\overset{+}{C}H_2 \qquad CH_3-\overset{+}{C}H-CH_3$$

the first is a primary carbonium ion, with only one alkyl group attached to the positively charged carbon atom, whereas the second is a secondary carbonium ion, with two alkyl groups attached to the positively charged carbon atom. Since an alkyl group is electron-releasing relative to a hydrogen atom (4.8), it reduces the density of positive charge on the neighbouring carbon atom, and this makes the ion more stable. Two alkyl groups are more effective than one, so that the secondary carbonium ion is more stable than the primary carbonium ion. Consequently, the secondary ion is formed the more rapidly, and the major product from propene is of the type $CH_3-CHX-CH_3$.

When three alkyl groups are attached to a positively charged carbon atom, the stabilising effect is increased further; i.e. the order of stability of carbonium ions is tertiary > secondary > primary. Thus,

$$R_2C=CH_2 \text{ forms } R_2\overset{+}{C}-CH_3 \text{ faster than } R_2CH-\overset{+}{C}H_2$$
$$\qquad\qquad\qquad \text{tertiary} \qquad\qquad\qquad \text{primary}$$

$$R_2C=CHR \text{ forms } R_2\overset{+}{C}-CH_2R \text{ faster than } R_2CH-\overset{+}{C}HR$$
$$\qquad\qquad\qquad \text{tertiary} \qquad\qquad\qquad \text{secondary}$$

Free-radical addition to propene

When propene reacts with hydrogen bromide in the presence of an organic peroxide, the product is mainly 1-bromopropane:

$$CH_3-CH=CH_2 + HBr \xrightarrow{\text{peroxide}} CH_3-CH_2-CH_2-Br$$

whereas in the absence of a peroxide the main product is 2-bromopropane.

This is because, in the presence of a peroxide, a radical chain reaction occurs which is more rapid than the electrophilic addition which gives 2-bromopropane. The peroxide, containing the relatively weak O—O bond (p. 51), breaks down to give two radicals:

$$R-O-O-R \rightarrow 2R-O\cdot$$

These radicals react with hydrogen bromide to give a bromine atom:

$$R-O\cdot + H-Br \rightarrow R-O-H + Br\cdot$$

The bromine atom adds to the alkene mainly at the unsubstituted carbon atom, giving the radical $CH_3-\dot{C}H-CH_2-Br$ (which is more stable than the alternative radical, $CH_3-CHBr-\dot{C}H_2$):

$$CH_3-CH=CH_2 + Br\cdot \rightarrow CH_3-\dot{C}H-CH_2-Br$$

and this radical abstracts a hydrogen atom from another molecule of hydrogen bromide:

$$CH_3-\dot{C}H-CH_2-Br + H-Br \rightarrow CH_3-CH_2-CH_2-Br + Br\cdot$$

The bromine atom formed can add to another molecule of propene, so that a chain reaction is propagated; the chain ends when two radicals meet and combine. In summary:

Initiation
$$\begin{cases} RO-OR \rightarrow 2RO\cdot \\ RO\cdot + HBr \rightarrow ROH + Br\cdot \end{cases}$$

Propagation
$$\begin{cases} Br\cdot + CH_3-CH=CH_2 \rightarrow CH_3-\dot{C}H-CH_2Br \\ CH_3-\dot{C}H-CH_2Br + HBr \rightarrow CH_3-CH_2-CH_2Br + Br\cdot \end{cases}$$

Termination
$$\begin{cases} 2Br\cdot \longrightarrow Br_2 \\ Br\cdot + CH_3-\dot{C}H-CH_2Br \longrightarrow CH_3-CHBr-CH_2Br \\ 2CH_3-\dot{C}H-CH_2Br \longrightarrow \begin{array}{c} CH_3 \\ \diagdown \\ BrCH_2 \end{array} CH-CH \begin{array}{c} CH_3 \\ \diagup \\ CH_2Br \end{array} \end{cases}$$

Hydrogen chloride and hydrogen iodide do not react in this way. This is because, although halogen atoms are generated in the initiation step in each case, one or other of the propagating steps is so slow that the overall rate of the chain reaction is less than that of the electrophilic addition. With hydrogen chloride, the slow step is

$$CH_3-\dot{C}H-CH_2Cl + HCl \rightarrow CH_3-CH_2-CH_2Cl + Cl\cdot$$

and with hydrogen iodide it is

$$I\cdot + CH_3-CH=CH_2 \rightarrow CH_3-\dot{C}H-CH_2I$$

Since alkenes form peroxides slowly when exposed to air, it is not always necessary to add a peroxide in order to bring about the chain reaction with hydrogen bromide. Indeed, if the product of the electrophilic addition is required, it is necessary for the alkene to be a freshly prepared (or freshly distilled) sample.

ALKENES

Uses

As with ethene, the principal use of propene is as a raw material to make plastics, the most important being poly(propene), poly(propenonitrile) (*via* propenonitrile), perspex (*via* propanone) and the alkyd resins (*via* propane-1,2,3-triol). These are discussed in more detail in Chapter 21.

The single most important chemical made from propene is propanone (12.3). The uses of propene are summarised in Section 20.4.

6.5 Practical work

Small-scale preparation of cyclohexene

To 10 cm^3 of cyclohexanol in a flask, add, with a dropping pipette, 4 cm^3 of concentrated phosphoric acid, shaking the flask.

Assemble the apparatus (Fig. 6.1), and heat the flask gently, distilling over the liquid.

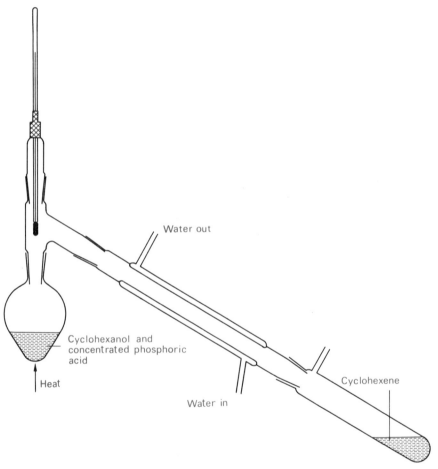

FIG. 6.1. *Preparation of cyclohexene*

Pour the distillate into a separating funnel and add 2 cm^3 of a saturated solution of sodium chloride. Shake the mixture and allow the two layers to separate. Run off the lower layer and then run the top layer, containing cyclohexene, into a small flask. Add 2 or 3 pieces of anhydrous calcium chloride, stopper the flask and shake until the liquid is clear.

Decant the liquid into a clean distillation flask and distil it, collecting the liquid boiling at 81–85°C.

ALKENES

Test-tube preparation of ethene

Place ethanol in a test-tube, to a depth of 2·5 cm. Add Rocksil until the ethanol has been soaked up. Place about 1 g of aluminium oxide half-way along the tube (Fig. 6.2). Fit a cork and delivery tube to the test-tube

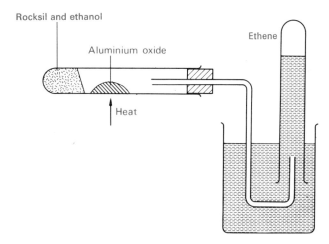

FIG. 6.2. *Test-tube preparation of ethene*

and heat the aluminium oxide with a gentle flame. Collect 4 or 5 test-tubes of ethene by displacement of water, placing corks in the test-tubes when they have been filled.

Cracking of oil to form alkenes

For details, see Section 19.10.

Reactions of alkenes

Carry out the following reactions with a gaesous alkene (e.g. ethene) and a liquid alkene (e.g. cyclohexene), comparing the results of experiments 1–3 with those obtained with an alkane (5.7).

1. (a) Ignite the ethene by applying a lighted splint to the mouth of the test-tube.
(b) Place a few drops of cyclohexene on a watch-glass and ignite.
Note the colour of the flame.

2. (a) Add a few drops of a solution of bromine in tetrachloromethane to a test-tube of ethene and shake the mixture.
(b) To a few drops of cyclohexene in a test-tube add a few drops of bromine in tetrachloromethane and shake the mixture.

3. (a) Add 3–4 drops of an alkaline solution of potassium permanganate (made by dissolving about 0·1 g of anhydrous sodium carbonate in 1 cm^3 of a 1 per cent solution of potassium permanganate) to a test-tube of ethene and shake the mixture.
Note the colour of the solution and whether any precipitate is formed.
(b) Repeat this experiment using 2 or 3 drops of cyclohexene instead of ethene.

ALKENES

4. *Polymerisation of alkenes.* The polymerisation of an aromatic alkene, phenylethene, C_6H_5—CH=CH$_2$, is described on p. 325.

5. *Mechanism of the addition of bromine to an alkene.*
(a) Make up a solution of 5 cm³ of pent-1-ene in 10 cm³ of ethanol in a 100 cm³ conical flask. Add 1 cm³ of a saturated solution of sodium nitrate in water. Some of this salt recrystallises out of solution. Add water dropwise, with shaking, until the sodium nitrate redissolves.

Shake the solution vigorously, so that the two layers are thoroughly mixed, and add bromine, from a dropping pipette, until the mixture is yellow.

Using a dropping pipette, transfer a few drops of the bottom layer to an ignition tube and add enough powdered anhydrous sodium carbonate to absorb the liquid. Carry out the Lassaigne sodium test for nitrogen (p. 43).

(b) Make up a solution of 5 cm³ of pent-1-ene in 10 cm³ of ethanol in a 100 cm³ conical flask. Shake vigorously and add bromine dropwise until the mixture is pale yellow. Add, with shaking, 1 cm³ of a saturated solution of sodium nitrate in water and then add water until the sodium nitrate redissolves. Allow the mixture to stand for a minute and transfer a small amount of the bottom layer to an ignition tube and add enough powdered anhydrous sodium carbonate to absorb the liquid. Carry out the Lassaigne sodium test for nitrogen (p. 43).

(c) Answer the following questions:
  (i) Write an equation for the reaction between the alkene and bromine.
  (ii) Was nitrogen present in the compounds formed in these experiments?
  (iii) From these results suggest a mechanism for the reaction between an alkene and bromine.

6.6 Questions

1 Describe in detail two test-tube reactions to show whether a given compound contains an unsaturated carbon–carbon link.

To 30 cm³ of a gaseous mixture of butadiene CH_2=CH—CH=CH_2 and but-1-ene C_2H_5CH=CH_2, 100 cm³ of hydrogen were added and the mixture was passed repeatedly over a hydrogenation catalyst in a closed system until no further reduction in volume occurred. The total volume was then 90 cm³. What was the composition by volume of the original mixture? (All volumes were measured at the same temperature and pressure.)

Give two sets of reactions by which chloroethene may be prepared.

Indicate the structure of poly(chloroethene). Give one industrial application of this material and mention the properties upon which this use depends. (JMB)

2 Describe how you would prepare a pure specimen of ethene from ethanol. By what reactions can the following be obtained from ethene: (a) ethanol, (b) ethyne, (c) ethane-1,2-diol.

3 Write an account of the laboratory preparation and of the properties of ethene, indicating how this substance may be distinguished from methane and from benzene.

4 (a) Name the organic products of the reaction of but-2-ene, CH_3CH=$CHCH_3$, with each of the following reagents and write a balanced equation for each reaction:
  (i) hydrogen bromide,
  (ii) bromine dissolved in trichloromethane,
  (iii) a dilute alkaline solution of potassium permanganate.

ALKENES

(b) When a hydrocarbon C_xH_y is completely oxidised in oxygen the reaction which occurs can be represented by the following equation:

$$C_xH_y + (x + y/4)O_2 \rightarrow xCO_2 + (y/2)H_2O$$

20·0 cm³ of a gaseous, unsaturated hydrocarbon A were exploded with 200 cm³ of oxygen and, after cooling to the original room temperature, 140 cm³ of gas remained. This volume was reduced to 20·0 cm³ on shaking the gas with potassium hydroxide solution. Deduce the molecular formula of A.

When the gas undergoes oxidation at the double bond, one mole of A gives one mole each of B and C, both of which have the molecular formula C_3H_6O. Both B and C give an orange precipitate with 2,4-dinitrophenylhydrazine and B reduces Fehling's solution while C does not.

Write structural formulae for A, B and C and explain the reactions involved.

(L(X))

5 Draw a labelled diagram of an apparatus suitable for the preparation of a sample of trioxygen. Describe a method for determining the percentage by volume of trioxygen in the sample.

Write structural formula (excluding cyclic structures) for the different isomers corresponding to the molecular formula C_4H_8 and explain how trioxygen may be used to distinguish between the isomers. (W)

6 Reaction of compound A (C_9H_{18}) with trioxygen and then water gave neutral compounds B (C_3H_6O) and C ($C_6H_{12}O$). B did not reduce Fehling's solution, but C did. Reduction of C with hydrogen and a catalyst gave D ($C_6H_{14}O$), which, when heated with concentrated hydrobromic acid, gave E ($C_6H_{13}Br$). E was heated with a concentrated solution of potassium hydroxide in ethanol, and gave F (C_6H_{12}). After being heated with alkaline potassium permanganate, F gave on acidification an acid G ($C_5H_{10}O_2$); treatment of G with d-morphine gave two products, separated by fractional crystallisation.

Identify the compounds A–G, explaining your reasoning. C(N, S))

7 Reaction of compound A ($C_{11}H_{14}$) with trioxygen and then water gave B (C_8H_8O) and C (C_3H_6O) in equimolar amounts. C was unaffected by ammoniacal silver nitrate, but B reacted with this reagent to give D ($C_8H_8O_2$). When D was treated successively with sulphur dichloride oxide (or phosphorus trichloride) and ammonia, E (C_8H_9NO) was formed. E reacted with bromine and sodium hydroxide to give F(C_7H_9N), a much weaker base than methylamine.

Identify the compounds A–F so far as is possible and comment on the reactions. Suggest further experiments (a) to confirm the identification of C, and (b) to be carried out on compound D to complete the identification of compounds D–F.

(C(T, S))

8 Describe in outline the preparation of propene starting from propanone.

Write down the structural formulae of the compounds obtained when propene is treated with the following reagents.

(a) Concentrated sulphuric acid.
(b) Bromine.
(c) Cold potassium permanganate solution.
(d) Trioxygen followed by water. (O and C)

9 A gaseous hydrocarbon (10 cm³) was mixed with oxygen (80 cm³) in a eudiometer. After sparking the mixture the volume of the gases remaining was 60 cm³. Sodium hydroxide solution was then added and the volume of the gas decreased to 20 cm³. All measurements were made at room temperature and atmospheric pressure. Deduce the molecular formula of the gas.

Write down two of the possible structures for the gas and indicate how you would distinguish between these two structures. (O and C)

ALKENES

10 Reaction of a hydrocarbon, A, of molecular formula, $C_{10}H_{20}$, with trioxygen and then water gives two isomeric products, B and C, $C_5H_{10}O$.

Oxidation of B leads to an optically active acid, D, $C_5H_{10}O_2$, whereas oxidation of C leads to an acid E, $C_4H_8O_2$, which cannot be resolved into optically active forms.

Reduction of C gives F, $C_5H_{12}O$, dehydration of which, followed by reaction with trioxygen and then water, gives propanone as one of the products.

Deduce the structure of A and elucidate the reaction sequence.

11 A compound A reacts with bromine giving B ($C_5H_{10}Br_2$). Oxidation of A leads to two compounds C ($C_2H_4O_2$) and D (C_3H_6O). The silver salt of C contains 64·6 per cent Ag while D with 2,4-dinitrophenylhydrazine forms a derivative having the composition $C_9H_{10}N_4O_4$. D also reacts with iodine and sodium hydroxide forming tri-iodomethane. Deduce a structure for A.

Chapter 7

Alkynes

General formula

$$C_nH_{2n-2}$$

7.1 Nomenclature

The compounds are named as for the alkenes but with the suffix **-yne**. For example:

$$\overset{4}{C}H_3—\overset{3}{C}H_2—\overset{2}{C}\equiv\overset{1}{C}—H$$

is but-1-yne. However, the first member, ethyne (CH≡CH) is often described by its original name, acetylene, and its simple derivatives are sometimes described as substituted acetylenes: e.g. propyne (CH_3—C≡CH) is methylacetylene.

7.2 Physical properties of alkynes

The melting points and boiling points of the alkynes are similar to those of the alkanes with the same number of carbon atoms (cf. Table 5.1).

Table 7.1. Some alkynes

NAME	FORMULA	B.P. (°C)
Ethyne	CH≡CH	−83
Propyne	CH_3—C≡C—H	−23
But-1-yne	C_2H_5—C≡C—H	9
But-2-yne	CH_3—C≡C—CH_3	27
Pent-1-yne	C_3H_7—C≡C—H	40
Phenylethyne	C_6H_5—C≡C—H	143

7.3 Ethyne

Structural formula

$$H—C\equiv C—H$$

Manufacture

1. Calcium dicarbide is obtained by heating coke with calcium oxide in an electric furnace above 2000°C:

$$CaO + 3C \rightarrow CaC_2 + CO$$

Ethyne is produced by the treatment of calcium dicarbide with water:

$$CaC_2 + 2H_2O \rightarrow CH\equiv CH + Ca(OH)_2$$

2. Ethyne is now being increasingly made by the pyrolysis of methane at 1500°C (5.4).

Chemical properties

(a) Electrophilic addition reactions

Ethyne resembles ethene in undergoing addition reactions with electrophilic reagents. Examples are:

1. It reacts readily with the halogens. Chlorine reacts explosively:

$$CH\equiv CH + Cl_2 \rightarrow 2C + 2HCl$$

To prevent explosions, ethyne and chlorine are generally mixed in retorts filled with Kieselguhr and iron filings to absorb the heat of the reaction. Addition compounds are then formed:

$$CH\equiv CH \xrightarrow{Cl_2} \underset{\text{1,2-Dichloro-ethene}}{ClCH=CHCl} \xrightarrow{Cl_2} \underset{\text{1,1,2,2-Tetra-chloroethane}}{Cl_2CH-CHCl_2}$$

Bromine reacts less violently, giving 1,2-dibromoethene and, with an excess of bromine, 1,1,2,2-tetrabromoethane.

2. It reacts with the halogen acids, for example:

$$CH\equiv CH \xrightarrow{HCl} \underset{\text{Chloroethene}}{CH_2=CHCl} \xrightarrow{HCl} \underset{\text{1,1-Dichloro-ethane}}{CH_3-CHCl_2}$$

(b) Other reactions

1. Ethyne resembles ethene in being reduced by hydrogen on certain metal surfaces (e.g. nickel at 150°C or platinum or palladium at room temperature). First ethene and then ethane are formed:

$$CH\equiv CH \xrightarrow{H_2} CH_2=CH_2 \xrightarrow{H_2} CH_3-CH_3$$

2. Ethyne differs from ethene in reacting with water in the presence of dilute sulphuric acid and mercury(II) sulphate ($HgSO_4$) at 60°C:

$$CH\equiv CH + H_2O \rightarrow \underset{\text{Ethanal}}{CH_3CHO}$$

This is the basis of a method for manufacturing ethanal, but it is now being superseded by one which starts with ethene, a cheaper raw material than ethyne (12.3).

3. Ethyne also differs from ethene in forming salts with electropositive metals; for example, if ethyne is passed through an ammoniacal solution of copper(I) chloride, a red precipitate of copper(I) dicarbide is formed:

$$CH\equiv CH + Cu_2Cl_2 + 2NH_3 \rightarrow \underset{\substack{\text{Copper(I)}\\\text{dicarbide}}}{Cu_2C_2} + 2NH_4Cl$$

Similarly, if ethyne is passed through an ammoniacal solution of silver nitrate, a white precipitate of silver dicarbide, Ag_2C_2, is produced.

Uses

Many of the industrial uses of ethyne are now losing importance as methods for making the same compounds from ethene are being developed, since the alkene is cheaper to produce; examples are the production of ethanal mentioned above and the formation of chloroethene by addition of hydrogen chloride to ethyne (p. 90), which is being superseded by the chlorination of ethene (6.3.)

One small-scale use is in oxyacetylene welding, which is based on the very high temperatures (*ca.* 3000°C) attained when ethyne burns in oxygen:

$$2CH\equiv CH + 5O_2 \rightarrow 4CO_2 + 2H_2O$$

7.4 Other alkynes

The properties of other alkynes resemble those of ethyne except that only those alkynes in which the triple bond is at the end of the chain form metallic compounds. For example, propyne forms a silver salt, $CH_3C\equiv CAg$, but but-2-yne ($CH_3-C\equiv C-CH_3$) does not. This provides a method of distinguishing ethyne and alkynes of the type $RC\equiv CH$, which are acidic and form metal salts, from other alkynes and also from alkenes.

7.5 Practical work

Small-scale preparation of ethyne

Place 2 or 3 small pieces of calcium dicarbide in a test-tube and arrange the apparatus for collection of ethyne (Fig. 7.1). Add water dropwise and collect 4 or 5 test-tubes of gas, putting a cork in the test-tube when it is full of gas.

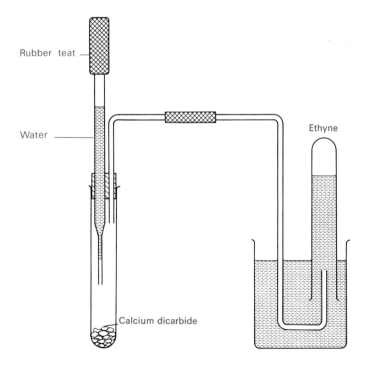

FIG. 7.1. *Test-tube preparation of ethyne*

ALKYNES

Reactions of alkynes

Carry out the following experiments with a gaseous alkyne (e.g. ethyne) and a liquid alkyne (e.g. phenylethyne). Compare the results of experiments 1–3 with those obtained with an alkane (5.7) and an alkene (6.5).

1. (a) Ignite the gas by applying a lighted splint to the mouth of the test-tube.

(b) Place a small sample (a few drops) of the liquid alkyne on a watch-glass and ignite it.

Note the colour of the flames.

2. (a) Add a few drops of a solution of bromine in tetrachloromethane to a test-tube of gas and shake the mixture.

(b) To a few drops of the liquid alkyne in a test-tube, add a few drops of bromine in tetrachloromethane and shake the mixture.

3. (a) Add 3–4 drops of an alkaline solution of potassium permanganate (made by dissolving about 0·1 g of anhydrous sodium carbonate in 1 cm^3 of a 1 per cent solution of potassium permanganate) to a test-tube of gas and shake the mixture. Note the colour of the solution and whether any precipitate is formed.

(b) Repeat the experiment with 2 or 3 drops of the liquid alkyne instead of the gas, warming the mixture gently.

4. (a) Prepare an ammoniacal solution of copper(I) chloride by dissolving about 0·1 g of copper(I) chloride in 1 cm^3 of a dilute solution of ammonia. Generate some ethyne, and pass the gas through the solution (Fig. 7.2).

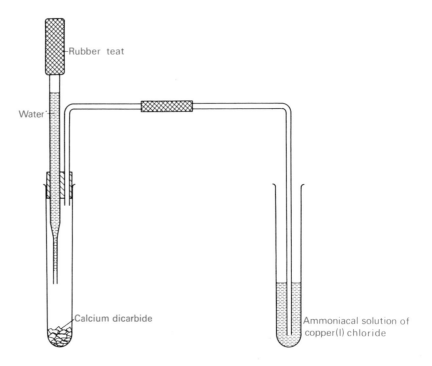

FIG. 7.2. *Reaction between ethyne and a copper(I) salt*

ALKYNES

(b) Will this reaction occur with (i) propyne (CH_3—C≡C—H), (ii) but-2-yne (CH_3—C≡C—CH_3)?

(c) Try the reaction with phenylethyne but shaking 2 or 3 drops of the liquid alkyne with 1 cm^3 of an ammoniacal solution of copper(I) chloride.

N.B. *If any solids are formed in 4(a) or 4(c), do not allow them to become dry. Wash them down the sink with plenty of water.*

7.6 Questions

1 With the aid of equations describe how and under what conditions ethyne (acetylene) reacts with (a) hydrogen, (b) chlorine, (c) hydrogen cyanide and (d) a solution of copper(I) chloride (cuprous chloride) in ammonia.

Explain briefly how you would demonstrate the presence of carbon and hydrogen in ethyne. Describe an experiment you would carry out to determine its formula. (AEB)

2 Discuss the meaning of the term *unsaturation*, illustrating your answer with reference to ethene and ethyne.

Outline the reactions by which ethyne may be converted to (a) ethanamide, (b) chloroethene, CH_2=CHCl, (c) ethane-1,2-diol, (d) benzene. (C(N))

3 Describe in outline **one** commercial preparation of ethyne.

State the reactions of ethyne with (a) chlorine and (b) aqueous mercury(II) sulphate (mercuric sulphate).

One mole of a hydrocarbon S (C_6H_{10}) gives 2 moles of ethanal with trioxygen. What are the possible structural formulae for S? (O and C)

4 (a) When 20 cm^3 of a gaseous hydrocarbon A was exploded with 150 cm^3 of oxygen, the residual gases occupied 110 cm^3. After shaking these gases with aqueous sodium hydroxide, the final volume was 30 cm^3 (all volumes at the same temperature and pressure). Calculate the molecular formula of A.

(b) Another hydrocarbon B, of molecular formula C_4H_6, formed a red precipitate with ammoniacal copper(I) chloride (cuprous chloride) and reacted with water in the presence of sulphuric acid and mercury(II) sulphate (mercuric sulphate) to give a compound C. C was unaffected by potassium permanganate but with iodine and warm aqueous sodium hydroxide gave a yellow crystalline precipitate with a characteristic odour. Deduce the structural formulae of B and C and explain the above reactions. (C(T))

Chapter 8

Aromatic compounds

8.1 Introduction

The term 'aromatic' was first used to describe a group of compounds which have a pleasant smell (aroma). These compounds include the cyclic compound, benzene, and its derivatives. The name aromatic has been retained since it is useful to classify these compounds separately; this is because their properties are so different from those of the aliphatic and alicyclic compounds.

8.2 Structure of benzene

It has been known for over 100 years that benzene is a cyclic compound, with a six-membered ring of six carbon atoms and with one hydrogen atom attached to each carbon atom. Bearing in mind that carbon and hydrogen form four bonds and one bond respectively, it was natural to represent its structure as

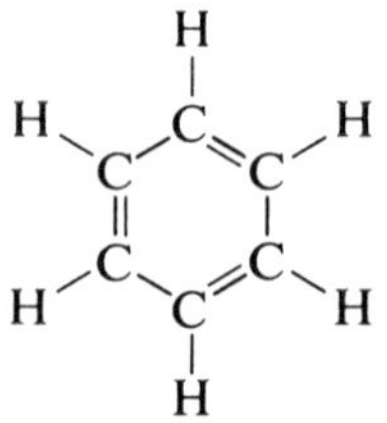

in which single and double bonds alternate round the ring. However, the following evidence shows that this representation does not adequately describe the structure of benzene.

1. If the structure was correct, we should expect to be able to isolate two isomers of any disubstituted benzene in which the substituents were adjacent, for example:

However, in no case has more than one isomer of any such compound been obtained.

2. As we shall see (8.3), benzene does not undergo the addition reactions which are characteristic of ethene (6.3) and other compounds which contain C=C bonds.

3. Whereas the length of a carbon–carbon single bond is 1·54 Å and that of a carbon–carbon double bond is 1·34 Å, the six carbon–carbon bonds in benzene are of equal length (1·39 Å), somewhat longer than double bonds but shorter than single bonds.

4. Benzene is a more stable compound than would be expected by comparison of its structure with that of an alkene, as shown by thermochemical

measurements. When benzene is reduced to cyclohexane:

benzene + 3H$_2$ ⟶ cyclohexane

the heat evolved is 207 kJ per mol. When cyclohexene is reduced to cyclohexane:

cyclohexene + H$_2$ ⟶ cyclohexane

the heat evolved is 119 kJ per mol. Now, the heat change in the latter reaction is associated with the conversion of a double bond into a single bond and the breaking of the H—H bond, each of which requires energy, and the formation of two C—H bonds, which releases energy. If benzene contained three alkene-like double bonds, we should expect that the heat evolved during its reduction would be three times that for cyclohexene, i.e. 357 kJ per mol, since the bond changes associated with the reduction of one molecule of cyclohexene take place three times over on the reduction of one molecule of benzene. The experimental value is (357 − 207) = 150 kJ per mol less than this, from which we infer that benzene is more stable by this amount of bonding energy than would be expected if it possessed alkene-like double bonds.

The differences described above between benzene and the alkenes can all be understood by consideration of the nature of the bonding in benzene. Each carbon atom is *sp*^{2}-hybridised, so that it forms three coplanar bonds, two with other carbon atoms and one with a hydrogen atom. These bonds are at an angle of 120° to each other, just as in an alkene, and this means that the six carbon atoms are in a plane and that there is no strain in the ring (compare cyclohexane, p. 73, where a planar structure would be strained because the normal bond angle for *sp*^{3}-hybridised carbon is 109° 28′). The remaining electron is in a *p* orbital perpendicular to the plane of this ring, just as in ethene each of the carbon atoms has a *p* orbital perpendicular to the plane containing the atoms. Now, in benzene, each of these *p* orbitals can overlap with *both* the neighbouring *p* orbitals, Fig. 4.15 (p. 57). Since bond energies are determined by the extent of orbital-overlap (4.3), there is a greater degree of bonding in benzene than would be expected by comparison with ethene. This underlies the thermochemical stability of benzene and also its resistance to addition reactions, since a reaction such as:

benzene + Br$_2$ ⟶ dibromocyclohexadiene

would result in the loss of the extensive *p*-orbital interactions. It also enables us to understand why the bond lengths in the ring are equal, because, as Fig. 4.15 shows, there is no difference in the type of bonding between one pair of adjacent carbon atoms and any other pair.

The bonds associated with the *p*-orbital interactions in benzene are termed **delocalised π-bonds**, in contrast to the **localised π-bond** in an alkene which is confined to only two atoms. The extra thermochemical stability of benzene as compared with what would be expected if it possessed alkene-like double bonds is termed its **delocalisation energy** or **stabilisation energy**.

As described in Chapter 4, there is an alternative way of representing the delocalised bonds in benzene, that is, by describing benzene as a resonance hybrid of two structures:

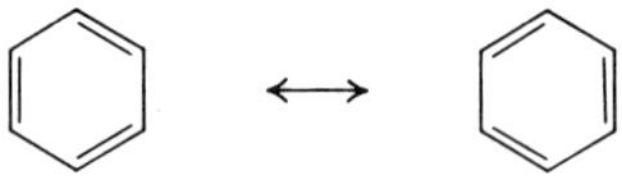

Although delocalisation in this book is often described in terms of molecular orbitals, it is sometimes easier to discuss the properties of aromatic compounds in terms of resonance hybrids. In equations, benzene is represented as:

8.3 Benzene

Manufacture

1. *Based on petroleum*

(a) If the gasoline and naphtha fractions from the distillation of petroleum (19.4) are passed over a catalyst (either platinum or molybdenum(VI) oxide, suspended on alumina) in presence of an excess of hydrogen, the straight-chain alkanes undergo cyclisation and dehydrogenation. For example, hexane is converted into benzene,

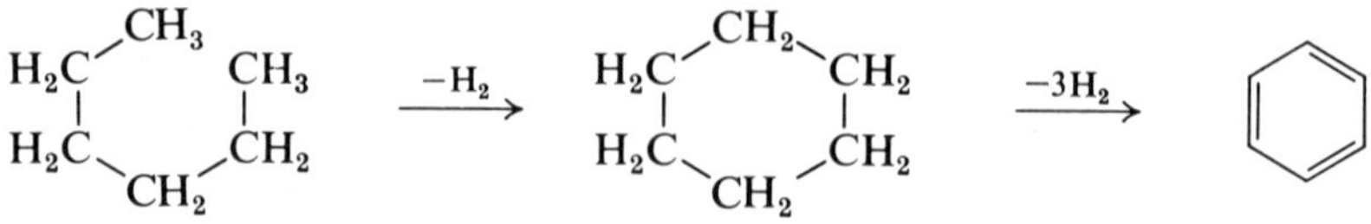

and heptane into methylbenzene (p. 295). Benzene is separated from the other aromatic hydrocarbons which have been formed by solvent extraction and by fractional distillation and crystallisation (20.5). The process is known as **reforming**.

(b) Benzene is one of the products when the light gas-oil fraction undergoes **catalytic cracking** (19.6).

(c) It is also formed from other aromatic hydrocarbons by **hydrodealkylation** (20.5), for example:

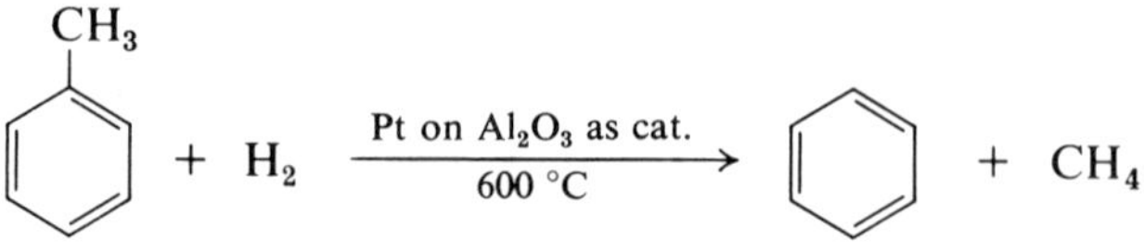

2. *Based on coal*

Benzene is obtained by the fractional distillation of coal tar, which is itself obtained by the destructive distillation of coal.

Plate 8.1. *A reforming unit. Naphtha is heated in a furnace (A), mixed with hydrogen, compressed to a high pressure (B) and passed over a heated platinum catalyst. The reactors are hidden in the photograph (C). Low boiling hydrocarbons are removed by fractional distillation, for example ethane, propane, butane (D). The residue, which is fractionated (E), contains principally the aromatic hydrocarbons, benzene, methylbenzene and dimethylbenzenes (Esso Petroleum Co. Ltd.)*

Physical properties

Benzene is a colourless liquid with a characteristic odour. It is insoluble in water but soluble in all organic solvents, and it is itself a very good solvent for organic compounds. It freezes at 5°C and boils at 80°C. Both the liquid and the vapour are highly poisonous, so that benzene must be used with care.

Chemical properties

(a) Substitution reactions

Benzene takes part in a variety of substitution reactions with electrophilic reagents.

1. When treated with a mixture of concentrated nitric acid and concentrated sulphuric acid at room temperature, nitrobenzene is formed:

$$\text{C}_6\text{H}_6 + \text{HNO}_3 \longrightarrow \text{C}_6\text{H}_5\text{NO}_2 + \text{H}_2\text{O}$$

Nitrobenzene

AROMATIC COMPOUNDS

This reaction occurs in several stages. First, sulphuric acid is so strong an acid that it transfers a proton to nitric acid:

$$H_2SO_4 + H-O-NO_2 \rightleftharpoons HSO_4^- + \overset{H}{\underset{H}{>}}\overset{+}{O}-NO_2$$

The protonated nitric acid then undergoes a spontaneous heterolysis to form the **nitronium ion**, NO_2^+:

$$\overset{H}{\underset{H}{>}}\overset{+}{O}-NO_2 \longrightarrow H_2O + O=\overset{+}{N}=O$$

The nitronium ion is the electrophilic reagent with which benzene reacts. The first step is an addition:

[benzene + NO_2^+ → cyclohexadienyl cation with H and NO_2]

The positive charge in this adduct is actually delocalised over three of the carbon atoms; the ion can be described as a resonance hybrid of three structures:

[three resonance structures of the arenium ion with H and NO_2 on sp3 carbon and + charge delocalised]

In the final step of the reaction, a proton is removed from the adduct by the hydrogen sulphate anion:

[arenium ion + HSO_4^- → nitrobenzene (NO_2) + H_2SO_4]

There is strong evidence for the existence of the nitronium ion. For example, compounds such as nitronium perchlorate, NO_2ClO_4, have been prepared and shown, by X-ray analysis, to contain the ion NO_2^+. In addition, the depression of the freezing point of sulphuric acid by dissolved nitric acid is four times greater than expected; evidently each molecule of nitric acid provides four particles in the solution, consistent with the ionisation:

$$HNO_3 + 2H_2SO_4 \rightarrow NO_2^+ + H_3O^+ + 2HSO_4^-$$

The mechanism described above is typical of the electrophilic substitution reactions of benzene, and the others will not be described in such detail.

2. If chlorine is passed through benzene *at room temperature* and in the presence of a catalyst, substitution takes place:

$$C_6H_6 + Cl_2 \rightarrow C_6H_5\text{---}Cl + HCl$$
$$\text{Chlorobenzene}$$

Suitable catalysts are iron filings and aluminium trichloride; they are referred to as **halogen carriers**.

The function of the catalyst is to withdraw the electrons from the bond between the chlorine atoms, a process represented for aluminium trichloride as:

$$Cl\text{---}Cl \quad AlCl_3$$

where the curved arrow represents the tendency for two electrons to move into the vacant $2p$ orbital of the aluminium atom. As this happens, the benzene ring provides two electrons to make good the deficiency of electrons on one of the chlorine atoms, so that the whole process can be represented as:

$$C_6H_6 + Cl\text{---}Cl + AlCl_3 \longrightarrow [C_6H_6Cl]^+ + AlCl_4^-$$

The carbonium ion then reacts with the $AlCl_4^-$ ion:

$$[C_6H_6Cl]^+ + AlCl_4^- \longrightarrow C_6H_5\text{---}Cl + HCl + AlCl_3$$

so that the catalyst is regenerated.

When iron is used as the halogen carrier, it first reacts with chlorine to give iron(III) chloride which then catalyses the reaction in the same way as aluminium trichloride.

In the presence of iron filings or aluminium tribromide, benzene reacts with bromine to give bromobenzene:

$$C_6H_6 + Br_2 \rightarrow C_6H_5\text{---}Br + HBr$$
$$\text{Bromobenzene}$$

The mechanism of the reaction is the same as that in chlorination.

The iodination of benzene is usually brought about by refluxing benzene, iodine and concentrated nitric acid:

$$C_6H_6 + I_2 \rightarrow C_6H_5I + HI$$

It was once thought that the function of the nitric acid was to oxidise the hydrogen iodide as it was formed and thereby prevent the reverse reaction. However, this cannot be the correct explanation, for hydrogen iodide does not react with iodobenzene to give benzene and iodine. It is now thought that the acid serves to provide the active iodinating species, possibly

$$I\text{---}\overset{+}{N}\begin{smallmatrix}\diagup O \\ \diagdown OH\end{smallmatrix}$$

This is one example of the way in which our interpretations of the mechanisms of organic reactions change in the light of experimental evidence.

3. Benzene reacts with an alkyl halide in the presence of an aluminium halide to give an alkylbenzene, for example:

$$C_6H_6 + CH_3CH_2\text{—Br} \xrightarrow{\text{AlBr}_3 \text{ as cat.}} C_6H_5\text{—}C_2H_5 + HBr$$
$$\text{Ethylbenzene}$$

The reaction, which is described as **alkylation**, occurs in a similar way to chlorination.

The catalyst withdraws the pair of electrons from the C—Br bond, the benzene ring provides a pair of electrons to form a bond to the alkyl group:

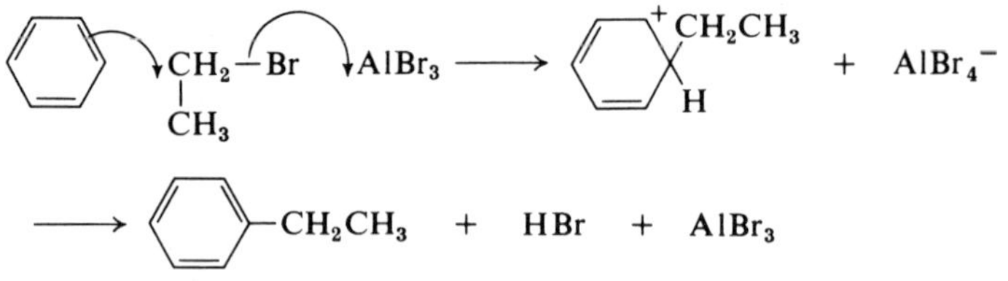

[reaction mechanism diagram showing benzene attacking CH₂(CH₃)—Br with AlBr₃, forming the arenium ion intermediate with +CH₂CH₃ and H, plus AlBr₄⁻]

$$\longrightarrow \bigcirc\text{—}CH_2CH_3 + HBr + AlBr_3$$

Methylbenzene is formed from benzene and iodomethane (the iodide being used as it is the only methyl halide which is a liquid at room temperature):

$$C_6H_6 + CH_3I \xrightarrow{\text{AlCl}_3 \text{ as cat.}} C_6H_5\text{—}CH_3 + HI$$
$$\text{Methylbenzene}$$

A similar reaction takes place with acid halides to give ketones, for example:

$$C_6H_6 + CH_3\text{—CO—Cl} \xrightarrow{\text{AlCl}_3 \text{ as cat.}} C_6H_5\text{—CO—}CH_3 + HCl$$
$$\text{Phenylethanone}$$

The process is known as **acylation**.

The reactions of aromatic compounds with alkyl halides and acid halides are known as **Friedel-Crafts reactions**, after the names of their two discoverers.

4. Benzene reacts with an alkene in the presence of an acid to give an alkylbenzene, for example:

$$C_6H_6 + CH_3\text{—CH=}CH_2 \xrightarrow{\text{H}^+ \text{ as cat.}} C_6H_5\text{—CH}(CH_3)_2$$
$$\text{(1-Methylethyl)benzene}$$

Reaction occurs by addition of a proton to the alkene to give a carbonium ion which then reacts with the aromatic compound:

$$CH_3\text{—CH=}CH_2 + H^+ \rightarrow CH_3\text{—}\overset{+}{C}H\text{—}CH_3$$

$$C_6H_6 + CH_3\text{—}\overset{+}{C}H\text{—}CH_3 \rightarrow C_6H_5\text{—CH}(CH_3)_2 + H^+$$

5. If a mixture of benzene and concentrated sulphuric acid is heated for about 8 hours, benzenesulphonic acid is formed:

$$C_6H_6 + H_2SO_4 \rightarrow C_6H_5\text{—}SO_2OH + H_2O$$
$$\text{Benzenesulphonic acid}$$

AROMATIC COMPOUNDS

The reaction is another example of an electrophilic substitution. The reagent is sulphur trioxide, which is present in a solution of concentrated sulphuric acid and accepts a pair of electrons from benzene:

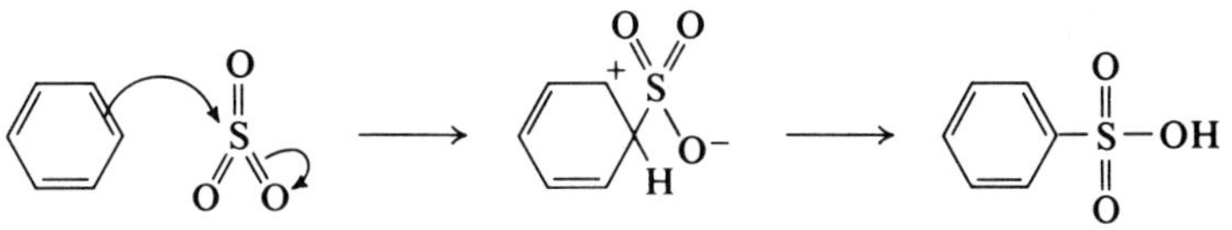

Note that in this organic sulphur compound, carbon is linked to sulphur, whereas in ethyl hydrogen sulphate (p. 78) it is linked to oxygen ($CH_3-CH_2-O-SO_2-OH$).

(b) Other reactions

1. If a mixture of hydrogen and benzene vapour is passed over nickel at 150°C, cyclohexane is formed:

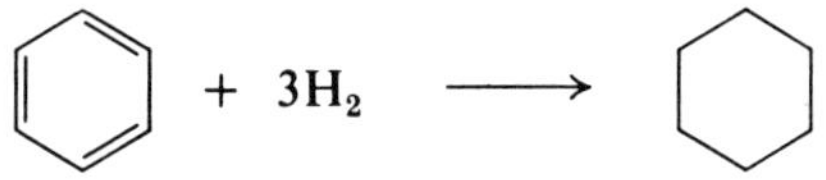

$$\text{C}_6\text{H}_6 + 3H_2 \longrightarrow \text{C}_6\text{H}_{12}$$

2. In the presence of ultraviolet light, benzene reacts with chlorine by addition:

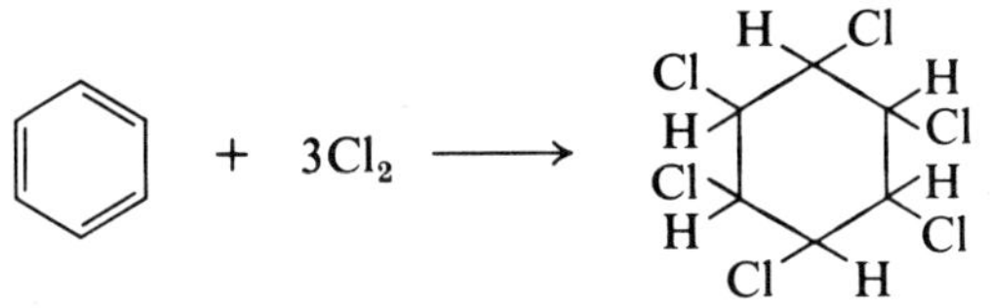

$$\text{C}_6\text{H}_6 + 3Cl_2 \longrightarrow \text{C}_6\text{H}_6\text{Cl}_6$$

The resulting product, hexachlorocyclohexane, is a mixture of geometrical isomers, one of which is an important insecticide, sold as Gammexane.

Uses

The uses of benzene are discussed in Section 20.5.

8.4 Methylbenzene

Structural formula

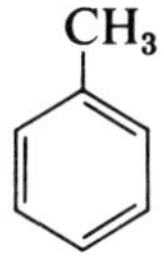

Manufacture

Methylbenzene (toluene) is obtained both from coal and from petroleum in the same way as benzene.

Physical properties

Like benzene, methylbenzene is a colourless liquid which is insoluble in water but soluble in organic solvents. It melts at −95°C and boils at 111°C. Note that the melting point is lower than that of benzene although methylbenzene has the higher formula weight. This is because the planar molecules of benzene can pack closely together in the crystal and the cohesive forces are strong, whereas the methyl group in methylbenzene prevents such close packing.

Chemical properties

Methylbenzene undergoes three types of reactions: (a) electrophilic substitution in the ring, (b) addition to the ring and (c) substitution in the methyl group.

(a) Electrophilic substitution reactions

Methylbenzene reacts with all the electrophilic reagents which attack benzene, and in each reaction it is more reactive than benzene. There are three possible isomeric products in each case; for example, for nitration:

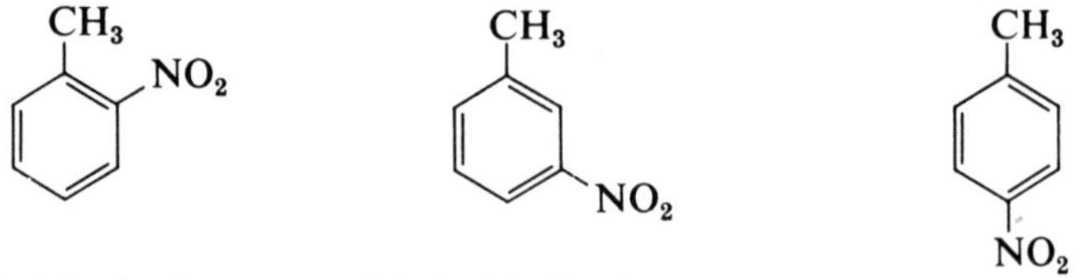

Methyl-2-nitrobenzene   Methyl-3-nitrobenzene   Methyl-4-nitrobenzene

The prefixes *ortho*, *meta* and *para*, usually abbreviated to *o*, *m* and *p*, are often used instead of the numbers 2, 3 and 4, respectively, for describing the relative positions of the substituents in a disubstituted benzene; for example, methyl-2-nitrobenzene can be called *o*-methylnitrobenzene.

Of these three products, the principal ones from methylbenzene are always the 2- and 4-isomers; for example, in nitration, with a mixture of concentrated nitric and sulphuric acids, the relative amounts of the three products, expressed as percentages, are 2-, 59; 3-, 4; 4-, 37. The methyl group in methylbenzene is described as **2-,4-directing** (*ortho*, *para*-directing).

The reasons for both the greater reactivity of methylbenzene than benzene and the predominance of 2- and 4-substitution in methylbenzene can be understood by considering the first step in the reaction. Thus, in nitration:

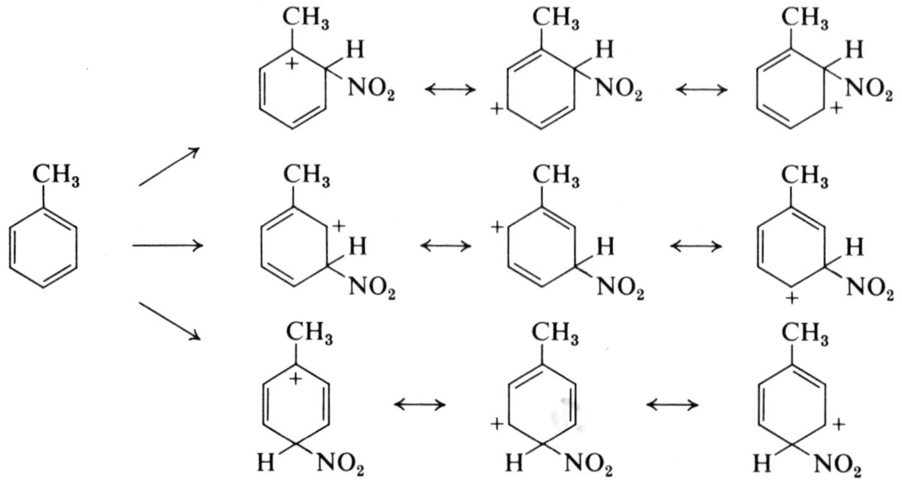

In each case a carbonium ion is formed which is a resonance hybrid of three structures; the positive charge is shared by three of the ring-carbon atoms.

Since the methyl group is electron-releasing ($+I$) and serves to stabilise a positive charge (4.9), each of the three adducts is more stable than that from benzene and is formed faster; thus, methylbenzene is more reactive than benzene. Of the three adducts, those formed by reaction at the 2- and 4-positions have positive charge adjacent to the methyl group, so that they are more stable than the adduct formed by reaction at the 3-position in which a carbon atom is interposed between the positive charge and the methyl group; consequently, reaction occurs faster at the 2- and 4-positions than at the 3-position, that is, methylbenzene is 2-,4-directing.

Other examples of electrophilic substitutions in methylbenzene are:

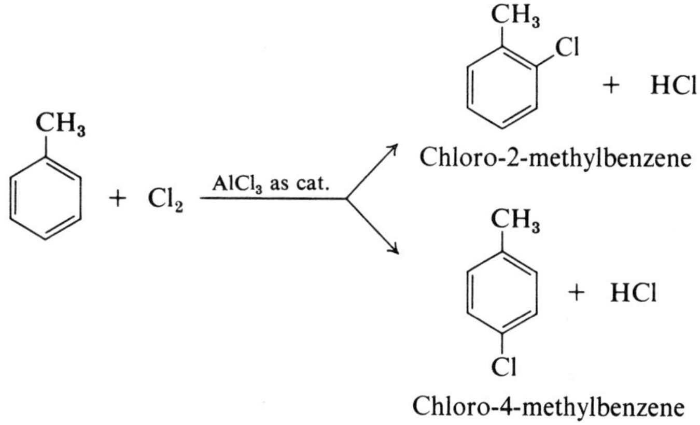

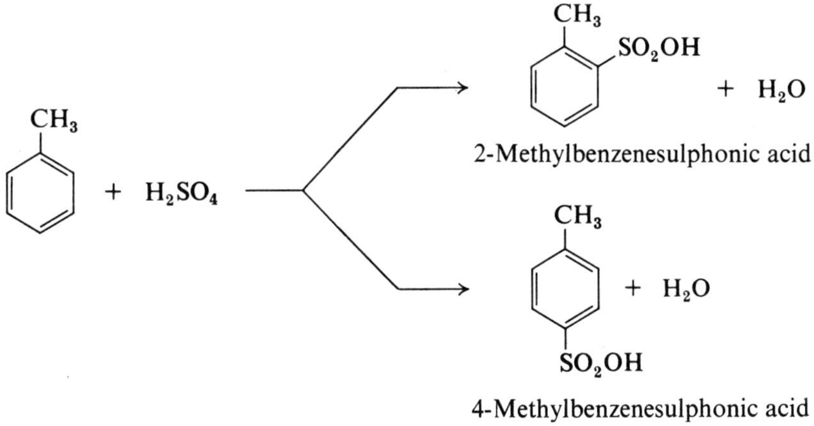

$C_6H_5CH_3 + Cl_2 \xrightarrow{AlCl_3 \text{ as cat.}}$ Chloro-2-methylbenzene + HCl
and Chloro-4-methylbenzene + HCl

$C_6H_5CH_3 + H_2SO_4 \longrightarrow$ 2-Methylbenzenesulphonic acid + H_2O
and 4-Methylbenzenesulphonic acid + H_2O

(b) Addition reactions

Methylbenzene resembles benzene in reacting with hydrogen on nickel at 150°C:

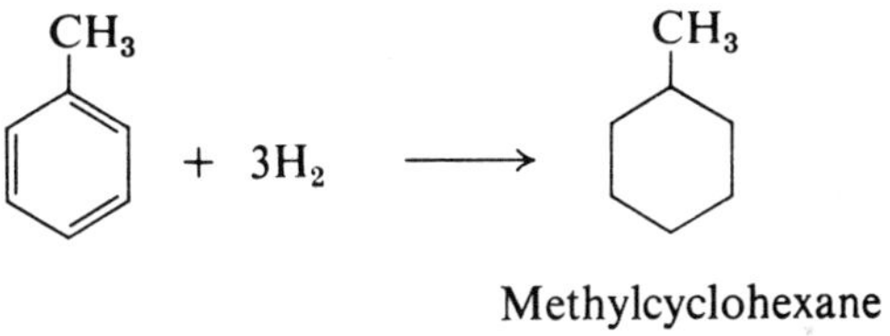

$C_6H_5CH_3 + 3H_2 \longrightarrow$ Methylcyclohexane

However, when treated with chlorine in the presence of ultraviolet light, it preferentially reacts in the methyl group (p. 104).

AROMATIC COMPOUNDS

(c) Substitution in the methyl group

1. When chlorine is passed through boiling methylbenzene which is exposed to ultraviolet light, substitution of chlorine for hydrogen occurs:

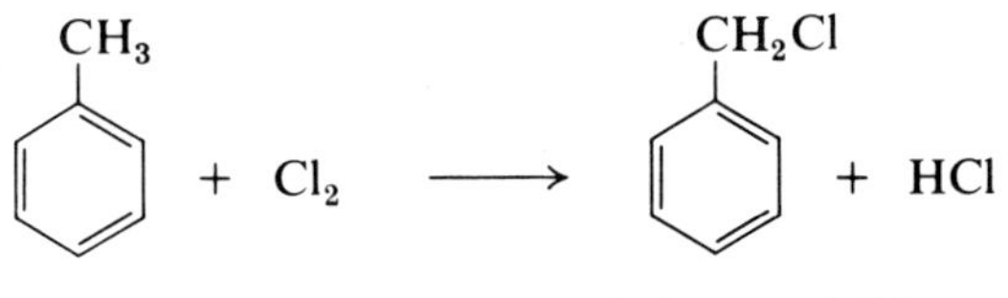

$$C_6H_5\text{—}CH_3 + Cl_2 \longrightarrow C_6H_5\text{—}CH_2Cl + HCl$$

(Chloromethyl)benzene

$$C_6H_5\text{—}CH_2Cl + Cl_2 \longrightarrow C_6H_5\text{—}CHCl_2 + HCl$$

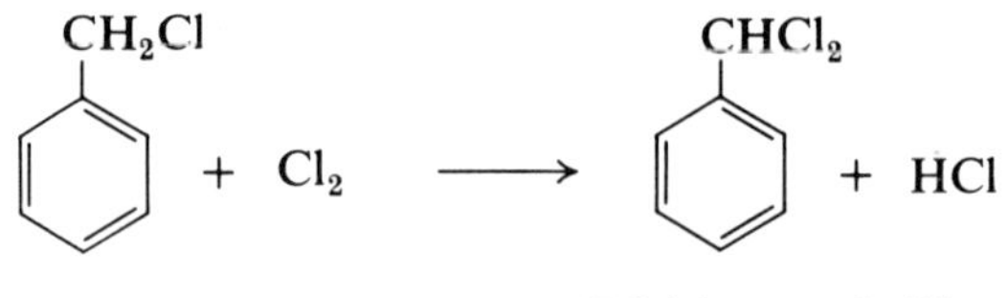

(Dichloromethyl)benzene

$$C_6H_5\text{—}CHCl_2 + Cl_2 \longrightarrow C_6H_5\text{—}CCl_3 + HCl$$

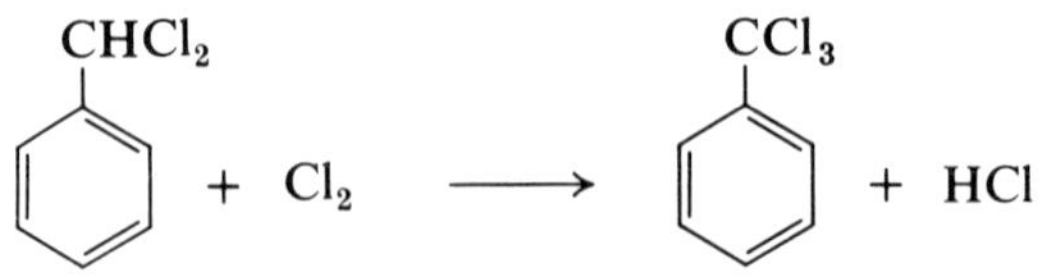

(Trichloromethyl)benzene

These reactions occur *via* free radicals, as in the chlorination of methane (p. 69). They are in contrast to the ionic reactions which occur when methylbenzene is treated with chlorine at room temperature in the presence of a halogen carrier, which give mainly 2- and 4-chloro-derivatives in the same way as benzene gives chlorobenzene (p. 99).

2. The methyl group in methylbenzene can be oxidised to the aldehyde group, —CH=O, or to the carboxylic acid group, —CO$_2$H, depending on the oxidising agent:

$$C_6H_5\text{—}CH_3 \xrightarrow{CrO_2Cl_2} C_6H_5\text{—}CHO$$
Benzaldehyde

$$C_6H_5\text{—}CH_3 \xrightarrow[\text{alkaline}]{KMnO_4} C_6H_5\text{—}CO_2H$$
Benzoic acid

8.5 Substitution in other aromatic compounds

As we have seen, the methyl group in methylbenzene is 2-,4-directing in electrophilic substitutions because it is an electron-releasing group and preferentially stabilises the adducts formed by the reaction at the 2- and 4-positions. Other alkyl groups have the same effect; for example, chlorination of ethylbenzene in the presence of a halogen carrier gives mainly chloro-2- and chloro-4-ethylbenzene:

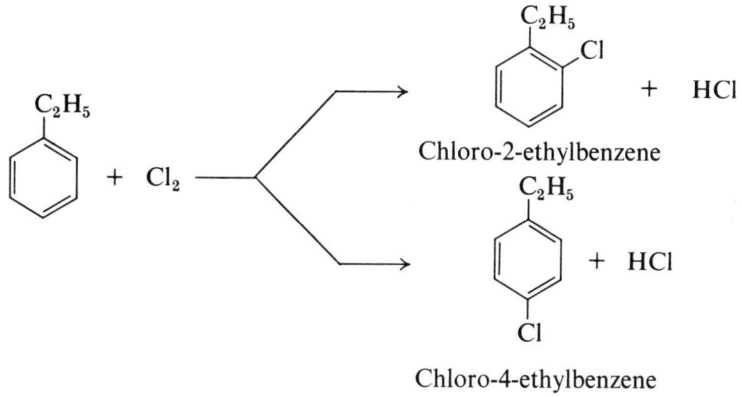

AROMATIC COMPOUNDS

C₂H₅–C₆H₅ + Cl₂ →
- Chloro-2-ethylbenzene + HCl
- Chloro-4-ethylbenzene + HCl

When an electron-attracting group such as —NO$_2$ or —CN (p. 61) is attached to the benzene ring, it reduces the stability of the intermediate cation in an electrophilic substitution reaction. Consequently, compounds such as nitrobenzene (C$_6$H$_5$NO$_2$) and benzonitrile (C$_6$H$_5$CN) react less rapidly than benzene. This destabilising effect is greatest for reaction at the 2- or 4-position, for the electron-attracting substituent is then adjacent to one of the carbon atoms which shares the positive charge, for example,

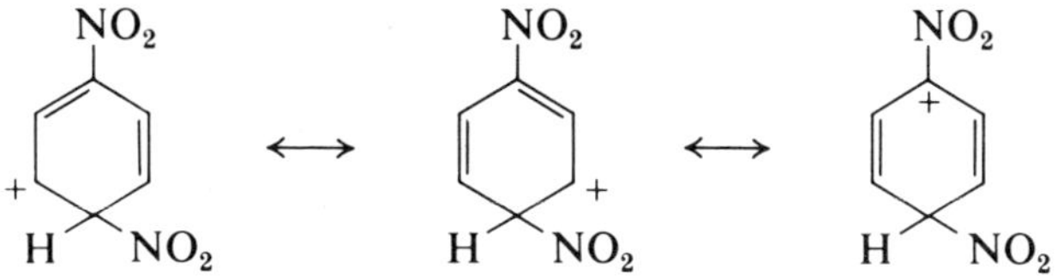

whereas for reaction at the 3-position an extra carbon atom is interposed between the substituent and the charge, for example:

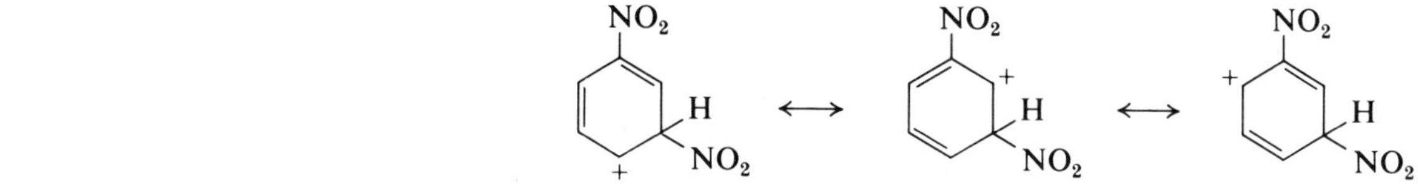

Hence these electron-attracting substituents are 3-directing.

An illustration of the inhibiting effect which the nitro group has on electrophilic substitution, as well as its 3-directing capacity, is provided by the nitration of nitrobenzene: when benzene is treated with a mixture of concentrated nitric and concentrated sulphuric acids at room temperature, nitrobenzene is formed, but the temperature has to be raised to about 50°C for further reaction to occur to give 1,3-dinitrobenzene.

In contrast, those electron-attracting groups bonded to benzene which contain an atom with an unshared pair of electrons directly attached to the aromatic ring are 2-,4-directing in electrophilic substitution; examples are —Cl, —OH and —NH$_2$.

The reason is again apparent when the intermediate adducts in the substitution are considered. For example, when chlorobenzene is nitrated, the possible adducts are:

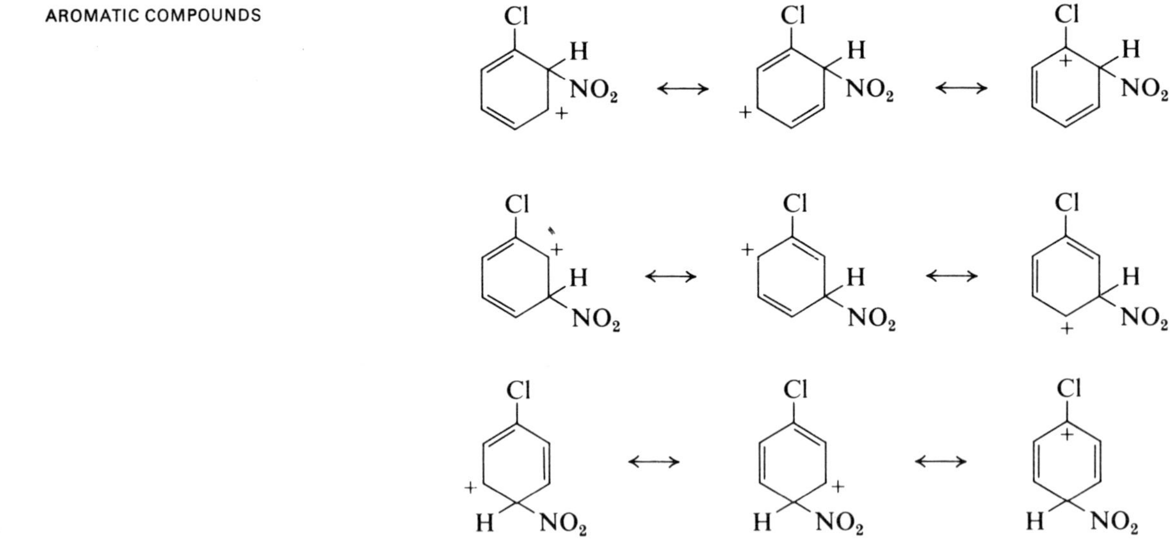

[Resonance structures of chloroarene-nitro adducts at 2-, 3-, and 4-positions]

In the cases of the 2- and 4-adducts, the positive charges are delocalised not only on to three carbon atoms but also on to the chlorine atom; that is, a p orbital on chlorine can overlap with the adjacent carbon p orbital, the result being that the pair of electrons in the chlorine p orbital is partly donated to the carbon p orbital, so reducing the deficiency of electrons in the latter. This can be represented as follows:

[Structures showing Cl lone pair donation into ring for 2- and 4-adducts]

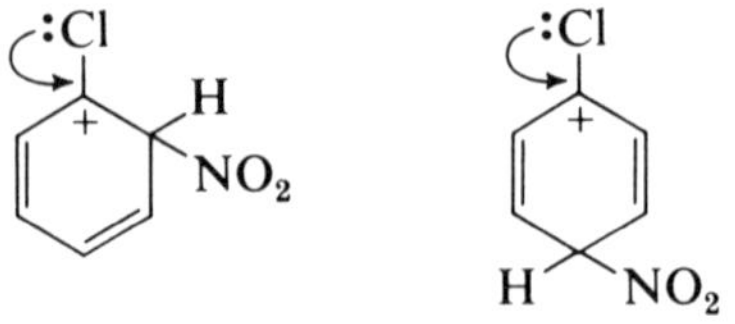

This corresponds to the sharing of the positive charge by the chlorine atom in addition to the carbon atoms, and could be represented alternatively by the structures:

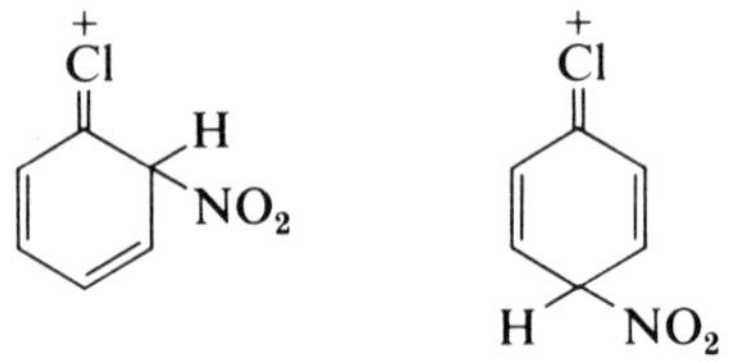

[Structures with Cl^+ double-bonded to ring for 2- and 4-adducts]

This extra delocalisation of the charge, which cannot occur when the reagent adds to the 3-position, makes the adducts formed by reaction at the 2- and 4-positions more stable than that formed by reaction at the 3-position; hence the chlorine substituent is 2-,4-directing.

The hydroxyl (—OH) and amino (—NH_2) substituents, each of which contains unshared pairs of electrons, act in the same way as the chlorine substituent; their ability to stabilise the 2- and 4-adducts formed during electrophilic substitution can be represented as follows:

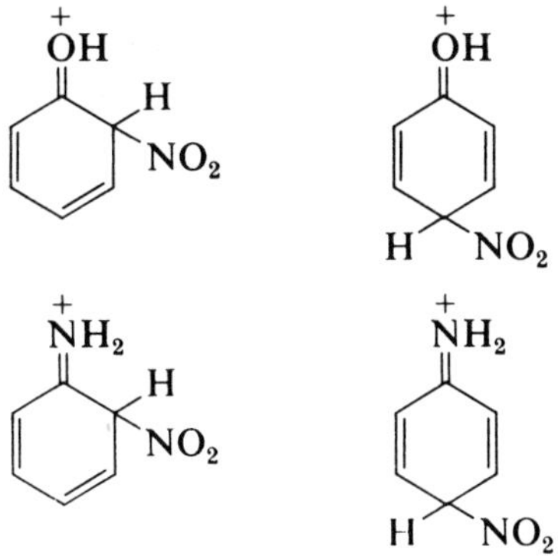

AROMATIC COMPOUNDS

[Resonance structures showing protonated OH and NH₂ intermediates with H and NO₂ at ortho and para positions]

Oxygen and nitrogen release p electrons in this way more readily than does chlorine, and as a result phenol (C_6H_5—OH) (p. 153) and phenylamine (C_6H_5—NH_2) (p. 249) are much more reactive than chlorobenzene or benzene. For instance, they are so reactive towards chlorine and bromine that it is impossible to stop the reaction before all the activated positions have been substituted, even in the absence of a halogen carrier, for example:

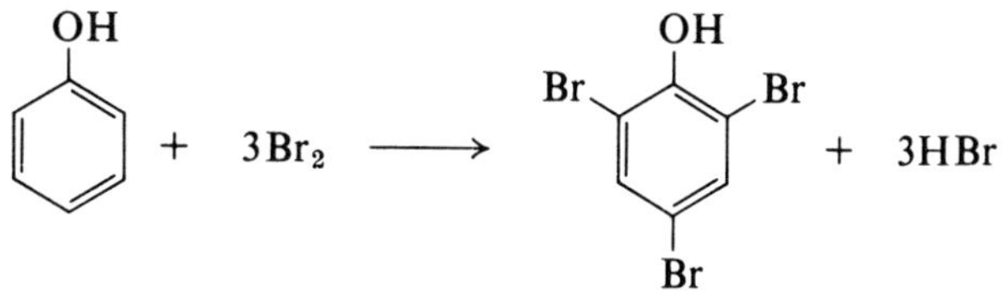

$$C_6H_5OH + 3Br_2 \longrightarrow \text{2,4,6-Tribromophenol} + 3HBr$$

Summary of directing power of substituents

	2-,4-DIRECTING (ORTHO,PARA-DIRECTING)		3-DIRECTING (META-DIRECTING)
	Substituent is electron-releasing (+I effect)	Substituent is electron-attracting (−I effect) but possesses an unshared pair of electrons	Substituent is electron-attracting (−I effect)
	—Alkyl, as in methylbenzene C_6H_5—CH_3	—Halogen (—F, —Cl, —Br, —I), as in chlorobenzene, C_6H_5—Cl —OH, as in phenol, C_6H_5—OH —NH_2, as in phenylamine, C_6H_5—NH_2	—NO_2, as in nitrobenzene C_6H_5—NO_2 —CHO, as in benzaldehyde, C_6H_5—CHO —COR, as in phenylethanone C_6H_5—$COCH_3$ —CO_2H, as in benzoic acid, C_6H_5—CO_2H —SO_2OH, as in benzenesulphonic acid, C_6H_5—SO_2OH

8.6 Practical work

Reactions of methylbenzene

Compare your results with those obtained with an alkane (p. 74) and an alkene (p. 85).

1. To 5 drops of methylbenzene in a test-tube, add 1 cm^3 of a solution of bromine in tetrachloromethane and shake.

2. Place 5 drops of methylbenzene in each of two test-tubes, and to one of them add a few iron filings. Add 3 drops of bromine to both test-tubes and note whether there is evolution of gas from either. It may take a few minutes to see whether a gas is evolved. Test the gas with moist blue litmus paper.

3. To 10 drops of *concentrated* nitric acid in a test-tube, carefully add 10 drops of *concentrated* sulphuric acid, shaking the mixture and cooling the test-tube under a stream of cold water. Add the mixture of acids to 5 drops of methylbenzene in another test-tube, and shake this mixture under a stream of cold water; then pour into a beaker containing about 10 cm^3 of cold water. Observe whether a new liquid is formed and note its smell.

4. To 4 drops of methylbenzene, add 10 drops of *concentrated* sulphuric acid in a test-tube. Warm until the methylbenzene has dissolved into the acid layer. Pour the mixture into 10 cm^3 of a cold saturated solution of sodium chloride. White crystals of a mixture of sodium methylbenzenesulphonates are formed.

5. Make up about 25 cm^3 of an alkaline solution of potassium permanganate (add 0.1 g of sodium carbonate to a 10 per cent solution of potassium permanganate).
To 10 drops of methylbenzene, add the alkaline solution of potassium permanganate in a flask. Heat the mixture under reflux (Fig. 2.1) until the purple colour disappears.
Cool the mixture and add dilute sulphuric acid until the mixture is acid to litmus. Add solid sodium metabisulphite until the brown solid (manganese(IV) oxide) has dissolved. Filter the white crystals of benzoic acid and recrystallise them from hot water. Find the m.p. of benzoic acid.

Preparation of hexachlorocyclohexane

The experiment must be carried out in a fume-cupboard and take great care not to breathe in the vapour of chlorine or benzene. Benzene is highly toxic and must only be used under supervision.

The apparatus must be in a fume-cupboard, covered with black paper or cloth so that no harm will be done to eyes by the ultraviolet light. The chlorine escaping should be washed in 2M sodium hydroxide solution and passed into a vent of the fume cupboard.

Chlorine (if a cylinder is not available, the gas may be generated by the action of concentrated hydrochloric acid on potassium permanganate) must be washed with water and dried by passage through concentrated sulphuric acid. Allow it to pass through 20 cm^3 of benzene at a slow rate for about an hour, keeping the benzene at about 45°C on a water-bath and irradiating with an ultraviolet lamp (Fig. 8.1).

Steam distil the resulting solution (Fig. 2.6) until no more benzene passes over, and filter, dry and recrystallise the solid residue from benzene. One isomer of hexachlorocyclohexane is an insecticide (Gammexane).

FIG. 8.1. *Preparation of hexachlorocyclohexane*

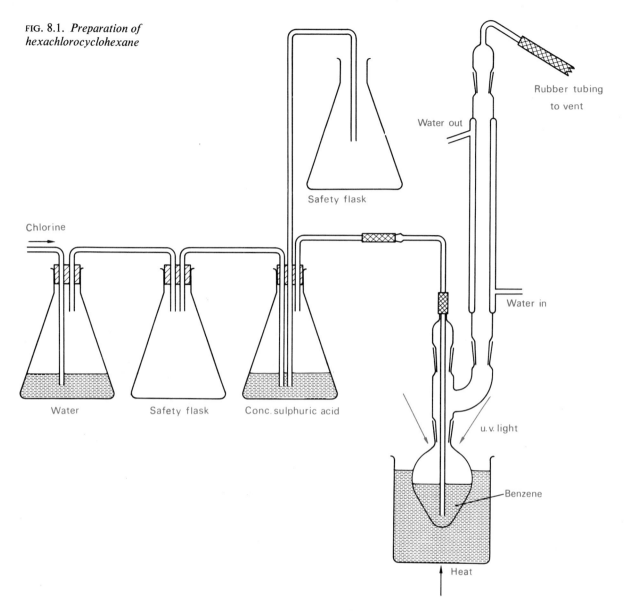

8.7 Questions

1. Describe **two** reactions which show that ethene (an alkene) and benzene are different and **two** reactions which show them to be similar.
   Show by diagrams how you consider benzene to be structurally related to ethene and then briefly explain why it differs in its behaviour. (JMB)

2. How, and under what conditions, does (i) chlorine, (ii) nitric acid, react with (a) benzene, (b) methylbenzene?
   Outline how benzene may be converted into methylbenzene and vice versa. (C(N))

3. Distinguish between an aliphatic and an aromatic compound.
   Give one reaction to illustrate the saturated nature and one reaction to illustrate the unsaturated nature of the benzene molecule.
   Explain how you would convert benzene into (a) benzene carboxylic acid (benzoic acid), (b) methylbenzene (toluene), (c) hydroxybenzene (phenol), and also how each of these products may be reconverted into benzene. (AEB)

AROMATIC COMPOUNDS

4 The average bond energies associated with the C—H bond, the C—C bond and the C=C bond are respectively 98·7, 82·6 and 146·0 kcal mole^{-1}. Use these values to calculate the theoretical enthalpy of formation of the molecule:

[Structure of benzene with alternating single and double bonds, all C—H bonds shown]

In practice, the enthalpy of formation of benzene from atoms is 1317 kcal mole^{-1}. How do you account for the difference in these quantities? Discuss the action of bromine on benzene in the light of your answer.

The normal length of a C—C bond is 1·54 Å, and for a C=C bond is 1·33 Å. Suggest values, which might be observed experimentally, for the lengths of the various bonds in

(a) benzene,
(b) graphite,
(c) diamond, and
(d) buta-1,3-diene (CH$_2$=CH—CH=CH$_2$).

Indicate briefly the reasons for your answers. (L(X))

5 There are **three** typical ways in which methylbenzene might react with chlorine. State what these are and indicate the conditions necessary for two of them to take place.

How can methylbenzene or its chlorination products be converted into benzaldehyde?

6 (a) Write equations to show how, given supplies of benzene and methylbenzene, you would prepare the following compounds. (Full practical details are not required, but reagents and conditions should be indicated.)

(i) C$_6$H$_5$I,   (ii) C$_6$H$_5$NHCOCH$_3$,   (iii) C$_6$H$_5$CHO.

(b) A compound is believed to have structure X.

[Benzene ring with OCH$_2$·CO·OC$_2$H$_5$ and NH$_2$ substituents] X

By what chemical reactions would you show the presence of:
(i) the benzene ring;
(ii) the amino group;
(iii) the ester group? (C(N, S))

7 How is methylbenzene obtained industrially and how may it be prepared from benzene? By what reactions may (a) benzaldehyde, (b) benzoic acid, (c) a chloromethylbenzene, and (d) benzene be obtained from methylbenzene?

8 What are the main differences between aliphatic and aromatic hydrocarbons? How can these types of hydrocarbons be converted into (a) acids, and (b) bases?

9 How would you obtain a pure sample of methylbenzene from benzene? Suggest feasible routes for the conversion of methylbenzene into 4-methylphenol, phenylethanoic acid, 4-nitrobenzoic acid, and 3-nitrobenzoic acid. (O Schol.)

Chapter 9 — Halogen compounds

9.1 Introduction

The four halogens (fluorine, chlorine, bromine and iodine) are contained in several types of organic compound:

Alkyl halides, in which the halogen atom is attached to a saturated carbon atom (e.g. bromoethane, CH_3—CH_2—Br). These can be subdivided into three classes, according to how many alkyl groups are attached to the carbon atom which is bonded to the halogen:

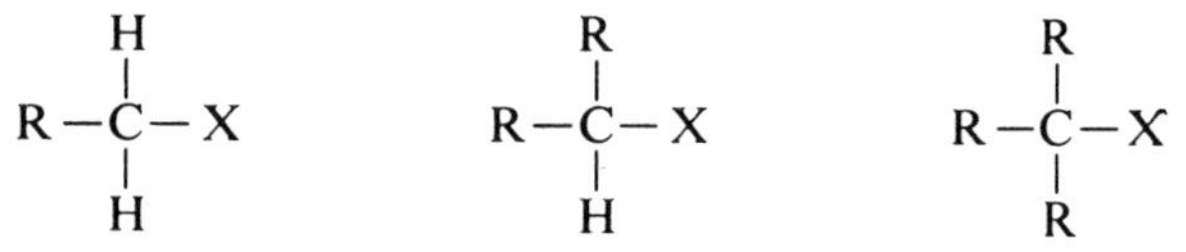

$$
\begin{array}{ccc}
\text{H} & \text{R} & \text{R} \\
| & | & | \\
\text{R}-\text{C}-\text{X} & \text{R}-\text{C}-\text{X} & \text{R}-\text{C}-\text{X} \\
| & | & | \\
\text{H} & \text{H} & \text{R} \\
\text{Primary} & \text{Secondary} & \text{Tertiary}
\end{array}
$$

These descriptions correspond to those of the related alcohols (10.2).

Unsaturated halides, in which the halogen atom is attached to a carbon atom which forms a double or triple bond (e.g. chloroethene, CH_2=CH—Cl).

Aryl halides, in which the halogen atom is attached to an aromatic ring (e.g. chlorobenzene, C_6H_5—Cl).

There are also **polyhalides**, which contain more than one halogen atom (e.g. 1,2-dichloroethane, CH_2Cl—CH_2Cl; trichloromethane, $CHCl_3$).

9.2 Nomenclature

Alkyl halides are named as derivatives of the corresponding alkane, a number being inserted to indicate the position of the halogen atom in the carbon chain where there would otherwise be ambiguity. The simpler ones can also be named by combining the names of the appropriate alkyl group and the halide; for example, CH_3Cl is chloromethane or methyl chloride. Unsaturated halides are generally named as derivatives of the corresponding alkene or alkyne. Examples are in Table 9.1. Aryl halides are described in Section 9.6.

9.3 Alkyl halides

This section considers chlorides, bromides and iodides. Fluorides are discussed separately (9.7).

Physical properties

Chloromethane, chloroethane and bromomethane are colourless gases at room temperature. The other lower members are colourless liquids with a sweet smell. Iodides have a higher boiling point than bromides, which, in turn, boil at higher temperatures than the chlorides (Table 9.1).

The alkyl halides are insoluble in water, but are soluble in organic solvents.

Alkyl chlorides are less dense than water, but the bromides and iodides are denser.

HALOGEN COMPOUNDS

Table 9.1. Some aliphatic halides

NAME	FORMULA	B.P. (°C)
Chloromethane	CH_3Cl	−24
Bromomethane	CH_3Br	4
Iodomethane	CH_3I	42
Chloroethane	CH_3CH_2Cl	12
Bromoethane	CH_3CH_2Br	38
Iodoethane	CH_3CH_2I	72
1-Chloropropane	$CH_3CH_2CH_2Cl$	47
2-Chloropropane	$CH_3\overset{\underset{\mid}{Cl}}{C}HCH_3$	36
1-Chlorobutane	$CH_3CH_2CH_2CH_2Cl$	78
2-Chlorobutane	$CH_3CH_2\overset{\underset{\mid}{Cl}}{C}HCH_3$	68
2-Chloro-2-methylpropane	$(CH_3)_3CCl$	51
1-Bromobutane	$CH_3CH_2CH_2CH_2Br$	102
Dichloromethane	CH_2Cl_2	40
1,1-Dichloroethane	CH_3CHCl_2	57
1,2-Dichloroethane	$ClCH_2CH_2Cl$	84
1,2-Dibromoethane	$BrCH_2CH_2Br$	131
Trichloromethane	$CHCl_3$	61
Tri-iodomethane	CHI_3	m.p. 119
Tetrachloromethane	CCl_4	77
Chloroethene	$CH_2{=}CHCl$	−14

Laboratory preparations

1. From alcohols

(a) *Alkyl chlorides*

(i) By treating the alcohol with sulphur dichloride oxide (a liquid, b.p. 77°C). Sulphur dioxide and hydrogen chloride are evolved:

$$R-OH + SOCl_2 \rightarrow R-Cl + SO_2 + HCl$$

An organic base (for example, pyridine) is often added to neutralise the hydrogen chloride.

(ii) By treating the alcohol with hydrogen chloride:

$$R-OH + HCl \rightarrow R-Cl + H_2O$$

To obtain good yields from primary and secondary alcohols, anhydrous conditions are necessary and even then primary alcohols react slowly unless anhydrous zinc chloride is used as a catalyst with the hydrogen chloride. However, tertiary alcohols react readily even in the presence of water and concentrated hydrochloric acid can be used.

(b) *Alkyl bromides*

(i) By treating the alcohol with a mixture of red phosphorus and bromine:

HALOGEN COMPOUNDS

$$2P + 3Br_2 \rightarrow 2PBr_3$$

$$3R{-}OH + PBr_3 \rightarrow 3R{-}Br + H_3PO_3$$

(ii) By treating the alcohol with hydrogen bromide. It is convenient to generate the hydrogen bromide *in situ* by the reaction of potassium bromide with concentrated sulphuric acid:

$$KBr + H_2SO_4 \rightarrow KHSO_4 + HBr$$

$$R{-}OH + HBr \rightarrow R{-}Br + H_2O$$

(c) *Alkyl iodides*

(i) By treating the alcohol with a mixture of red phosphorus and iodine:

$$2P + 3I_2 \rightarrow 2PI_3$$

$$3R{-}OH + PI_3 \rightarrow 3R{-}I + H_3PO_3$$

(ii) By treating the alcohol with a concentrated solution of hydriodic acid:

$$R{-}OH + HI \rightarrow R{-}I + H_2O$$

The hydrogen iodide can be prepared *in situ* from potassium iodide and phosphoric acid:

$$3KI + H_3PO_4 \rightarrow 3HI + K_3PO_4$$

(Sulphuric acid is not used because it oxidises hydrogen iodide.)

2. From alkenes

Hydrogen halides react with alkenes to form alkyl halides. The orientation in the addition reaction is described by Markownikoff's rule (6.4), for example:

$$CH_3{-}CH{=}CH_2 + HCl \rightarrow CH_3{-}CHCl{-}CH_3$$
$$\text{2-chloropropane}$$

Manufacture of alkyl chlorides

1. By the direct chlorination of alkanes (5.5), for example:

$$CH_3{-}CH_3 + Cl_2 \rightarrow CH_3{-}CH_2Cl + HCl$$

In order to minimise the extent of further chlorination of the product, an excess of the alkane is used.

2. By the addition of hydrogen chloride to an alkene (6.3).

Chemical properties

(a) Substitution reactions

The most important reactions of alkyl halides are those in which the halogen atom, X, is replaced by another group. They can be represented by the general equation:

$$R{-}X + Y{-}Z \rightarrow R{-}Z + Y{-}X$$

HALOGEN COMPOUNDS

An example is the hydrolysis of an alkyl halide with sodium hydroxide:

$$R{-}X + NaOH \to R{-}OH + NaX$$

The mechanisms of these reactions have been studied in great detail and are well understood. There are two general mechanisms. The first applies to *primary* halides, RCH_2X. The reagent (e.g. hydroxide ion) approaches the carbon atom of the C—X bond in the halide from the side opposite to the halogen atom. As it does so, it begins to form a bond to the carbon atom, while the bond between the carbon atom and the halogen begins to break (Fig. 9.1). The pair of electrons which forms the new bond is supplied

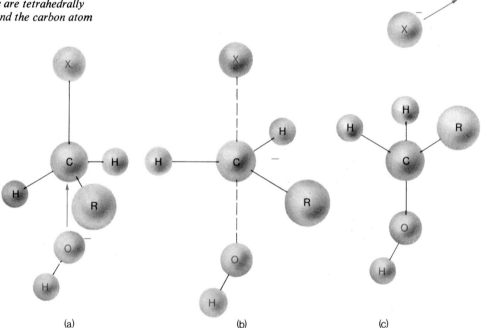

FIG. 9.1. *The mechanism of an S_N2 reaction of an alkyl halide, RCH_2X. (a) The approach of a hydroxide ion; (b) The transition state, showing that the alkyl group, RCH_2, is planar; (c) The reaction product, RCH_2OH, is formed and the four bonds are tetrahedrally arranged around the carbon atom*

by the reagent, and the pair of electrons in the C—X bond is gradually acquired completely by the halogen atom. A convenient representation of these processes is:

$$H\overset{..}{\underset{..}{O}}{:}^{(-)} \curvearrowright \overset{R}{\underset{|}{C}H_2}{-}X \longrightarrow HO{-}\overset{R}{\underset{|}{C}H_2} + X^-$$

where each curved arrow represents the movement of a pair of electrons. Eventually, the new bond is fully formed and the C—X bond is completely broken.

The energy change during the reaction is shown in Fig. 9.2. At first, more energy is needed to break the C—X bond than is supplied by the formation of the new C—O bond, and the energy of the system increases. A peak is reached, corresponding approximately to the situation in which the C—X bond is 'half-broken' and the C—O bond is 'half-formed'; the system is

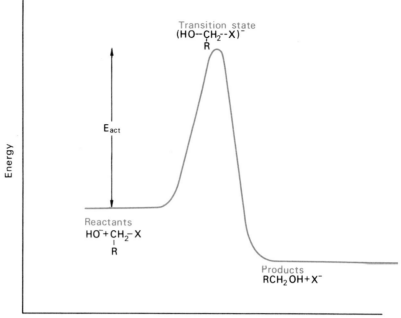

FIG. 9.2. *The variation in energy as the reactants $RCH_2X + OH^-$ are converted into the products $RCH_2OH + X^-$*

then described as being at its **transition state**. The energy then decreases, corresponding to the completion of the formation of the C—O bond. The difference in energy between the transition state and the reactants is the **activation energy** of the reaction. There is in effect a barrier to the reaction; a collision between the reactants can only lead to the formation of products when the reactant molecules possess between them enough excess of kinetic energy to surmount the barrier. In many cases, only a small proportion possesses sufficient energy, and reaction is slow. However, kinetic energy rises with temperature, and so the rate is increased by heating.

Other reagents which possess at least one unshared pair of electrons can replace the hydroxide ion in this reaction. Such reagents are described as **nucleophiles** ('nucleus-seeking') or **nucleophilic reagents**. Since two species —the reagent and the alkyl halide—are brought together in formation of the transition state, the reaction is **bimolecular**. The overall reaction is referred to as an S_N2 reaction (Substitution, nucleophilic, bimolecular).

The second mechanism applies to *tertiary* halides. These ionise spontaneously in solution, forming a carbonium ion and a halide ion, for example:

$$CH_3-\underset{\underset{CH_3}{|}}{\overset{\overset{CH_3}{|}}{C}}-Cl \rightleftharpoons CH_3\underset{CH_3}{\overset{\overset{CH_3}{|}}{C^+}}CH_3 + Cl^-$$

The rate of ionisation is fairly small, and the equilibrium lies well to the left-hand side. However, the carbonium ion is very reactive and is attacked by other nucleophiles which may be present. For example, if water is the solvent, water itself acts as the nucleophile and an alcohol is formed:

HALOGEN COMPOUNDS

$$\underset{H}{\overset{H}{>}}O: \quad \underset{CH_3}{\overset{CH_3}{>}}\overset{+}{C}\underset{CH_3}{\overset{|}{}} \longrightarrow \underset{H}{\overset{H}{>}}\overset{+}{O}-\underset{CH_3}{\overset{CH_3}{\underset{|}{C}}}-CH_3 \longrightarrow HO-\underset{CH_3}{\overset{CH_3}{\underset{|}{C}}}-CH_3 + H^+$$

If the solvent is an alcohol, an ether is formed:

$$\underset{H}{\overset{R}{>}}O: \quad \underset{CH_3}{\overset{CH_3}{>}}\overset{+}{C}\underset{CH_3}{\overset{|}{}} \longrightarrow \underset{H}{\overset{R}{>}}\overset{+}{O}-\underset{CH_3}{\overset{CH_3}{\underset{|}{C}}}-CH_3 \longrightarrow RO-\underset{CH_3}{\overset{CH_3}{\underset{|}{C}}}-CH_3 + H^+$$

These reactions can be represented as in Fig. 9.3. The slow step is the breaking of the C—Cl bond, and only one molecule is involved in the transition state; the process is therefore **unimolecular**, and the whole reaction is described as an S_N1 reaction.

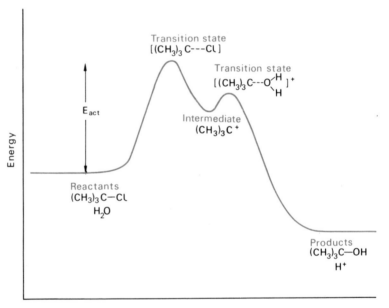

FIG. 9.3. *The variation in energy during the hydrolysis of $(CH_3)_3CCl$ by the S_N1 reaction*

The difference between tertiary and primary halides arises because a tertiary carbonium ion is a relatively more stable species than a primary carbonium ion (6.4) and is formed much faster. The activation energy for formation of a primary carbonium ion is so great that it is not formed at a significant rate under most conditions; instead, a nucleophilic reagent is needed to assist the breaking of the C—X bond.

Secondary alkyl halides (e.g. 2-chloropropane, $(CH_3)_2CH-Cl$) are intermediate in behaviour between primary and tertiary halides; they react partly by the bimolecular S_N2 reaction and partly by the unimolecular S_N1 reaction.

The following are examples of these nucleophilic substitution reactions with **primary** alkyl halides.

(i) *Preparation of an alcohol.* By treatment of the halide with an aqueous solution of sodium hydroxide, for example:

$$C_2H_5-Br + NaOH \rightarrow C_2H_5-OH + NaBr$$

(ii) *Preparation of an ether.* By treatment of the halide with a solution of a sodium alkoxide in the corresponding alcohol, for example:

$$C_2H_5-Br + C_2H_5-O^- Na^+ \rightarrow C_2H_5-O-C_2H_5 + NaBr$$
Diethyl ether

The nucleophile is the alkoxide ion, $C_2H_5O^-$. However, with higher homologues, an alternative reaction—elimination of the hydrogen halide—is also important and is often the major reaction (p. 118).

(iii) *Preparation of amines.* By heating the halide with concentrated ammonia in a sealed tube. A mixture of amines is formed, for example:

$$C_2H_5-I + NH_3 \rightarrow C_2H_5-NH_2 + HI$$
Ethylamine

$$C_2H_5-I + C_2H_5-NH_2 \rightarrow (C_2H_5)_2NH + HI$$
Diethylamine

$$C_2H_5-I + (C_2H_5)_2NH \rightarrow (C_2H_5)_3N + HI$$
Triethylamine

$$C_2H_5-I + (C_2H_5)_3N \rightarrow (C_2H_5)_4N^+ I^-$$
Tetraethylammonium iodide

The hydrogen iodide liberated in the first three reactions reacts with ammonia or an amine to form a salt.

Ammonia is a nucleophile:

$$H_3N: \curvearrowright CH_2-I \longrightarrow H_3\overset{+}{N}-CH_2 \quad I^- \longrightarrow C_2H_5-NH_2 + HI$$
$$CH_3 \phantom{-I \longrightarrow H_3\overset{+}{N}-}CH_3$$

Likewise, ethylamine, diethylamine and triethylamine are nucleophilic reagents.

(iv) *Preparation of an ester.* By treatment of the halide with the silver salt of a carboxylic acid, for example:

$$C_2H_5-Br + CH_3-CO_2^- Ag^+ \rightarrow C_2H_5-O-CO-CH_3 + AgBr$$
Silver ethanoate Ethyl ethanoate

The nucleophile is the ethanoate anion:

$$CH_3-\underset{\underset{O}{\|}}{C}-O^- \curvearrowright CH_2-Br \longrightarrow CH_3-\underset{\underset{O}{\|}}{C}-O-CH_2 + Br^-$$
$$CH_3 CH_3$$

(v) *Preparation of a nitrile.* By refluxing a solution of the halide and potassium cyanide in ethanol, for example:

$$C_2H_5-Br + KCN \rightarrow C_2H_5-CN + KBr$$
Propanonitrile

The nucleophile is the cyanide anion, $N\equiv C^-$.

If silver cyanide is used, an *isocyano-compound* is formed:

$$C_2H_5-Br + AgCN \rightarrow C_2H_5-\overset{+}{N}{\equiv}\overset{-}{C} + AgBr$$
$$\text{Isocyanoethane}$$

(vi) *Preparation of a nitroalkane.* By refluxing a solution of the halide in ethanol with silver nitrite. A mixture of a nitroalkane and an alkyl nitrite is formed and can be separated by fractional distillation, for example:

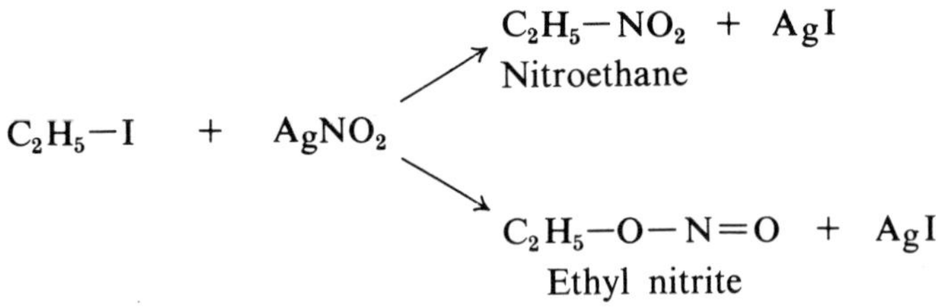

$$C_2H_5-I + AgNO_2 \nearrow C_2H_5-NO_2 + AgI \quad \text{Nitroethane}$$
$$\searrow C_2H_5-O-N=O + AgI \quad \text{Ethyl nitrite}$$

The nucleophile is the nitrite ion, NO_2^-. It can react at either an oxygen atom or the nitrogen atom:

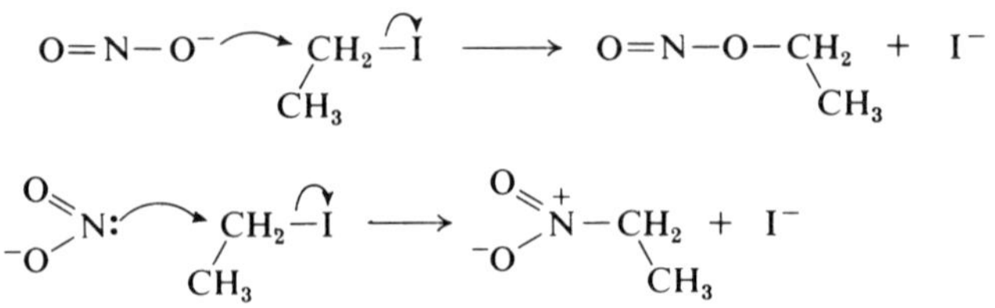

$$O=N-O^- \frown CH_2-I \longrightarrow O=N-O-CH_2 + I^-$$
$$\qquad\qquad\quad\; CH_3 \qquad\qquad\qquad\qquad\quad CH_3$$

$$\underset{-O}{\overset{O}{\diagdown}}N: \frown CH_2-I \longrightarrow \underset{-O}{\overset{O}{\diagdown}}\overset{+}{N}-CH_2 + I^-$$
$$\qquad\qquad CH_3 \qquad\qquad\qquad\qquad\; CH_3$$

(vii) *Preparation of alkylaromatic compounds.* By reaction of the halide with the aromatic compound in the presence of an aluminium halide (Friedel-Crafts reaction, p. 100), for example:

$$C_6H_6 + C_2H_5-Br \xrightarrow{AlBr_3 \text{ as cat.}} C_6H_5-C_2H_5 + HBr$$
$$\text{Ethylbenzene}$$

These reactions can be regarded either as electrophilic substitutions in the aromatic compound or nucleophilic substitutions by the aromatic compound on the alkyl halide (p. 100).

(b) Elimination reactions

When 2-bromopropane is refluxed with a solution of sodium hydroxide in ethanol, both an ether and propene are formed:

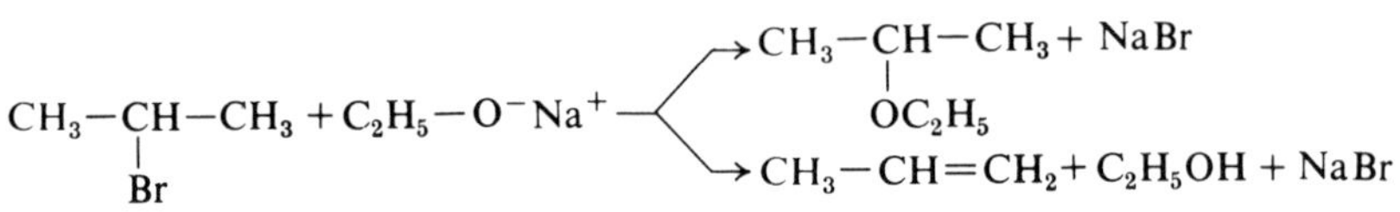

$$CH_3-\underset{Br}{CH}-CH_3 + C_2H_5-O^-Na^+ \nearrow CH_3-\underset{OC_2H_5}{CH}-CH_3 + NaBr$$
$$\searrow CH_3-CH=CH_2 + C_2H_5OH + NaBr$$

The ether is formed by the S_N2 reaction and propene is formed by an elimination reaction: the base, ethoxide ion, abstracts a proton from the halide at the same time as the halide ion breaks away. The movements of electron-pairs are represented as follows:

$$\text{C}_2\text{H}_5-\text{O}^-\overset{\curvearrowright}{\text{H}} \qquad\qquad \text{C}_2\text{H}_5-\text{OH}$$
$$\underset{\underset{\text{Br}}{|}}{\text{CH}_2}-\text{CH}-\text{CH}_3 \quad\longrightarrow\quad \text{CH}_2=\text{CH}-\text{CH}_3$$
$$\qquad\qquad\qquad\qquad\qquad\qquad\qquad\qquad \text{Br}^-$$

At the transition state, the C—H and C—Br bonds are partially broken and the C=C and O—H bonds are partially formed:

$$\text{C}_2\text{H}_5-\text{O}$$
$$\vdots$$
$$\text{H}$$
$$\vdots$$
$$\text{CH}_2\mathbin{\vcenter{\hbox{$=$}}}\text{CH}-\text{CH}_3$$
$$\vdots$$
$$\text{Br}$$

Since two molecules are involved in the formation of the transition state, the reaction is bimolecular; it is described as an $E2$ reaction (E, elimination; 2, bimolecular).

Competition between the S_N2 and $E2$ reactions occurs with other primary and secondary halides. The relative importance of each type of reaction depends on the solvent, the temperature and the structure of the halide; the ratio of elimination to substitution increases as:

(i) the solvent is changed from water (where the reagent is hydroxide ion) to an alcohol, ROH (where the reagent is the corresponding alkoxide ion, RO$^-$);

(ii) the temperature is increased;

(iii) the number of alkyl groups adjacent to the double bond in the resulting alkene is increased (for example, the reaction of bromoethane with a solution of sodium hydroxide in ethanol gives only 1 per cent of ethene, whereas 2-bromopropane, under the same conditions, gives 80 per cent of propene).

Tertiary alkyl halides, and to some extent secondary alkyl halides, undergo elimination by a different mechanism in which the first step is heterolysis of the carbon–halogen bond, for example:

$$\underset{\underset{\text{CH}_3}{|}}{\overset{\overset{\text{CH}_3}{|}}{\text{CH}_3-\text{C}-\text{Br}}} \quad\rightleftharpoons\quad \underset{\text{CH}_3}{\overset{\overset{\text{CH}_3}{|}}{\text{C}^+}}\text{CH}_3 \quad+\quad \text{Br}^-$$

$$\underset{\text{CH}_3}{\overset{\overset{\text{CH}_3}{|}}{\text{C}^+}}\text{CH}_3 \quad\longrightarrow\quad \underset{\text{CH}_3}{\overset{\text{CH}_3}{\text{C}}}=\text{CH}_2 \quad+\quad \text{H}^+$$

This is described as an $E1$ reaction (E, elimination; 1, unimolecular). The intermediate carbonium ion is the same as is involved in substitution reactions, so that the $E1$ elimination competes with the S_N1 substitution. Under most conditions, the $E1$ reaction is the major process.

(c) Reactions with metals

Alkyl halides react with sodium to give alkanes, for example:

$$2C_2H_5-I + 2Na \rightarrow CH_3CH_2CH_2CH_3 + 2NaI$$
$$\text{Butane}$$

This is the **Wurtz** reaction.

With magnesium, they form **Grignard reagents** (alkylmagnesium halides), for example:

$$C_2H_5-Br + Mg \rightarrow C_2H_5-MgBr$$
$$\text{Ethylmagnesium bromide}$$

These reagents are of especial value in synthesis (9.8).

Uses of alkyl halides

Alkyl halides are of great value in organic synthesis because of the variety of compounds which can be made from them by nucleophilic substitution and *via* Grignard reagents (9.8).

Tetraethyllead, which is the principal 'anti-knock' additive in petrol (19.4), is made by heating chloroethane with a lead–sodium alloy:

$$4C_2H_5-Cl + Pb + 4Na \rightarrow (C_2H_5)_4Pb + 4NaCl$$

9.4 Polyhalides

Compounds in which two halogen atoms are attached to adjacent carbon atoms are known as *vic*-dihalides (*vicinal*, adjacent). Compounds in which two halogen atoms are attached to one carbon atom are known as *gem*-dihalides (*gemini*, twins). For example:

```
    H  H              H  H
    |  |              |  |
H — C— C — Cl     H — C— C — H
    |  |              |  |
    H  Cl             Cl Cl
```

1,1-Dichloroethane   1,2-Dichloroethane
(a *gem*-dihalide)   (a *vic*-dihalide)

There are also compounds in which three or four halogen atoms are attached to one carbon atom [for example, trichloromethane ($CHCl_3$), tri-iodomethane (CHI_3), tetrachloromethane (CCl_4)].

vic-Dihalides are prepared by the addition of the halogen to an alkene, for example:

$$CH_2\!=\!CH_2 + Br_2 \rightarrow CH_2Br-CH_2Br$$

The properties of *vic*-dihalides are very similar to those of alkyl halides. Thus, they undergo both substitution and elimination reactions, for example:

1. They undergo hydrolysis when heated with aqueous sodium hydroxide:

HALOGEN COMPOUNDS

$$CH_2Cl-CH_2Cl + 2NaOH \rightarrow CH_2OH-CH_2OH + 2NaCl$$
$$\text{Ethane-1,2-diol}$$

2. They react with a solution of potassium cyanide in ethanol to give dinitriles:

$$CH_2Cl-CH_2Cl + 2KCN \rightarrow \underset{\underset{CN}{|}}{CH_2}-\underset{\underset{CN}{|}}{CH_2} + 2KCl$$

3. They undergo elimination with a hot solution of sodium hydroxide in ethanol:

$$CH_2Cl-CH_2Cl + C_2H_5-O^-Na^+ \rightarrow CH_2=CHCl + C_2H_5OH + NaCl$$
$$CH_2=CHCl + C_2H_5-O^-Na^+ \rightarrow CH\equiv CH + C_2H_5OH + NaCl$$

gem-Dihalides are prepared from aldehydes or ketones with phosphorus pentahalides, for example:

$$CH_3-CH=O + PCl_5 \rightarrow CH_3-CHCl_2 + POCl_3$$

Hydrolysis regenerates the aldehyde or ketone, for example:

$$CH_3-CHCl_2 + 2NaOH \rightarrow CH_3-CHO + H_2O + 2NaCl$$

Hence, they can readily be distinguished from *vic*-dihalides, which give diols (dihydric alcohols) on hydrolysis.

Trichloromethane (chloroform) can be made in the laboratory by heating ethanol with bleaching powder. The bleaching powder provides chlorine, and reaction occurs in three stages, which can be represented by the following equations:

Oxidation:
$$CH_3-CH_2OH + Cl_2 \rightarrow CH_3-CHO + 2HCl$$

Chlorination:
$$CH_3-CHO + 3Cl_2 \rightarrow CCl_3-CHO + 3HCl$$

Hydrolysis:
$$2CCl_3-CHO + Ca(OH)_2 \rightarrow 2CHCl_3 + (H-CO_2)_2Ca$$
$$\text{Calcium methanoate}$$

$$(Ca(OH)_2 + 2HCl \rightarrow CaCl_2 + 2H_2O)$$

Trichloromethane is a colourless liquid with a characteristic sickly smell. It is almost insoluble in water but is soluble in most organic solvents. Its chemical properties are as follows:

1. It is oxidised in the presence of light and air to carbonyl chloride:

$$CHCl_3 + \tfrac{1}{2}O_2 \rightarrow COCl_2 + HCl$$
$$\text{Carbonyl chloride}$$

Trichloromethane is stored in dark bottles to prevent the formation of carbonyl chloride, as it is intensely poisonous.

2. It is hydrolysed with alkali:

$$CHCl_3 + 4NaOH \rightarrow H\text{—}CO_2^- \, Na^+ + 3NaCl + 2H_2O$$
$$\text{Sodium methanoate}$$

This reaction occurs by a different mechanism from those in the hydrolysis of other aliphatic halides. Trichloromethane is weakly acidic, and in basic solution ionises to a small extent to form a **carbanion**, CCl_3^-:

$$CHCl_3 + OH^- \rightleftharpoons \bar{C}Cl_3 + H_2O$$

The carbanion eliminates a chloride ion to form a **carbene**:

$$\bar{C}Cl_3 \rightarrow :CCl_2 + Cl^-$$
$$\text{Dichloro-carbene}$$

Dichlorocarbene has two electrons available for bonding and is therefore very reactive; thus, it is attacked by water:

$$:CCl_2 + H_2O \rightarrow HO\text{—}CHCl_2$$

The resulting unstable species reacts further:

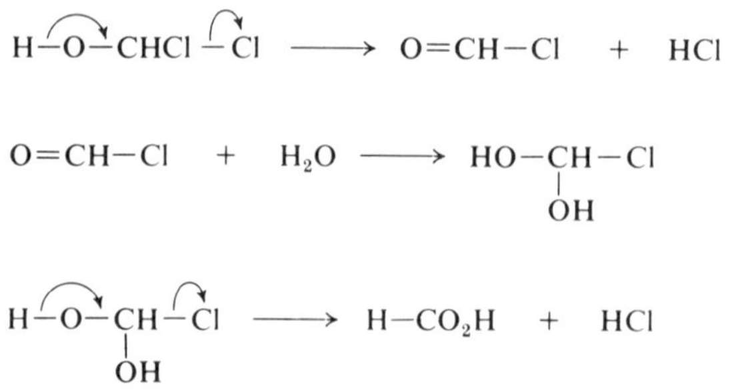

$$H\text{—}O\text{—}CHCl\text{—}Cl \longrightarrow O=CH\text{—}Cl + HCl$$

$$O=CH\text{—}Cl + H_2O \longrightarrow HO\text{—}\underset{OH}{CH}\text{—}Cl$$

$$H\text{—}O\text{—}\underset{OH}{CH}\text{—}Cl \longrightarrow H\text{—}CO_2H + HCl$$

The high reactivity of dichlorocarbene is also shown by its reaction with alkenes to give cyclopropane derivatives:

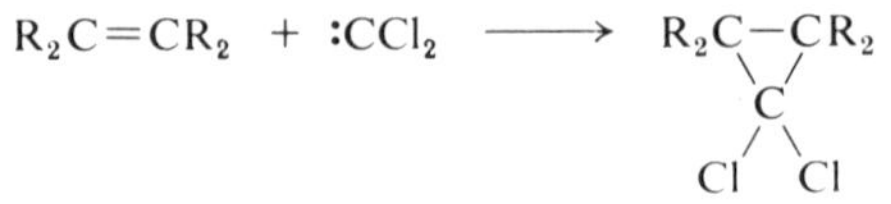

$$R_2C=CR_2 + :CCl_2 \longrightarrow \begin{array}{c} R_2C\text{—}CR_2 \\ \diagdown \diagup \\ C \\ \diagup \diagdown \\ Cl \quad Cl \end{array}$$

3. It reacts with a primary amine, in an ethanolic solution of potassium hydroxide, to form an isocyano-compound (**carbylamine reaction**):

$$CHCl_3 + 3KOH + R\text{—}NH_2 \rightarrow R\text{—}\overset{+}{N}\equiv\bar{C} + 3H_2O + 3KCl$$

Isocyano-compounds have characteristic powerful and unpleasant smells, and their formation has sometimes been used as a test for primary amines.

HALOGEN COMPOUNDS

This reaction also occurs *via* dichlorocarbene, which is formed in the basic solution:

$$CHCl_3 + OH^- \rightarrow \overline{C}Cl_3 + H_2O$$

$$\overline{C}Cl_3 \rightarrow :CCl_2 + Cl^-$$

$$R-NH_2 + :CCl_2 \rightarrow R-NH-CHCl_2$$

$$R-NH-CHCl_2 + 2OH^- \rightarrow R-\overset{+}{N}\equiv\overset{-}{C} + 2Cl^- + 2H_2O$$

Tetrachloromethane (carbon tetrachloride) is manufactured by the chlorination of carbon disulphide. Aluminium trichloride is used as a catalyst.

$$CS_2 + 3Cl_2 \xrightarrow{\text{AlCl}_3 \text{ as cat.}} CCl_4 + \underset{\substack{\text{Disulphur} \\ \text{dichloride}}}{S_2Cl_2}$$

It is a colourless liquid which is insoluble in water but soluble in all organic solvents. It is inert to most reagents; for example, it is not hydrolysed by alkali. It is used as a fire-extinguisher (Pyrene) because it is non-inflammable and its dense vapour prevents oxygen getting to the flame.

9.5 Unsaturated halides

Chloroethene (vinyl chloride), $CH_2=CHCl$, is the most important of the unsaturated halides.

It is manufactured from ethyne (7.2) and, to an increasing extent, from ethene (6.3). Its principal use is for the manufacture of PVC (poly(chloroethene)) (p. 314).

The reactions of chloroethene are strikingly different from those of alkyl chlorides in that it does not react with nucleophilic reagents; for example, it is not hydrolysed by sodium hydroxide.

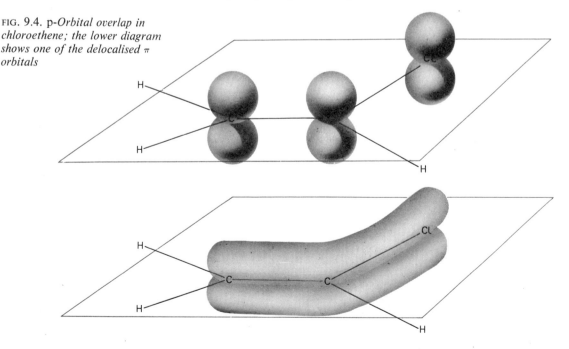

FIG. 9.4. p-*Orbital overlap in chloroethene; the lower diagram shows one of the delocalised π orbitals*

HALOGEN COMPOUNDS

The reason for this difference is that the C—Cl bond in chloroethene is stronger than one in an alkyl chloride such as chloroethane and is therefore less readily broken. This in turn is because, in chloroethene, a p orbital on chlorine interacts with the p orbital on the adjacent carbon atom (Fig. 9.4), providing additional bonding as compared with an alkyl chloride.

Trichloroethene, $CHCl=CCl_2$, is manufactured from ethyne:

$$CH\equiv CH \xrightarrow{2Cl_2} CHCl_2-CHCl_2 \xrightarrow[\text{over Ba(OH)}_2]{\text{Pass vapour}} CHCl=CCl_2 + HCl$$

It resembles chloroethene in being unreactive towards nucleophilic reagents, and because of its chemical stability and its property of dissolving oils and greases, it is used as a cleaning agent. Surfaces are cleaned by suspending the object (for example, an aeroplane wing or large engine) in the vapour of trichloroethene, which condenses on the surface and 'strips off' the grease and dirt.

9.6 Aromatic halides

There are two types of halides which contain an aromatic ring, examples of which are given in Table 9.2.

(a) **Aryl halides** have the halogen atom attached to the aromatic ring. They are named as the halogen derivatives of the aromatic compound, for example:

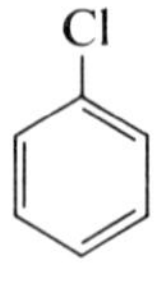

Chlorobenzene 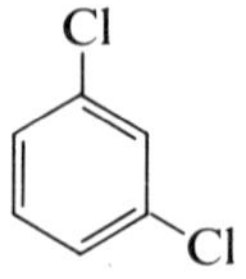   1,3-Dichloro-benzene   3,5-Dibromo-nitrobenzene

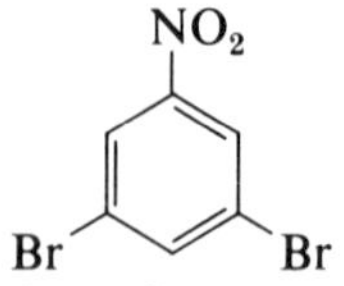

Aryl halides (aryl, from aromatic) have different properties from alkyl halides.

(b) Some aromatic halides have the halogen atom in a side-chain, for example:

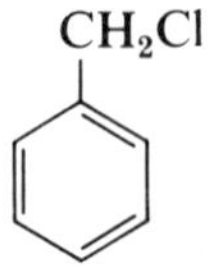

(Chloromethyl)-benzene 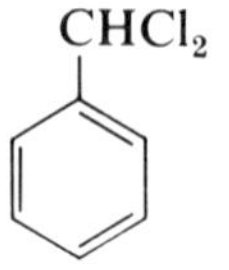   (Dichloromethyl)-benzene

These behave like aliphatic halides, and their reactions are compared with those of alkyl and aryl halides in the practical section (p. 132).

Preparation of aryl halides

1. Chlorobenzene and bromobenzene are prepared by the reaction of the halogen with benzene at room temperature in the presence of a 'halogen carrier' (8.3); iron, or aluminium trichloride or tribromide, are usually employed. The mechanism of the reaction is discussed on p. 99, and details of the laboratory preparation of bromobenzene are given on p. 131.

HALOGEN COMPOUNDS

Table 9.2. **Some aromatic halides**

NAME	FORMULA	B.P. (°C)
Fluorobenzene	C_6H_5—F	85
Chlorobenzene	C_6H_5—Cl	132
Bromobenzene	C_6H_5—Br	156
Iodobenzene	C_6H_5—I	189
Chloro-2-methylbenzene	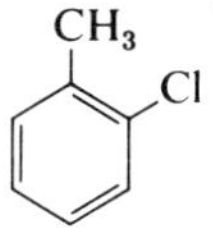 2-CH₃-C₆H₄-Cl	159
Chloro-3-methylbenzene	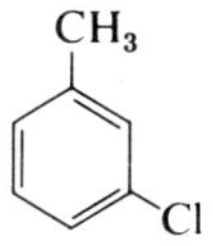 3-CH₃-C₆H₄-Cl	162
Chloro-4-methylbenzene	 4-CH₃-C₆H₄-Cl	162
(Chloromethyl)benzene	C_6H_5—CH_2Cl	179
(Dichloromethyl)benzene	C_6H_5—$CHCl_2$	206
(Trichloromethyl)benzene	C_6H_5—CCl_3	221

2. All aryl halides can be prepared from the corresponding aromatic amine *via* the diazonium salt (16.8).

Manufacture of chlorobenzene

Chlorobenzene is manufactured from benzene by the **Raschig process**, in which benzene, hydrogen chloride and oxygen are passed over copper(II) chloride at 250°C:

$$C_6H_6 + HCl + \tfrac{1}{2}O_2 \rightarrow C_6H_5\text{—}Cl + H_2O$$

Reaction occurs in two stages: the oxidation of hydrogen chloride to chlorine, followed by the reaction of benzene with chlorine.

Chemical properties of aryl halides

1. They are unreactive towards nucleophilic reagents under ordinary laboratory conditions. Thus, they resemble unsaturated halides and differ from alkyl halides. However, some reactions with nucleophiles can be effected under very vigorous conditions, and one which is of industrial importance is the hydrolysis of chlorobenzene with sodium hydroxide solution at 200°C under a pressure of 200 atmospheres:

$$C_6H_5\text{—}Cl + 2NaOH \rightarrow C_6H_5\text{—}O^- Na^+ + H_2O + NaCl$$

Phenol is liberated on the addition of acid:

$$C_6H_5\text{—}O^- Na^+ + H^+ \rightarrow C_6H_5\text{—}OH + Na^+$$

HALOGEN COMPOUNDS

2. Aryl halides undergo substitution in the aromatic ring with electrophilic reagents (8.5). The 2- and 4-derivatives are the major products, for example:

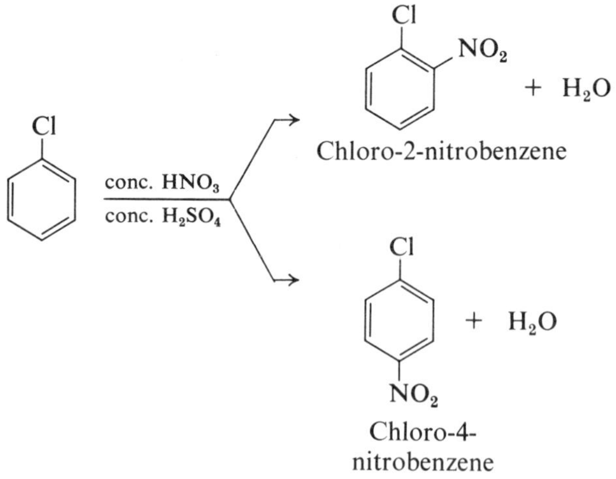

$$\text{C}_6\text{H}_5\text{Cl} \xrightarrow{\text{conc. HNO}_3,\ \text{conc. H}_2\text{SO}_4} \text{Chloro-2-nitrobenzene} + \text{H}_2\text{O}$$

$$\text{and}\ \text{Chloro-4-nitrobenzene} + \text{H}_2\text{O}$$

3. Bromobenzene and iodobenzene form Grignard reagents with magnesium (9.8). Details of the preparation of phenylmagnesium bromide are given on p. 196.

Uses of chlorobenzene

Chlorobenzene is used in the manufacture of phenol (10.8) and the important insecticide D.D.T., by reaction with trichloroethanal in the presence of concentrated sulphuric acid.

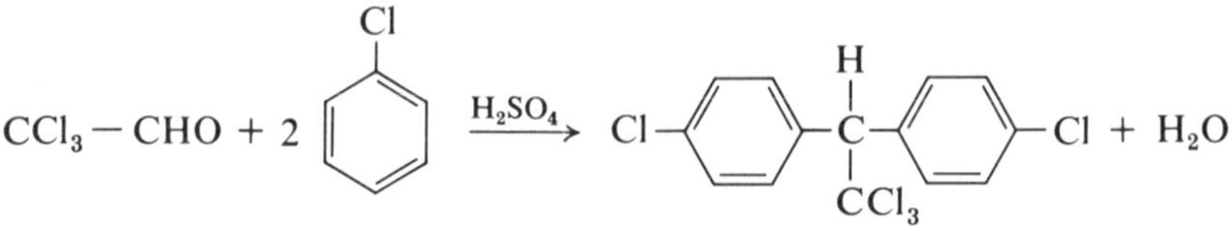

$$\text{CCl}_3-\text{CHO} + 2\ \text{C}_6\text{H}_5\text{Cl} \xrightarrow{\text{H}_2\text{SO}_4} \text{Cl}-\text{C}_6\text{H}_4-\text{CH}(\text{CCl}_3)-\text{C}_6\text{H}_4-\text{Cl} + \text{H}_2\text{O}$$

9.7 Fluorocarbons

For convenience, the alkyl fluorides, R—F, and other fluorine derivatives of the hydrocarbons are discussed together. They are much less reactive than the other halides because of the much greater strength of the C—F bond (485 kJ per mol) than C—Cl (339 kJ per mol), C—Br (284 kJ per mol) or C—I (213 kJ per mol). In general, they behave like alkanes (hydrocarbons), so that they are usually referred to as fluorocarbons. Many have become of industrial importance during the last 20 years.

Physical properties

Fluorine derivatives have physical properties (for example, boiling points) which are similar to those of the parent alkane:

	$n=1$	2	3	4
C_nH_{2n+2} B.p. °C	−162	−89	−42	−0·5
C_nF_{2n+2} B.p. °C	−128	−79	−38	−5

Preparations

1. *By fluorination of alkanes.* Fluorine reacts far more vigorously with the alkanes than do the other halogens, and it is necessary to use nitrogen as a diluent for the fluorine to help remove the heat evolved in the reaction. A complex mixture of products is formed and the carbon skeleton of the alkane is often broken down.

2. *By substitution of fluorine for other halogens.* (a) A halogen atom is replaced by fluorine when an alkyl halide is heated with anhydrous potassium fluoride in ethane-1,2-diol, for example:

$$CH_3-Cl + KF \rightarrow CH_3-F + KCl$$

(b) Antimony trifluoride, in the presence of antimony pentachloride, can also be used as a fluorinating agent, for example:

$$3CCl_4 + 2SbF_3 \rightarrow 3CF_2Cl_2 + 2SbCl_3$$

Addition of anhydrous hydrogen fluoride regenerates antimony trifluoride from the antimony trichloride.

3. *By substitution of fluorine for oxygen.* Sulphur tetrafluoride replaces oxygen by fluorine in alcohols ($R-OH \rightarrow R-F$), ketones ($R_2C=O \rightarrow R_2CF_2$) and acids ($R-CO_2H \rightarrow R-CF_3$).

4. *By electrolysis.* When an organic compound in pure hydrogen fluoride is electrolysed with nickel electrodes at 0°C, it undergoes fluorination at the anode. For example, diethyl ether gives $C_2F_5-O-C_2F_5$, and ethanoic acid gives CF_3-CO-F, which on hydrolysis gives trifluoroethanoic acid, CF_3-CO_2H.

Chemical properties

Unlike the other halogen derivatives of alkanes, the alkyl fluorides are chemically stable. They do not react with oxidising or reducing agents, or with strong acids and alkalis. They react slowly with sodium or potassium metal at elevated temperatures. However, Grignard reagents have been prepared, but the magnesium compound is only stable below −20°C.

The fluorocarbons are regarded as parents of a new branch of chemistry, similar to the organic compounds formed from hydrocarbons. Chains of $-CF_2-$ are stable, similar to $-CH_2-$ chains. Thus there is a wide range of fluorocarbon derivatives of the type R_FZ, where R_F is the fluorocarbon group (CF_3-, CHF_2-, CH_2F-, etc.), and the functional group Z can be $-CO_2H$, $-CHO$, $-CH_2OH$, $-OH$, etc., and a vast new series of compounds is now being developed.

Uses

1. Fluorocarbons are generally very stable. They are used as oils, sealing liquids and coolants.

2. Tetrafluoroethene, C_2F_4, is the fluorine analogue of ethene, C_2H_4. It is prepared by the fluorination of trichloromethane, generally by antimony trifluoride; reaction occurs in two stages:

$$2CHCl_3 \xrightarrow{SbF_3} 2CHF_2Cl \xrightarrow{700°C} CF_2=CF_2 + 2HCl$$

In the presence of a catalyst, tetrafluoroethene undergoes a free-radical polymerisation (21.2) to give a plastic, poly(tetrafluoroethene) (**PTFE**), marketed as Teflon, which is highly resistant to all chemicals.

3. The fluorochloro derivatives of methane and ethane, for example CF_2Cl_2 and $CFCl_3$, are used as refrigerants (sold under the name of Freons) and as aerosol propellants in dispensers for fly-killers, shaving cream, etc.

4. 1-Bromo-1-chloro-2,2,2-trifluoroethane is used as an anaesthetic (Fluothane).

9.8 Grignard reagents

The preparation of organic compounds containing magnesium was first described by Grignard in 1900. The importance of the compounds lies in their usefulness in organic synthesis, and Grignard was awarded the Nobel Prize for his work in 1912.

Preparation of Grignard reagents

A Grignard reagent is prepared by refluxing an alkyl or aryl bromide or iodide, dissolved in dry ether, with small magnesium turnings. A small crystal of iodine is sometimes added to initiate the reaction:

$$R-Br + Mg \rightarrow R-MgBr$$

Grignard reagents cannot be isolated. The ethereal solution of the reagent is used for further reaction.

Alkyl chlorides also form Grignard reagents, though less readily than the bromides or iodides, but aryl chlorides react too slowly. Iodides react more readily than bromides, and are used in the following examples.

Reactions of Grignard reagents

The Grignard reagent, dissolved in ether, is usually kept in the apparatus used for its preparation, and the other reagent, which is sometimes dissolved in ether, is added from a tap-funnel.

1. With water, to form an **alkane**:

$$R-MgI + H_2O \rightarrow R-H + Mg(OH)I$$

2. With methanal, to form a **primary alcohol**. Methanal gas is passed into the solution of the Grignard reagent, and the mixture is then hydrolysed with dilute acid:

$$R-MgI + CH_2=O \rightarrow R-CH_2-O-MgI$$
$$R-CH_2-O-MgI + H_2O \rightarrow R-CH_2-OH + Mg(OH)I$$

3. With other aldehydes, to form a **secondary alcohol**:

$$R-MgI + R'-CHO \longrightarrow \underset{R'}{\overset{R}{>}}CH-O-MgI$$

$$\xrightarrow{H_2O} \underset{R'}{\overset{R}{>}}CH-OH + Mg(OH)I$$

4. With ketones, to form a **tertiary alcohol**:

$$R-MgI + R'_2CO \longrightarrow R-\underset{R'}{\overset{R'}{\underset{|}{\overset{|}{C}}}}-O-MgI$$

$$\xrightarrow{H_2O} R-\underset{R'}{\overset{R'}{\underset{|}{\overset{|}{C}}}}-OH + Mg(OH)I$$

HALOGEN COMPOUNDS

5. When carbon dioxide is passed through the solution of the Grignard reagent and the mixture is then hydrolysed, a **carboxylic acid** is formed:

$$R-MgI + CO_2 \longrightarrow R-\underset{\underset{O}{\|}}{C}-O-MgI$$

$$\xrightarrow{H_2O} R-\underset{\underset{O}{\|}}{C}-OH + Mg(OH)I$$

Details of this reaction are given on p. 196.

6. With epoxyethane, to form a **primary alcohol**:

$$R-MgI + H_2C\underset{O}{-}CH_2 \longrightarrow R-CH_2-CH_2-O-MgI$$

$$\xrightarrow{H_2O} R-CH_2-CH_2-OH + Mg(OH)I$$

In this method for primary alcohols, *two* carbon atoms are introduced into the Grignard reagent, whereas when methanal is used *one* carbon atom is added.

The reactions of Grignard reagents described above share a common feature in their mechanisms. The C—Mg bond in the Grignard reagent is strongly polarised by the electropositive metal:

$$-\overset{|}{\underset{|}{C}} \twoheadleftarrow Mg-I$$

As a result, the carbon atom tends to break away with the bonding pair of electrons; that is, it behaves as a nucleophilic reagent, for example:

$$I-Mg-CH_3 \quad CH_2=O \longrightarrow I^- + Mg^{2+} + CH_3-CH_2-O^-$$

9.9 Practical work

Small-scale preparation of bromoethane

Place 6 cm³ of ethanol in the flask, immerse the flask in cold water and add, slowly, 7 cm³ of *concentrated* sulphuric acid. Shake the mixture gently while adding the acid. Add 6 g of potassium bromide and set up the apparatus (Fig. 9.5).

Heat the flask very gently until all the alkyl halide has distilled over into the receiver, which should be surrounded by ice.

Transfer the distillate to a separating funnel, add about 3 cm³ of a dilute solution of sodium hydrogen carbonate, fit the stopper and shake. Remove the stopper several times to relieve the pressure.

Allow the mixture to settle and run off the lower organic layer. After discarding the upper layer, wash the alkyl halide in the separating funnel with water and then run it into a test-tube. Add some anhydrous calcium chloride, stopper the test-tube and shake it.

Redistil the dry bromoethane, collecting the fraction boiling between 35 and 40°C (Fig. 2.2); heat the flask with a beaker of hot water.

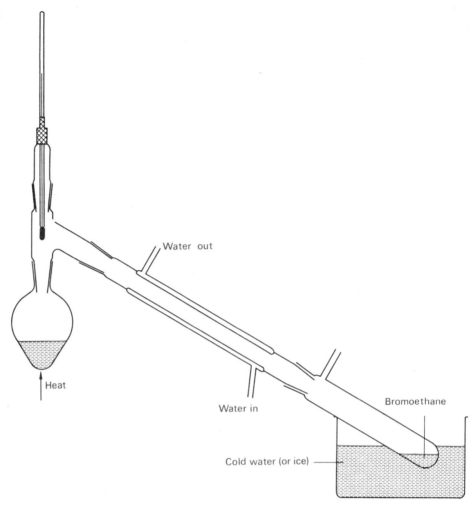

FIG. 9.5. *Preparation of bromoethane*

Small-scale preparation of 1-bromobutane

Place 6 cm³ of butan-1-ol in the flask, immerse the flask in cold water and add, slowly, 8 cm³ of *concentrated* sulphuric acid. Shake the mixture gently while adding the acid. Add 8 g of potassium bromide, set up the apparatus as in Fig. 2.1 and reflux the mixture using a small bunsen flame for 30 minutes.

Rearrange the apparatus as in Fig. 2.2 and distil over 1-bromobutane. Purify the alkyl halide as above, collecting the fraction boiling between 99 and 103°C.

Small-scale preparation of 2-chloro-2-methylbutane

Shake 5 cm³ of 2-methylbutan-2-ol with 20 cm³ of *concentrated* hydrochloric acid for 10 minutes in a separating funnel. Run off the lower layer (acid) and add slowly 10 cm³ of dilute sodium hydrogen carbonate solution. Shake, making sure that you do not allow an excess of pressure of carbon dioxide to build up.

Discard the lower aqueous layer, run the alkyl halide into a test-tube, add a few pieces of anhydrous calcium chloride, stopper the tube and shake.

Decant the dry alkyl halide into a distillation flask and purify it by distillation (Fig. 2.2). Collect the fraction boiling between 84 and 86°C.

Small-scale preparation of bromobenzene

The experiment must be carried out in a fume-cupboard, and take great care not to breath in the vapour of either benzene or bromine. Benzene is highly toxic and must only be used under supervision.

Set up the apparatus shown in Fig. 9.6, with 6 cm³ of benzene and 0·2 g of iron filings in the flask. Run 3 cm³ of bromine slowly into the flask, and *slowly* raise the temperature of the water-bath to 70°C. Maintain this temperature until no more hydrogen bromide is evolved.

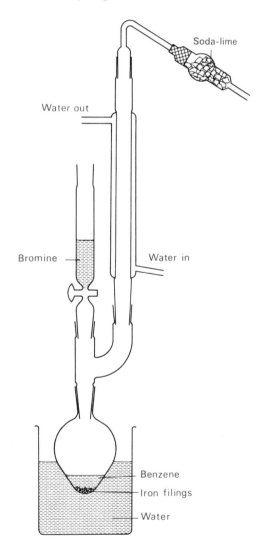

FIG. 9.6. *Preparation of bromobenzene*

Remove the flask, cool it in cold water and pour the mixture into a separating funnel. Purify the aryl halide in a similar way to that described for bromoethane (p. 129), collecting the fraction boiling between 152 and 158°C.

HALOGEN COMPOUNDS

Rates of hydrolysis of some halogen compounds

(a) *To compare the rates of hydrolysis of chloro-, bromo- and iodoalkanes*

To three separate test-tubes, add 2 cm³ of ethanol and place them in a beaker of water kept at about 60°C. When the ethanol has reached this temperature, using separate dropping pipettes, add 5 drops of 1-chlorobutane to one test-tube, 5 drops of 1-bromobutane to the second and 5 drops of 1-iodobutane to the third. Then, as quickly as possible, add 1 cm³ of 0·1M silver nitrate solution to each. Shake the test-tubes and observe (a) the order in which the precipitates appear, (b) the colour and density of the precipitates.

  (i) What are the precipitates?
  (ii) Test whether they occur when the silver nitrate is added to any of the alkyl halides by themselves.
  (iii) Deduce why the precipitates are formed. Write equations for the reactions.
  (iv) What is the effect on the rate of these reactions arising from changing the halogen in the alkyl halides?

(b) *To compare the rates of hydrolysis of some bromobutanes*

Repeat the experiment above, at room temperature, using 1-bromobutane, 2-bromobutane and 2-bromo-2-methylpropane. The test-tubes can be placed in a test-tube rack. Add the reagents as quickly as possible with shaking, and observe the test-tubes carefully for at least five minutes.

  (i) What are the precipitates?
  (ii) Write equations for the reactions.
  (iii) What effect does the alkyl group have on the rate of reaction? Suggest reasons.

(c) *To compare the reactivity of the halogen atoms in aliphatic and aromatic halogen compounds* (Carry out these experiments in a fume-cupboard. (Bromomethyl)benzene is a powerful lachrymator)

Repeat the experiments given in (a), first at room temperature and then in hot water (at 60°C), with 1-bromobutane, bromobenzene and (bromomethyl)benzene.

  (i) What are the precipitates?
  (ii) Write equations for the reactions.
  (iii) What effect does the phenyl group have on the reactivity of the halogen atom?

Take care that there is enough ethanol present to dissolve the aromatic halogen compounds. A slight turbidity on mixing may be due to an emulsion of the organic compound with water; to test for this, add a few drops of ethanol and shake.

Reactions of 1,1-dichloroethane and 1,2-dichloroethane

1. To 2–3 drops of 1,1-dichloroethane in a test-tube, add 2 cm³ of dilute sodium hydroxide solution. Shake and gently boil the mixture. Allow the mixture to cool, acidify with dilute nitric acid, then add silver nitrate solution.

  (a) Note the smell of the vapour evolved.
  (b) Note whether a precipitate is formed.

HALOGEN COMPOUNDS

Write an equation to explain your observations.

Repeat the experiments using 1,2-dichloroethane. Do you observe the same results? Write equations for the reactions.

2. Add a few pellets (about 0·5 g) of potassium hydroxide to 2 cm³ of ethanol in a test-tube. Warm the test-tube gently until the pellets have dissolved. Add 6 drops of 1,1-dichloroethane to the alcoholic solution of potassium hydroxide. Shake gently, then introduce a plug of Rocksil to absorb the solution. Fit the test-tube with a delivery tube dipping into 2 cm³ of an ammoniacal solution of copper(I) chloride. Warm the Rocksil plug gently and observe what occurs to the copper(I) chloride solution.

Repeat the experiment with 1,2-dichloroethane, writing equations for both reactions.

Preparation of a Grignard reagent

The preparation of phenylmagnesium bromide from bromobenzene is described on p. 196.

9.10 Questions

1 Outline how iodoethane (ethyl iodide) may be prepared from (a) ethene (ethylene), (b) ethanol. How from iodoethane would you prepare (i) ethane, (ii) ethene, (iii) butane, (iv) diethyl ether?

Give the structural formulae of the isomers of $C_2H_4Cl_2$ and describe the action of aqueous alkali on these isomers. (AEB)

2 Describe a laboratory method for the preparation of a named alkyl halide, giving a diagram and full practical details.

An alkyl bromide, A (0·615 g), was boiled under reflux with 100 cm³ of 0·125M sodium hydroxide solution. The mixture was allowed to cool and then titrated, using methyl orange as indicator, with 0·125M hydrochloric acid, of which 60 cm³ was required. Assuming that A contains only one bromine atom in its molecule, calculate its molecular weight.

The organic product of the hydrolysis of A is found to be easily oxidised to a ketone. Suggest the probable structural formula of A and indicate the reasoning by which you arrive at this result. (H = 1, C = 12, Br = 80.) (L(X))

3 Describe how you would prepare a pure specimen of bromoethane from ethanol (ethyl alcohol). Draw a sketch of the apparatus you would use to obtain the crude product.

How and under what conditions does bromoethane react with: (a) silver oxide, (b) potassium cyanide, (c) sodium hydroxide? Describe the experiments you would carry out to identify the organic product in one of these reactions. (O)

4 How would you prepare a pure sample of bromoethane from ethanol?

Outline the reactions by which bromoethane may be used in the preparation of (i) propanonitrile, (ii) ethane, (iii) ethylbenzene.

Chloroethane reacts with an alloy of sodium and lead to produce a liquid compound of the composition C, 29·7 per cent; H, 6·2 per cent; Pb, 64·1 per cent. Suggest a structural formula for this compound. (C(T))
(H = 1·0, C = 12, Pb = 207.)

5 Assume that the chemical properties of 1-bromopropane are the same as those of bromoethane. Deduce what products are formed when 1-bromopropane reacts with (a) hydrogen, (b) potassium hydroxide, (c) sodium, (d) sodium ethoxide. In each case give the essential conditions of reaction and name the chief product formed.

Give outline schemes of reactions for converting 1-bromopropane into (i) an

HALOGEN COMPOUNDS

acid, (ii) a primary amine, each containing the original number of carbon atoms per molecule. (AEB)

6 By means of equations with brief notes on reagents and experimental conditions show how an alkyl halide may be converted into (a) the corresponding hydrocarbon, (b) an alcohol, (c) an alkene, (d) a nitrile, and (e) a primary amine.
What prevents a good yield of the primary amine in (e)? (JMB)

7 Describe two methods of introducing a chlorine atom into (a) an aliphatic compound and (b) an aromatic compound. Give one example of each method.
By consideration of the reaction between methylbenzene and chlorine illustrate the importance of experimental conditions in determining the reaction products. Discuss the mechanism operating in each case.
Describe the reactivity of typical aliphatic monohalogen compounds towards (i) potassium hydroxide, (ii) potassium cyanide, and (iii) ammonia. Discuss the mechanisms of these reactions. (JMB (Syllabus B) Specimen question)

8 Give reasons for the items that are **underlined** in the following directions for the laboratory preparation of approximately 30 g of bromoethane.
Fit a 500 cm³ round-bottomed flask with a bent tube connected to a double-surface condenser set for downward distillation. To the lower end of the condenser attach an adapter. (These parts must be fitted together with tight-fitting joints.) Arrange for the end of the adapter to dip below the surface of about 50 cm³ of water contained in a 250 cm³ flask which is surrounded by an ice/water mixture.
Place 37 cm³ (30 g) of ethanol in the round-bottomed flask and add slowly 40 cm³ (74 g) of concentrated sulphuric acid. When the mixture has cooled add 50 g of powdered potassium bromide, reconnect the flask to the condenser and heat gently, at the same time ensuring that a copious supply of cold water is passing through the condenser. Continue heating until no more droplets pass from the end of the condenser.
Pour the contents of the receiving flask into a separating funnel and run off and retain the lower layer. Discard the upper layer. Return the lower layer to the separating funnel and wash it first with dilute aqueous sodium carbonate and then with water, retaining the lower layer each time. Add a few pieces of anhydrous calcium chloride to the lower layer and leave for 20 min in a stoppered flask.
Filter the solution through a fluted filter paper directly into a 50 cm³ distilling flask containing a few chips of unglazed porcelain. Fit the flask with a 100° thermometer and a double-surface condenser having as before a copious supply of cold water running through it. Collect the fraction b.p. 36–40°C.
The bromoethane obtained in this way contains approximately 15 per cent diethyl ether. How do you account for the presence of this impurity?
Given that the product contains 30 g of bromoethane, calculate the percentage yield for the overall reaction. (JMB)

9 Describe how you would prepare a pure specimen of 1,2-dibromoethane starting with ethanol. Draw a sketch of the apparatus you would use to obtain the crude product.
Mention one important use of 1,2-dibromoethane and state how and under what conditions it reacts with (a) potassium cyanide, (b) sodium hydroxide. (O)

10 Given a supply of ethanol and the usual laboratory reagents, describe how you would prepare a sample of 1,2-dibromoethane (ethylene dibromide).
Outline methods for the conversion of the dibromoethane into

(a) $\mathrm{CH_2OH}$ (b) $\mathrm{CH_2COOH}$ (c) $\mathrm{CH_2.O.CO.CH_3}$
    $|$ $|$ $|$
    $\mathrm{CH_2OH}$ $\mathrm{CH_2COOH}$ $\mathrm{CH_2.O.CO.CH_3}$ (L)

11 Give the structural formulae of the isomers represented by $C_2H_4Cl_2$ and outline how each may be prepared.
Describe how and under what conditions each isomer reacts with potassium hydroxide and name the chief products formed. (AEB)

HALOGEN COMPOUNDS

12 Outline schemes of reactions for the preparations of (chloromethyl)benzene and chlorobenzene from benzene, and for the preparation of chloroethane from methane.

How do these products react with (i) aqueous sodium hydroxide, (ii) alcoholic ammonia, (iii) lithium aluminium hydride? (AEB)

13 Compare and contrast the reactivities of the halogen atoms in chloroethane, ethanoyl chloride, chlorobenzene and tetrachloromethane.

Suggest a simple method for estimating chlorine in each of the first two compounds.

14 Write the names and structural formulae of the products of mono-nitration of bromobenzene. Indicate any necessary conditions for the nitration.

Compare the chemical reactivity of bromoethane and bromobenzene to aqueous sodium hydroxide.

Outline how bromobenzene may (a) be obtained from benzene and 4-bromobenzaldehyde respectively, (b) be converted to methylbenzene. (W)

15 Suggest possible structural formulae for aromatic compounds of empirical formula C_7H_7Cl. What (if any) would be the reactions of each of these compounds with sodium hydroxide?

Describe, giving essential experimental details, how you would detect the presence of chlorine in one of these isomers. How, and why, would you modify your method if the compound also contained nitrogen?

Which isomer would you use as a starting material for the preparation of benzoic acid? Describe briefly how you would perform the conversion in the laboratory. (AEB)

16 Describe how you would compare the rates of hydrolysis of chlorobenzene C_6H_5Cl and 1-phenyl-2-chloroethane C_6H_5—CH_2—CH_2Cl.

Predict the probable outcome of your investigation, explaining in detail the reasons for your prediction. (L (Nuffield trial) (S))

17 (a) Bromoethane may be prepared in the laboratory by distilling a mixture of ethanol, potassium bromide and concentrated sulphuric acid. Hydrogen bromide is generated and it reacts with the ethanol according to the equation:

$$C_2H_5OH + HBr \rightleftharpoons C_2H_5Br + H_2O$$

The bromoethane is collected under water in a suitable receiver. The distillate is transferred to a separating funnel and, after removal of the aqueous layer, it is shaken with dilute sodium carbonate solution. The bromoethane layer is then allowed to stand over anhydrous calcium chloride. The resulting liquid is distilled and the fraction boiling in the temperature range 35–40°C is collected.

(i) Name the organic compound(s), apart from ethanol and bromoethane, which may be present in the distillation flask during the preparation.

(ii) Explain why the reaction mixture usually turns brown on heating.

(iii) Explain why the condenser is fitted with an adaptor dipping below the surface of the water in the receiver.

(iv) Which impurities will be removed when the distillate is shaken with sodium carbonate solution?

(v) Why is the bromoethane layer allowed to stand over calcium chloride?

(vi) What is the main impurity in the product likely to be? (Boiling points: ethanol 78°C, diethyl ether 35°C, ethanal 21°C, ethanoic acid 118°C, methyl methanoate 32°C.)

(vii) Calculate the theoretical yield of bromoethane using the following quantities of reactants:

ethanol 4 g, concentrated sulphuric acid 12 g, potassium bromide 8 g.

($H = 1.0$, $C = 12$, $O = 16$, $S = 32$, $K = 39$, $Br = 80$)

HALOGEN COMPOUNDS

(b) The hydrolysis of an alkyl bromide (RBr) may be considered to proceed according to either a one-step (I) or a two-step (II) mechanism:

$$\text{I} \quad RBr + OH^- \rightarrow ROH + Br^-$$

or

$$\text{II} \quad RBr \rightarrow R^+ + Br^- \quad (slow)$$
$$R^+ + OH^- \rightarrow ROH \quad (fast)$$

For each of these possible mechanisms write an expression for the rate of reaction and comment on any points of interest in these expressions. (L(X))

18 The following results are those from an experiment in which equal volumes of equimolar solutions of a bromoalkane of formula C_4H_9Br and of potassium hydroxide were mixed and then 20 cm³ samples taken at intervals, the reaction quenched in an excess of ice cold water, and titrated against a standard acid solution.

Time (seconds × 10³)	0	0·45	0·9	1·8	2·7	3·6	4·5	5·4	6·3	7·2	8·1
Titre (cm³ of acid)	20	11·5	8·0	5·0	3·55	2·8	2·35	2·0	1·75	1·55	1·4

(a) Use these results to find the overall order of the reaction.

(b) From your result in (a), deduce the most probable mechanism for the hydrolysis, explaining your deduction.

(c) What is the most probable formula for the bromoalkane? Explain your reasoning carefully. (L (Nuffield trial) (S))

19 Devise experiments to determine:

(a) the structure of the product(s) obtained by addition of hydrogen bromide to propene $CH_3-CH=CH_2$;

(b) the structure of the product(s) obtained by elimination of hydrogen bromide from 2-bromobutane $CH_3-CHBr-CH_2-CH_3$. (O Schol.)

20 Discuss the differences in chemical properties between chloroethane and chlorobenzene. How would you expect the following compound to behave:

$$Br-\langle\bigcirc\rangle-CHBr_2$$

21 Write equations for some of the reactions of alkyl halides giving examples of as many different types of reaction as possible. How could you distinguish chemically between 1-chloropropane and 2-chloropropane?

When 1-chloropropane is further chlorinated the ratio of 1,2-dichloropropane to 1,3-dichloropropane in the product is 9:7. How does this result differ from that which might have been expected? (O Schol.)

Chapter 10

Alcohols and phenols

10.1 Introduction

General formula

$$R-OH$$

Alcohols are compounds containing one or more hydroxyl groups attached to saturated carbon atoms. Those with one hydroxyl group are known as **monohydric** alcohols; examples are ethanol, C_2H_5-OH (an aliphatic monohydric alcohol) and phenylmethanol, $C_6H_5-CH_2-OH$ (an aromatic monohydric alcohol). There are also **polyhydric** alcohols, which contain more than one hydroxyl group; examples are ethane-1,2-diol, $HOCH_2-CH_2OH$ (an aliphatic dihydric alcohol) and propane-1,2,3-triol, $HOCH_2-CH(OH)-CH_2OH$ (an aliphatic trihydric alcohol).

Phenols are compounds containing one or more hydroxyl groups attached to aromatic carbon atoms; the parent member of the series is phenol itself, C_6H_5-OH. Many of the properties of phenols are different from those of alcohols, and they are described separately (10.8).

10.2 Nomenclature of monohydric alcohols

Monohydric alcohols are named by replacing the final -e in the corresponding alkane by **-ol**. The position of the hydroxyl group in the carbon chain is given by numbering the carbon atoms as for alkanes. For example:

$$\overset{4}{C}H_3-\overset{3}{C}H_2-\overset{2}{\underset{OH}{C}H}-\overset{1}{C}H_3 \qquad \overset{4}{C}H_3-\overset{3}{C}H_2-\overset{2}{\underset{\underset{OH}{|}}{\overset{\overset{CH_3}{|}}{C}}}-\overset{1}{C}H_3$$

Butan-2-ol 2-Methylbutan-2-ol

There are three classes of alcohol. They differ in the number of alkyl groups attached to the hydroxyl-bearing carbon atom:

$$R-\underset{H}{\overset{H}{\underset{|}{\overset{|}{C}}}}-OH \qquad R-\underset{H}{\overset{R}{\underset{|}{\overset{|}{C}}}}-OH \qquad R-\underset{R}{\overset{R}{\underset{|}{\overset{|}{C}}}}-OH$$

Primary Secondary Tertiary

The three classes have many similar chemical properties, owing to the presence of the same functional group, —OH. However, there are also differences which are due to the different numbers of hydrogen atoms on the hydroxyl-bearing carbon atom.

The names and structural formulae of some alcohols are in Table 10.1.

There are two isomers with formula C_3H_7OH, one a primary and the other a secondary alcohol. There are four isomers of C_4H_9OH, two of which are primary alcohols, for 2-methylpropan-1-ol is a primary alcohol as it possesses a $-CH_2OH$ group, even though it contains a branched alkyl chain.

Table 10.1. The structural formulae, class and physical properties of some alcohols

NAME	FORMULA	STRUCTURAL FORMULA	CLASS	M.P. (°C)	B.P. (°C)
Methanol	CH_3OH	$H-CH_2-OH$	Primary	−97	64
Ethanol	C_2H_5OH	CH_3CH_2-OH	Primary	−117	78
Propan-1-ol	C_3H_7OH	$CH_3CH_2CH_2-OH$	Primary	−127	98
Propan-2-ol	C_3H_7OH	$(CH_3)_2CH-OH$	Secondary	−89	82
Butan-1-ol	C_4H_9OH	$CH_3CH_2CH_2CH_2-OH$	Primary	−90	118
2-Methylpropan-1-ol	C_4H_9OH	$(CH_3)_2CHCH_2-OH$	Primary	−108	108
Butan-2-ol	C_4H_9OH	$CH_3CH_2(CH_3)CH-OH$	Secondary		100
2-Methylpropan-2-ol	C_4H_9OH	$(CH_3)_3C-OH$	Tertiary	25	83
Pentan-1-ol	$C_5H_{11}OH$	$CH_3CH_2CH_2CH_2CH_2-OH$	Primary	−79	138
Hexan-1-ol	$C_6H_{13}OH$	$CH_3CH_2CH_2CH_2CH_2CH_2-OH$	Primary	−51	157
Phenylmethanol	$C_6H_5CH_2OH$	$C_6H_5CH_2-OH$	Primary	−15	205

10.3 Physical properties of monohydric alcohols

As with alkanes, the boiling points of the alcohols increase fairly regularly on the addition of each methylene ($-CH_2-$) group; the increment is about 20°C among the lower homologues (Table 10.1). Again, as with the alkanes, and for the same reason (5.2), increase in branching of the carbon chain is accompanied by decrease in boiling point; this is illustrated by the four isomeric alcohols, C_4H_9OH (Table 10.1).

However, the boiling points of alcohols are considerably higher than those of alkanes of approximately the same formula weight, as illustrated in Table 10.2.

Table 10.2. The boiling points of alkanes and alcohols of similar formula weight

ALKANE	F. WT.	B.P. (°C)	ALCOHOL	F. WT.	B.P. (°C)
Ethane	30	−89	Methanol	32	64
Propane	44	−42	Ethanol	46	78
Butane	58	−0·5	Propan-1-ol	60	98

This is because of the occurrence of strong attractive forces, known as **hydrogen-bonds**, between the molecules of the alcohol in the liquid phase. The bonds are electrostatic in nature; the proton in the hydroxyl group of one molecule is attracted by an unshared pair of electrons on the oxygen atom of another:

$$R-O \cdots H \cdots O-R$$
$$\quad H \quad H$$
$$\quad O$$
$$\quad R$$

ALCOHOLS AND PHENOLS

The hydrogen-bonds, represented by dotted lines, are longer than covalent bonds and are not as strong. A typical value for the strength of a hydrogen-bond is 20 kJ per mol, which is much larger than the usual attractive forces between molecules, which are generally of the order of 1–2 kJ per mol. These differences in the attractive forces are reflected in the higher temperatures needed to separate the molecules of an alcohol compared with the molecules of an alkane. Hydrogen-bonding in a liquid alcohol can be detected by infrared spectroscopy (Fig. 10.1).

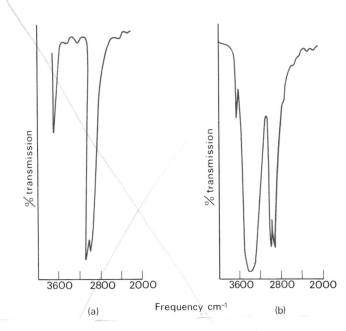

FIG. 10.1. *Part of the infrared absorption spectrum of ethanol. In (a), ethanol is in the vapour phase, and the —OH group is absorbing radiation at 3700 cm^{-1}. In (b), ethanol is dissolved in an inert solvent and the —OH group is absorbing radiation at 3300 cm^{-1}. The reduction in frequency indicates that the bond strength is reduced due to hydrogen-bonding between the molecules of ethanol*

The smaller alcohols are miscible with water, but as the number of carbon atoms increases, solubility decreases, for example:

Alcohol	Methanol	Ethanol	Propan-1-ol
Solubility (g/100 cm^3)	Miscible	Miscible	Miscible

Alcohol	Butan-1-ol	Pentan-1-ol	Hexan-1-ol
Solubility (g/100 cm^3)	8·3	2·6	1·0

Alcohols are more soluble in water than are alkanes of similar formula weight because of the attractive forces (hydrogen-bonds) between molecules of the alcohol and molecules of water:

$$\begin{array}{c} R \\ \diagdown \\ O-H \cdots O \\ \diagup \diagdown \\ H \quad H \end{array}$$

10.4 Methods of preparation of monohydric alcohols

General methods

1. By the hydration of an alkene (6.3), for example:

$$CH_2=CH_2 + H_2SO_4 \rightarrow CH_3-CH_2-O-SO_2-OH$$

$$CH_3-CH_2-O-SO_2-OH + H_2O \rightarrow CH_3-CH_2-OH + H_2SO_4$$

2. By the hydrolysis of an alkyl halide with an aqueous solution of an alkali (9.3), for example:

$$(CH_3)_2CH-Cl + NaOH \rightarrow (CH_3)_2CH-OH + NaCl$$

3. By the reaction of a Grignard reagent with an aldehyde or ketone (9.8). Methanal yields a primary alcohol, other aldehydes yield secondary alcohols and ketones yield tertiary alcohols.

4. By the reduction of an aldehyde (giving a primary alcohol) or a ketone (giving a secondary alcohol) (12.5).

5. By the reduction of a carboxylic acid with lithium aluminium hydride, giving a primary alcohol (13.5).

Manufacture of methanol

1. *From coal.* Steam is passed through white-hot coke to form **water gas** (a mixture of carbon monoxide and hydrogen):

$$C + H_2O \xrightarrow{1300°C} CO + H_2$$

Water gas is then mixed with half its volume of hydrogen and passed over a mixture of zinc and chromium(III) oxides at 300°C under pressure (300 atmospheres):

$$\underbrace{CO + H_2}_{\text{Water gas}} + H_2 \xrightarrow{ZnO + Cr_2O_3 \text{ as cat.}} CH_3OH$$

2. *From natural gas.* Methane, obtained from natural gas (19.2), is passed with steam over nickel at about 900°C and under pressure:

$$CH_4 + H_2O \xrightarrow{\text{Ni as cat.}} CO + 3H_2$$

This mixture of gases, known as **synthesis gas**, is then converted into methanol as described above (p. 308).

Manufacture of ethanol

1. Ethene is passed under pressure through concentrated sulphuric acid at 80°C to form ethyl hydrogen sulphate:

$$CH_2=CH_2 + H_2SO_4 \rightarrow CH_3-CH_2-O-SO_2-OH$$

The mixture is then diluted with water and distilled to give an aqueous solution of ethanol:

$$CH_3-CH_2-O-SO_2-OH + H_2O \rightarrow CH_3CH_2OH + H_2SO_4$$

2. A more recent method is to hydrate ethene directly by passing a mixture of the alkene and steam over a solid acid catalyst (phosphoric acid on silica) at 300°C and a pressure of about 70 atmospheres:

$$CH_2=CH_2 + H_2O \rightarrow CH_3CH_2OH$$

Only about 4 per cent of the ethene reacts, but the remaining ethene and steam are recirculated over the catalyst many times to obtain a good yield of ethanol.

3. An older method is by the fermentation of starch (18.3).

Absolute ethanol

Regardless of the method of manufacture, all aqueous solutions of ethanol yield, on fractional distillation, a 'constant boiling mixture' of 96 per cent ethanol and 4 per cent water, known as **rectified spirit**. This means that further fractionation would remove no more water, since the distillate has the same composition as the liquid.

In the laboratory, the rectified spirit is stored over quicklime (freshly prepared by heating calcium carbonate). Subsequently the mixture is refluxed over quicklime for about 6 hours, and then allowed to stand overnight. The pure product, known as **absolute ethanol**, is then distilled off, precautions being taken to prevent absorption of water vapour by the hygroscopic alcohol.

In industry, benzene is added to the rectified spirit. Distillation yields three fractions:

At 65°C, a constant boiling mixture of ethanol, benzene and water (a 'ternary azeotrope').

At 68°C, a constant boiling mixture of ethanol and benzene (a 'binary azeotrope').

At 78°C, pure ethanol.

Manufacture of propan-2-ol

Propan-2-ol is made from propene (obtained from petroleum (20.4)) by methods analogous to those for the manufacture of ethanol from ethene:

1. $CH_3-CH=CH_2 \xrightarrow[30°C]{H_2SO_4} CH_3-CH(OSO_2OH)-CH_3 \xrightarrow{H_2O} CH_3-CH(OH)-CH_3 + H_2SO_4$

2. $CH_3-CH=CH_2 + H_2O \xrightarrow[250°C, 200 \text{ atm.}]{\text{Tungsten(VI) oxide as cat.}} CH_3-CH(OH)-CH_3$

10.5 Chemical properties of monohydric alcohols

The reactions of alcohols can be grouped in three classes:

(a) Reactions of the —OH group.

(b) Oxidation reactions, which depend on whether the alcohol is primary, secondary or tertiary.

(c) Elimination reactions, which depend on whether or not the alcohol contains at least one hydrogen atom attached to the carbon atom next to the C—OH group.

(a) Reactions of the —OH group

1. Alcohols react with sodium to form salts (sodium alkoxides) and hydrogen, for example:

$$2C_2H_5OH + 2Na \rightarrow 2C_2H_5O^-Na^+ + H_2$$
$$\text{Sodium ethoxide}$$

2. Alcohols react with the hydrogen halides to form alkyl halides:

$$R-OH + HCl \rightarrow R-Cl + H_2O$$
$$R-OH + HBr \rightarrow R-Br + H_2O$$
$$R-OH + HI \rightarrow R-I + H_2O$$

The reaction with hydrogen chloride is catalysed by zinc chloride. A test (Lucas test) to distinguish between simple primary, secondary and tertiary alcohols is based on this reaction and the fact that tertiary alcohols react faster than secondary alcohols which in turn react faster than primary alcohols. The alcohol is shaken with a solution of zinc chloride in concentrated hydrochloric acid. Immediate cloudiness (due to formation of the alkyl chloride) indicates a tertiary alcohol; if the solution turns cloudy within about five minutes, a secondary alcohol is indicated; while primary alcohols show no cloudiness at room temperature.

The mechanism of the reaction with a hydrogen halide depends on whether the the alcohol is primary, secondary or tertiary. In each case, the first step is the same: the alcohol reacts reversibly with the hydrogen halide:

$$R-OH + HX \rightleftharpoons R-\overset{+}{\underset{H}{O}}-H + X^-$$

That is, in the presence of a strong acid, the oxygen atom of the alcohol is acting as a base, accepting a proton from the hydrogen halide. Alcohols are very weak bases and the equilibrium lies to the left.

With *primary* alcohols, the next step is displacement by the halide ion of a molecule of water. This is an S_N2 reaction, analogous to the reaction of a primary alkyl halide with hydroxide ion (9.3):

$$X^- \curvearrowright CH_2-\overset{+}{\underset{R}{O}}\underset{H}{\overset{H}{\diagdown}} \longrightarrow R-CH_2-X + H_2O$$

With *tertiary* alcohols, the second step is heterolysis:

$$R-\underset{R}{\overset{R}{\underset{|}{C}}}-\overset{+}{\underset{H}{O}}\overset{H}{\diagdown} \longrightarrow \underset{R}{\overset{R}{\underset{\diagup}{C^+}}}{\diagdown}_R + H_2O$$

and this is followed by reaction of the carbonium ion with the halide ion:

$$\underset{R}{\overset{R}{\underset{\diagup}{C^+}}}{\diagdown}_R + X^- \longrightarrow R-\underset{R}{\overset{R}{\underset{|}{C}}}-X$$

Tertiary alcohols differ from primary alcohols because a tertiary carbonium ion, R_3C^+, is relatively more stable, and is formed more rapidly, than a primary carbonium ion, RCH_2^+ (9.3).

Secondary alcohols are intermediate in behaviour between primary and tertiary alcohols.

3. Alcohols react with phosphorus tribromide and phosphorus tri-iodide to form alkyl bromides and alkyl iodides:

$$3R\text{—}OH + PBr_3 \rightarrow 3R\text{—}Br + H_3PO_3$$

$$3R\text{—}OH + PI_3 \rightarrow 3R\text{—}I + H_3PO_3$$

These phosphorus trihalides are conveniently prepared *in situ* from red phosphorus and the halogen.

Alkyl chlorides can be obtained from the alcohol with phosphorus pentachloride or with sulphur dichloride oxide at room temperature:

$$R\text{—}OH + PCl_5 \rightarrow R\text{—}Cl + POCl_3 + HCl$$

$$R\text{—}OH + SOCl_2 \rightarrow R\text{—}Cl + SO_2 + HCl$$

The liberation of hydrogen chloride from an alcohol and phosphorus pentachloride is typical of the behaviour of organic compounds containing the hydroxyl group.

4. Alcohols react with concentrated sulphuric acid to form products which depend on the nature of the alcohol and the reaction conditions.

At 0°C, an alkyl hydrogen sulphate is formed:

$$R\text{—}OH + H_2SO_4 \rightarrow R\text{—}O\text{—}SO_2\text{—}OH + H_2O$$

When the acid is added to an excess of a primary alcohol and the mixture is heated to about 140°C, an ether is formed, for example:

$$2C_2H_5\text{—}OH \xrightarrow[140°C]{H_2SO_4} C_2H_5\text{—}O\text{—}C_2H_5 + H_2O$$
$$\text{Diethyl ether}$$

The reaction occurs by displacement of the hydrogen sulphate group by the excess of the alcohol, as in the S_N2 substitution of an alkyl halide (9.3):

$$\underset{H}{\overset{C_2H_5}{>}}O: \curvearrowright CH_2\text{—}O\text{—}SO_2\text{—}OH \longrightarrow \underset{H}{\overset{C_2H_5}{>}}\overset{+}{O}\text{—}CH_2$$
$$ CH_3 CH_3$$
$$ + ^-O\text{—}SO_2\text{—}OH$$

$$\longrightarrow C_2H_5\text{—}O\text{—}C_2H_5 + H_2SO_4$$

However, when there is an excess of the acid and a still higher temperature is used, elimination can occur and an alkene is formed (p. 145).

5. Alcohols react with organic acids to form esters. The reaction is slow unless an acid catalyst is used; hydrogen chloride or concentrated sulphuric acid are suitable catalysts. For example:

$$\underset{\underset{O}{\|}}{CH_3\text{—}C\text{—}OH} + CH_3\text{—}OH \underset{}{\overset{H^+}{\rightleftharpoons}} \underset{\underset{O}{\|}}{CH_3\text{—}C\text{—}O\text{—}CH_3} + H_2O$$

Ethanoic acid Methyl ethanoate

This type of reaction (esterification) is discussed on p. 204.

Alcohols also form esters when treated with acid halides or acid anhydrides (14.3, 14.4), for example:

$$CH_3-\underset{\underset{O}{\|}}{C}-Cl + CH_3-OH \longrightarrow CH_3-\underset{\underset{O}{\|}}{C}-O-CH_3 + HCl$$

Ethanoyl chloride   Methyl ethanoate

(b) Oxidation reactions

Primary and secondary alcohols are readily oxidised to aldehydes and ketones, respectively. Tertiary alcohols are resistant to oxidation.

Oxidation in solution can be brought about with acidified sodium or potassium dichromate:

$$RCH_2-OH \xrightarrow[H_2SO_4]{Na_2Cr_2O_7} RCH=O$$

$$R_2CH-OH \xrightarrow[H_2SO_4]{Na_2Cr_2O_7} R_2C=O$$

In the case of the primary alcohol, the resulting aldehyde undergoes further oxidation to the carboxylic acid unless precautions are taken to prevent it (12.3).

Potassium permanganate in acid solution also effects these oxidations.

Oxidation in the gas phase can be brought about either by passing the vapour of the alcohol, together with oxygen, over silver at about 500°C, for example,

$$CH_3OH + \tfrac{1}{2}O_2 \xrightarrow[500°C]{Ag\ as\ cat.} HCHO + H_2O$$

or by passing the vapour of the alcohol alone over heated copper, for example:

$$CH_3CH_2OH \xrightarrow[500°C]{Cu\ as\ cat.} CH_3CHO + H_2$$

$$(CH_3)_2CHOH \xrightarrow[500°C]{Cu\ as\ cat.} (CH_3)_2CO + H_2$$

Secondary alcohols which contain the grouping $CH_3-CH(OH)-$ are oxidised by iodine in the presence of sodium hydroxide to tri-iodomethane:

$$CH_3-CH(OH)-R + I_2 + 2NaOH \rightarrow CH_3-CO-R + 2NaI + 2H_2O$$

$$CH_3-CO-R + 3I_2 + 4NaOH \rightarrow CHI_3 + R-CO_2^-Na^+ + 3NaI + 3H_2O$$

One primary alcohol, ethanol, also undergoes this reaction:

$$CH_3-CH_2-OH + 4I_2 + 6NaOH \rightarrow$$
$$CHI_3 + HCO_2^-Na^+ + 5NaI + 5H_2O$$

Other primary and secondary alcohols are also oxidised and iodinated but do not then give tri-iodomethane, for example:

$$CH_3-CH_2-CH_2-OH + I_2 + 2NaOH \rightarrow$$
$$CH_3-CH_2-CH=O + 2NaI + 2H_2O$$

$$CH_3-CH_2-CH=O + 2I_2 + 2NaOH \rightarrow$$
$$CH_3-CI_2-CH=O + 2NaI + 2H_2O$$

(c) Elimination reactions

Alcohols which possess at least one hydrogen atom on the carbon atom next but one to the hydroxyl group undergo **dehydration** (elimination of water) when heated with concentrated sulphuric acid:

$$H-\overset{|}{\underset{|}{C}}-\overset{|}{\underset{|}{C}}-OH \xrightarrow{H_2SO_4} {>}C=C{<} + H_2O$$

The reaction occurs via the alkyl hydrogen sulphate, for example:

$$C_2H_5OH + H_2SO_4 \longrightarrow CH_3-CH_2-O-SO_2-OH + H_2O$$

$$CH_3-CH_2-O-SO_2-OH \xrightarrow{170°C} CH_2=CH_2 + H_2SO_4$$

This elimination reaction competes with the substitution reaction which occurs (p. 143), for example:

$$CH_3-CH_2-O-SO_2-OH + C_2H_5OH \rightarrow$$
$$CH_3CH_2-O-CH_2CH_3 + H_2SO_4$$

The substitution reaction is promoted by the presence of an excess of the alcohol, and the elimination reaction is promoted by the use of higher temperatures.

An alternative dehydrating agent is phosphoric acid. A practical example, the dehydration of cyclohexanol, is given on p. 84.

Alcohols can also be dehydrated by passing their vapour over aluminium oxide at about 300°C, for example:

$$CH_3-CH_2-OH \xrightarrow{Al_2O_3 \text{ as cat.}} CH_2=CH_2 + H_2O$$

Summary of the differences between primary, secondary and tertiary alcohols

1. *Oxidation.* Primary alcohols give aldehydes, secondary alcohols give ketones and tertiary alcohols are resistant to oxidation by mild oxidising agents in solution such as acidified potassium dichromate.

2. *Elimination.* All three types of alcohols containing the grouping $H-\overset{|}{\underset{|}{C}}-\overset{|}{\underset{|}{C}}-OH$ are dehydrated by sulphuric acid to alkenes, but tertiary alcohols react far more readily than secondary alcohols which in turn react more readily than primary alcohols.

3. *Reaction with halogen acids.* All three types of alcohol react, but tertiary alcohols react the most readily (e.g. with concentrated hydrochloric acid, p. 112) and primary alcohols the least readily (anhydrous conditions, with zinc chloride as a catalyst, are necessary).

10.6 Uses of monohydric alcohols

Methanol

1. In the manufacture of methanal, which is used to make thermosetting plastics such as Bakelite (p. 316).
2. To make methyl 2-methylpropenoate, which is used in the manufacture of Perspex (p. 315).
3. As a solvent for varnishes and paints.

Ethanol

1. In the manufacture of ethanal (12.3).
2. As a solvent for many organic compounds.

Propan-2-ol

1. In the manufacture of propanone (20.4) and of hydrogen peroxide (20.4).
2. As a solvent for spirit polishes and varnishes.

10.7 Polyhydric alcohols

Dihydric alcohols

According to the I.U.P.A.C. nomenclature, these are known as **diols**, for example:

		B.P. (°C)
Ethane-1,2-diol	HO—CH_2—CH_2—OH	197
Propane-1,2-diol	HO—CH_2—CH(OH)—CH_3	189
Propane-1,3-diol	HO—CH_2—CH_2—CH_2—OH	215

The older name is **glycol**; for example, ethane-1,2-diol is often referred to as ethylene glycol or simply as glycol.

Physical properties of diols

The lower members are viscous, colourless liquids which are soluble in water. Their boiling points are very much higher not only than those of alkanes of similar formula weight but also than those of monohydric alcohols of similar formula weight; this is because the presence of two hydroxyl groups gives rise to very extensive hydrogen-bonding. This is also the reason for their high viscosity, since neighbouring molecules in the liquid, being bonded by hydrogen-bonds, cannot move freely relative to each other. Finally, their solubility in water stems from their forming hydrogen-bonds with water molecules.

Manufacture of ethane-1,2-diol

Ethane-1,2-diol is manufactured by hydration of epoxyethane (11.6):

$$H_2C\underset{O}{\overset{}{\diagdown\!\!\diagup}}CH_2 + H_2O \longrightarrow \underset{HO\ \ OH}{H_2C\!-\!CH_2}$$

This is carried out in acid solution at about 60°C or with water at 200°C under pressure.

ALCOHOLS AND PHENOLS

Chemical properties of diols

Ethane-1,2-diol (ethylene glycol) is taken as a typical example. Its two primary alcohol groups behave in the same way as the one such group in a monohydric primary alcohol, except that more vigorous conditions are sometimes needed for reaction of the second of the two groups. For example:

1. It reacts with sodium to form a monoalkoxide and, at higher temperatures, a dialkoxide:

$$\begin{array}{c} CH_2OH \\ | \\ CH_2OH \end{array} \xrightarrow[50\,°C]{Na} \begin{array}{c} CH_2O^-Na^+ \\ | \\ CH_2OH \end{array} + \tfrac{1}{2}H_2$$

$$\begin{array}{c} CH_2O^-Na^+ \\ | \\ CH_2OH \end{array} \xrightarrow[150\,°C]{Na} \begin{array}{c} CH_2O^-Na^+ \\ | \\ CH_2O^-Na^+ \end{array} + \tfrac{1}{2}H_2$$

2. It reacts with phosphorus halides:

$$3\begin{array}{c} CH_2-OH \\ | \\ CH_2-OH \end{array} + 2PBr_3 \longrightarrow 3\begin{array}{c} CH_2-Br \\ | \\ CH_2-Br \end{array} + 2H_3PO_3$$

3. It reacts with carboxylic acids to form esters:

$$\begin{array}{c} CH_2-OH \\ | \\ CH_2-OH \end{array} \xrightarrow[H^+]{CH_3-CO_2H} \begin{array}{c} CH_2-O-CO-CH_3 \\ | \\ CH_2-OH \end{array} \xrightarrow[H^+]{\text{excess of } CH_3-CO_2H} \begin{array}{c} CH_2-O-CO-CH_3 \\ | \\ CH_2-O-CO-CH_3 \end{array}$$

When esterified with a dibasic acid, it forms polymers, for example:

$$n(HO_2C\text{—}\langle\;\rangle\text{—}CO_2H) + n(HO-CH_2-CH_2-OH) \longrightarrow$$

Benzene-1,4-dicarboxylic acid

$$H-O-(CO\text{—}\langle\;\rangle\text{—}CO-O-CH_2-CH_2-O)_n-H + (2n-1)H_2O$$

4. On oxidation, with nitric acid, both primary alcohol groups are oxidised, first to aldehyde and then to carboxyl groups. Ethanedioic acid is then oxidised to carbon dioxide and water:

$$\begin{array}{c} CH_2-OH \\ | \\ CH_2-OH \end{array} \longrightarrow \begin{array}{c} CH=O \\ | \\ CH_2-OH \end{array} \nearrow \begin{array}{c} CH=O \\ | \\ CH=O \end{array} \searrow \begin{array}{c} CO_2H \\ | \\ CHO \end{array} \longrightarrow \begin{array}{c} CO_2H \\ | \\ CO_2H \end{array}$$

$$\searrow \begin{array}{c} CO_2H \\ | \\ CH_2-OH \end{array} \nearrow$$

$$\downarrow$$

$$2CO_2 + H_2O$$

Uses of ethane-1,2-diol

1. In the manufacture of Terylene (p. 321).
2. As an anti-freeze for car radiators and as a de-icing fluid for aeroplane wings. Other chemicals (anti-oxidants) are added to inhibit the formation of acids, by oxidation of the diol, which would cause corrosion.

Propane-1,2,3-triol (glycerol)

Propane-1,2,3-triol is the simplest trihydric alcohol (triol):

$$\begin{array}{ccc} CH_2 & -CH- & CH_2 \\ | & | & | \\ OH & OH & OH \end{array}$$

It is a colourless, very viscous liquid which is soluble in water and ethanol. Its chemical properties are similar to those of monohydric alcohols.

Manufacture of propane-1,2,3-triol

From propene (obtained from petroleum; 20.4) in two ways:

1. *via* 3-Chloropropene:

$$CH_3-CH=CH_2 \xrightarrow[400-600\,°C]{Cl_2} \underset{\underset{Cl}{|}}{CH_2}-CH=CH_2 \xrightarrow{HOCl}$$

$$\underset{\underset{Cl}{|}\;\underset{Cl}{|}\;\underset{OH}{|}}{CH_2-CH-CH_2} \xrightarrow{Ca(OH)_2} \underset{\underset{Cl}{|}\quad\;\;\diagdown\;\diagup}{CH_2-CH-CH_2}\; O$$

$$\xrightarrow[150\,°C]{NaOH/H_2O}$$

$$\underset{\underset{OH}{|}\;\underset{OH}{|}\;\underset{OH}{|}}{CH_2-CH-CH_2}$$

2. *via* Propenal:

$$CH_3-CH=CH_2 \xrightarrow[\substack{CuO\;as\;cat. \\ 350\,°C}]{O_2} \underset{Propenal}{O=CH-CH=CH_2} \xrightarrow[\substack{MgO\;+\;ZnO\;cat. \\ 400\,°C}]{(CH_3)_2CH-OH}$$

$$HO-CH_2-CH=CH_2 \xrightarrow[WO_3\;as\;cat.]{H_2O_2} \underset{\underset{OH}{|}\;\underset{OH}{|}\;\underset{OH}{|}}{CH_2-CH-CH_2}$$

3. It is a by-product in the manufacture of soap (13.10).

Uses of propane-1,2,3-triol

1. In the manufacture of nitroglycerin, a constituent of several explosives:

$$\begin{array}{l} CH_2-OH \\ | \\ CH-OH \\ | \\ CH_2-OH \end{array} + 3HNO_3 \longrightarrow \begin{array}{l} CH_2-O-NO_2 \\ | \\ CH-O-NO_2 \\ | \\ CH_2-O-NO_2 \\ \text{Nitroglycerin} \end{array} + 3H_2O$$

ALCOHOLS AND PHENOLS

It should be noted that nitroglycerin is not a nitro-compound as its name may suggest. It is a nitrate ester (propane-1,2,3-triyl trinitrate).

Nitroglycerin is a colourless, oily liquid which is violently detonated on slight shock. Oxygen is present in the molecule, and carbon dioxide, water vapour and nitrogen are liberated to produce a very large pressure.

Dynamite, invented by the Swedish chemist, Nobel, is made by allowing kieselguhr to absorb nitroglycerin. Although it retains its explosive properties, the nitroglycerin is less sensitive to shock. Nobel also introduced gun-cotton (cellulose trinitrate), and blasting gelatin, a mixture of 90 per cent nitroglycerin and 10 per cent gun-cotton.

Cordite is a slower burning powder (30 per cent nitroglycerin and 65 per cent gun-cotton) and is used in shells and bullets.

2. In the manufacture of glyptal plastics. (p. 319).

10.8 Phenols

Phenols are the hydroxy-derivatives of aromatic compounds. Examples of some monohydric phenols:

NAME	FORMULA	M.P. (°C)	B.P. (°C)
Phenol	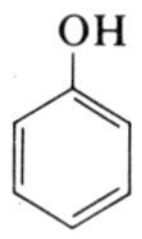 C_6H_5OH	43	181
2-Methylphenol	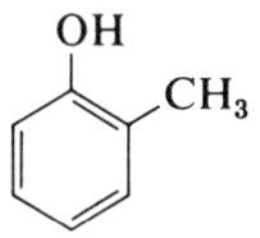 2-$CH_3C_6H_4OH$	30	191
3-Methylphenol	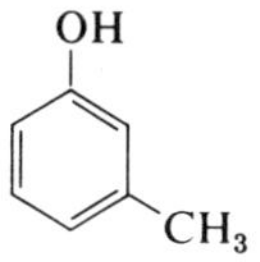 3-$CH_3C_6H_4OH$	11	201
4-Methylphenol	 4-$CH_3C_6H_4OH$	36	201
4-Nitrophenol	 4-$NO_2C_6H_4OH$	114	279

Phenol itself is chosen as a typical member of the group.

Physical properties of phenol

Phenol is a colourless, crystalline solid which becomes discoloured on exposure to air and light. It is only slightly soluble in water but is very soluble in organic solvents.

Preparation of phenol

1. From benzenesulphonic acid (obtained by the sulphonation of benzene; 8.3). The sodium salt of the acid is fused with sodium hydroxide at 300°C:

$$C_6H_5-SO_2O^-Na^+ + NaOH \rightarrow C_6H_5-OH + Na_2SO_3$$

$$C_6H_5-OH + NaOH \rightarrow C_6H_5-O^-Na^+ + H_2O$$
$$\text{Sodium phenoxide}$$

Phenol is released from sodium phenoxide with dilute acid:

$$C_6H_5-O^-Na^+ + H_2SO_4 \rightarrow C_6H_5-OH + NaHSO_4$$
$$\text{Phenol}$$

2. By warming an aqueous solution of benzenediazonium chloride (16.8).

Manufacture of phenol

The process *via* (1-methylethyl)benzene (cumene) now accounts for about 80 per cent of the total phenol produced, the older processes *via* benzenesulphonic acid and chlorobenzene having been largely superseded.

1. *The Cumene process.* Benzene is alkylated with propene, either in the liquid phase with aluminium trichloride as catalyst or in the gas phase with phosphoric acid on an inert solid as catalyst:

$$C_6H_6 + CH_3-CH=CH_2 \rightarrow C_6H_5-CH(CH_3)_2$$
$$\text{(1-Methylethyl)benzene}$$

Air is passed through (1-methylethyl)benzene (often called cumene) to form its hydroperoxide which is then decomposed with warm, dilute sulphuric acid:

$$C_6H_5-CH(CH_3)_2 + O_2 \longrightarrow C_6H_5-\underset{\underset{O-OH}{|}}{\overset{\overset{CH_3}{|}}{C}}-CH_3$$

$$\xrightarrow{H^+} C_6H_5-OH + (CH_3)_2C=O$$

The final reaction involves a rearrangement: as the O—O bond in the protonated hydroperoxide breaks, so the phenyl group migrates from carbon to oxygen:

$$C_6H_5-\underset{\underset{OH}{\overset{|}{O}}}{\overset{\overset{CH_3}{|}}{C}}-CH_3 \xrightleftharpoons{H^+} C_6H_5-\underset{\underset{\underset{H}{|}}{\overset{+}{O}}}{\overset{\overset{CH_3}{|}}{C}}-CH_3 \longrightarrow \underset{C_6H_5-O}{\overset{CH_3\diagdown\overset{+}{C}\diagup CH_3}{}} + H_2O$$

The carbonium ion reacts with water to form the products:

$$C_6H_5-O-\overset{+}{C}(CH_3)_2 + H_2O \longrightarrow C_6H_5-O-\underset{OH}{\overset{CH_3}{\underset{|}{\overset{|}{C}}}}-CH_3 + H^+$$

$$C_6H_5-O-\underset{OH}{\overset{CH_3}{\underset{|}{\overset{|}{C}}}}-CH_3 \longrightarrow C_6H_5-OH + (CH_3)_2C=O$$

This process is particularly valuable because of the formation of propanone as well as phenol.

2. *Chlorobenzene process.*

$$C_6H_6 \xrightarrow[40°C]{Cl_2} C_6H_5-Cl \xrightarrow[330°C]{NaOH} C_6H_5-O^-Na^+ \xrightarrow{H^+} C_6H_5-OH$$

3. *Benzenesulphonic acid process* (p. 150).

Chemical properties of phenol

Like alcohols, phenol reacts with acid chlorides to form esters, for example:

$$C_6H_5-OH + CH_3-CO-Cl \rightarrow C_6H_5-O-CO-CH_3 + HCl$$
Phenyl ethanoate

In the following respects, phenol behaves differently from both aliphatic and aromatic alcohols. In Section 10.9 experiments are suggested which compare and contrast the reactions of phenol with those of ethanol and phenylmethanol.

1. Phenol does not react with hydrogen halides to form aryl halides.
2. Phenol does not react with phosphorus tribromide or tri-iodide to form bromobenzene or iodobenzene.
3. Phenol is not oxidised in the same way as alcohols. Oxidation occurs readily, but the products are generally complex, polymeric materials.
4. Phenol does not undergo elimination reactions.
5. Phenol is a considerably stronger acid than an alcohol; thus, K_a is 1.3×10^{-10} for phenol and 10^{-16} for methanol.

This can be understood by considering the bonding in the phenoxide ion. One of the p orbitals on the oxygen atom, which contains two electrons, interacts with the singly occupied p orbital on the adjacent carbon atom. The latter also takes part in the p-orbital interactions which are characteristic of the benzene ring (p. 57), so that a total of seven π molecular orbitals is formed. The eight p electrons—six from carbon atoms and two from the oxygen atom—fill four of these; one is shown in Fig. 10.2. Consequently, the charge on the phenoxide ion is not confined to oxygen but is delocalised and therefore stabilised; in contrast, the charge on an alkoxide ion is confined to the oxygen atom. Thus, phenol has a greater tendency to dissociate than an alcohol.

Phenol is not as strong an acid as carbonic acid or a carboxylic acid. This affords a method for distinguishing phenol from a carboxylic acid, for phenol does not react with an aqueous solution of sodium carbonate, whereas carboxylic acids react to liberate carbon dioxide. The separation of a mixture of phenol and a carboxylic acid is based on the same principle (13.5).

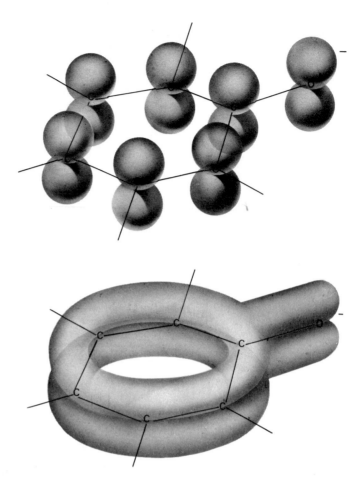

FIG. 10.2. π-*Bonding in the phenoxide ion; the lower diagram shows a delocalised π orbital formed by overlap of the p orbitals shown in the upper diagram*

The acidity of a phenol is increased when electron-attracting substituents are introduced into the aromatic ring. For example, 2,4,6-trinitrophenol is quite a strong acid ($K_a = 10^{-1}$), which liberates carbon dioxide from sodium carbonate solution; it is often described as an acid (picric acid) for this reason.

2,4,6-Trinitrophenol
(Picric acid)

6. Phenol reacts with a neutral solution of iron(III) chloride to give a violet colour. This is a characteristic reaction for compounds containing a hydroxyl group adjacent to an unsaturated carbon atom:

This is known as an **enol** group. Another example of a compound containing an enol group is ethyl 3-oxobutanoate (p. 208).

Reactions of the aromatic ring

1. Phenol is reduced to cyclohexanol when passed over nickel at about 200°C:

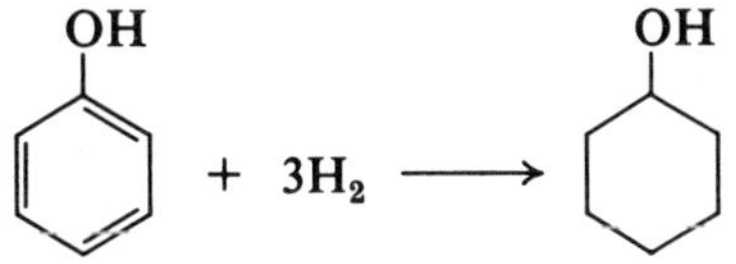

$$\text{C}_6\text{H}_5\text{OH} + 3\text{H}_2 \longrightarrow \text{Cyclohexanol}$$

Cyclohexanol is used in the manufacture of nylon (p. 307).

2. Phenol is very reactive towards electrophilic reagents, undergoing substitution at its 2- and 4-positions under mild conditions. The reason for the greater reactivity of phenol than benzene in these reactions has been described earlier (8.5). For example:

(a) With bromine, 2,4,6-tribromophenol is formed:

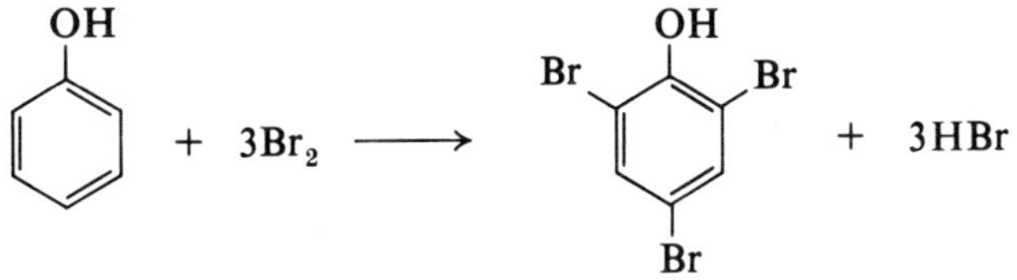

$$\text{C}_6\text{H}_5\text{OH} + 3\text{Br}_2 \longrightarrow \text{2,4,6-tribromophenol} + 3\text{HBr}$$

Likewise, chlorine gives 2,4,6-trichlorophenol.

(b) With dilute nitric acid, a mixture of 2- and 4-nitrophenol is formed:

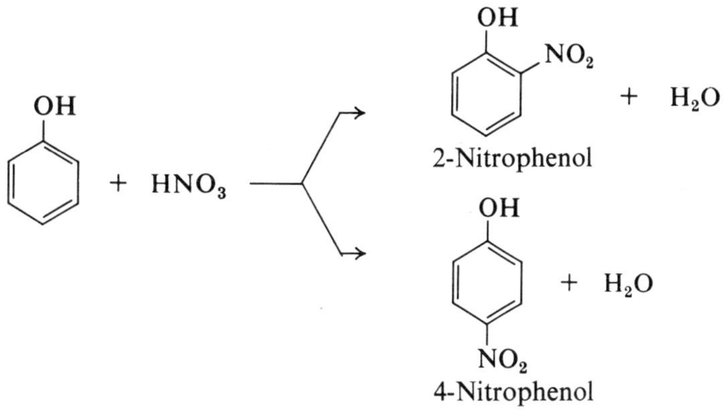

$$\text{C}_6\text{H}_5\text{OH} + \text{HNO}_3 \longrightarrow \text{2-Nitrophenol} + \text{H}_2\text{O}$$
$$\longrightarrow \text{4-Nitrophenol} + \text{H}_2\text{O}$$

2-Nitrophenol possesses an internal hydrogen-bond:

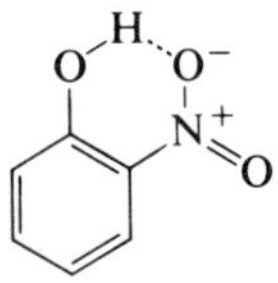

In contrast, the 4-isomer forms hydrogen-bonds by attraction of the hydroxyl hydrogen atom of one molecule to the nitro group of another:

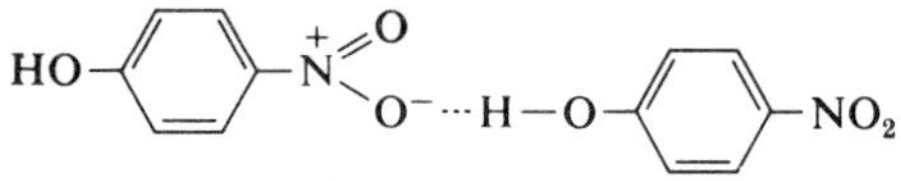

$$HO-\text{C}_6H_4-\overset{+}{N}(=O)(O^-\cdots H-O-\text{C}_6H_4-NO_2)$$

Consequently, more energy is needed to separate the molecules of 4-nitrophenol from each other, so that its boiling point (279°C) is higher than that of the 2-isomer (216°C). This makes it possible to separate the two compounds; a convenient method is by distillation in steam (2.3), the 2-compound having the higher vapour pressure and therefore being the more volatile in the steam.

With excess of nitric acid, 2,4,6-trinitrophenol (picric acid) is formed. The dry compound is explosive, but solutions have been used as yellow dyes, and it was probably the first artificial dye (1849).

Uses of phenol

1. In the manufacture of phenol-methanal plastics, e.g. Bakelite (p. 316).
2. As a starting material for the production of cyclohexanol, which is used in the manufacture of nylon (p. 320).
3. To make substituted phenols, which are used to make epoxy resins (p. 319).
4. To make 2,4-dichlorophenol which is used to make 2,4-dichlorophenoxyethanoic acid (known as 2,4-D), a selective weed killer:

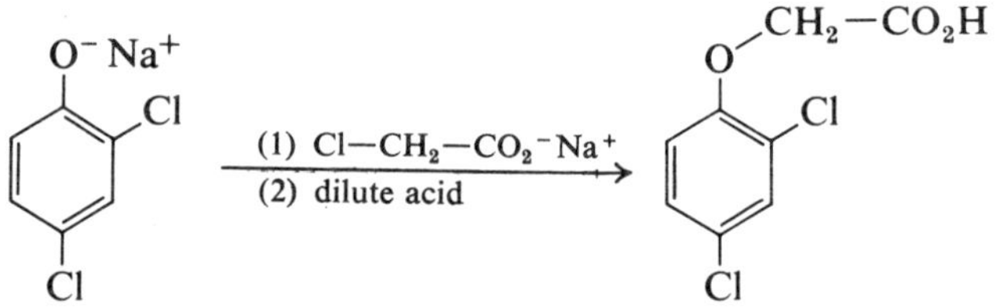

[2,4-dichlorophenol sodium salt] $\xrightarrow[\text{(2) dilute acid}]{\text{(1) Cl}-CH_2-CO_2^-Na^+}$ [2,4-D structure with $O-CH_2-CO_2H$]

5. To make 2,4-dichloro-3,5-dimethylphenol, a powerful antiseptic ('Dettol'):

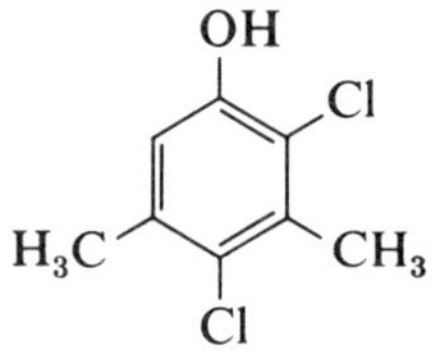

[Structure: phenol ring with OH, Cl at 2, CH₃ at 3, Cl at 4, CH₃ at 5]

10.9 Practical work

Reactions of alcohols and phenols

For reactions with alcohols, unless stated, use either ethanol or phenylmethanol (which must be freshly distilled), and use phenol itself as an example of a phenol.

Reactions of the —OH group

1. Dissolve 5 drops of ethanol in 5 cm³ of water in one test-tube and 0·5 g of phenol in 5 cm³ of water in another tube. To each solution, add 1 drop of blue litmus solution.

2. (a) To 1 cm³ of an alcohol in a test-tube, add a small pellet of sodium. Note the effervescence and test the gas evolved with a lighted splint.

(b) When all the sodium has reacted, evaporate the solution to dryness to obtain a white residue. Add 3 drops of water to the residue and test the solution with litmus solution.

Comment on whether these reactions would occur with phenol.

3. Warm a mixture of 5 drops of an alcohol and 5 drops of ethanoic acid with 1 drop of concentrated sulphuric acid. Note the characteristic smell of the product.

Comment on whether this reaction would occur with phenol.

4. To 5 drops of an alcohol in a test-tube, add 2 or 3 drops of ethanoyl chloride. Repeat the experiment with a few crystals of phenol.

5. *Schotten-Baumann reaction.* To 0·5 g of phenol, add 5 cm³ of 10 per cent sodium hydroxide. Add 5 drops of benzoyl chloride and shake. Filter the precipitate of phenyl benzoate, wash with water and recrystallise from hot ethanol; the product should melt at 69°C:

$$C_6H_5-OH + NaOH \rightarrow C_6H_5-O^-Na^+ + H_2O$$

$$C_6H_5-O^-Na^+ + C_6H_5-CO-Cl \rightarrow C_6H_5-O-CO-C_6H_5 + NaCl$$
Phenyl benzoate

Would you expect alcohols to undergo this reaction?

6. To 1 cm³ of an alcohol in a test-tube, add about 0·1 g of phosphorus pentachloride. Test the fumes evolved by (a) moist blue litmus paper, (b) breathing upon them.

Oxidation reactions

7. To 5 drops of ethanol, add 10 drops of dilute sulphuric acid and 2 drops of potassium dichromate solution. Warm gently, noting (a) the colour of the solution and (b) the smell of the product.

Repeat the experiment with (i) propan-2-ol, (ii) 2-methylpropan-2-ol, (iii) phenylmethanol.

8. Introduce about 10 cm³ of methanol into a 100-cm³ beaker. Introduce a red-hot spiral of platinum wire above the alcohol as shown in Fig. 10.3.

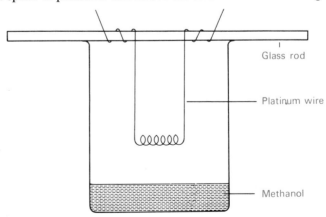

FIG. 10.3. *Catalytic oxidation of methanol*

The spiral continues to glow and the pungent odour of methanal is noticed:

$$2CH_3OH + O_2 \xrightarrow{\text{Pt as cat.}} 2HCHO + 2H_2O$$

ALCOHOLS AND PHENOLS

9. *The iodoform test.* To 5 drops of ethanol, add 5 drops of iodine solution (Appendix II) and then add dilute sodium hydroxide solution dropwise until the colour of the iodine is discharged. A yellow precipitate of tri-iodomethane (iodoform) is obtained. Filter, dry and view the crystals under a microscope and note their characteristic shape (hexagonal plates).

Repeat this experiment with (i) methanol and (ii) propan-2-ol.

Elimination reactions

10. The test-tube preparation of ethene in Section 6.5 is an example.

Would you expect (i) methanol, (ii) propan-2-ol, (iii) 2-methylpropan-2-ol, (iv) phenylmethanol, to undergo this reaction?

Substitution in the aromatic ring of phenol

11. To a few crystals of phenol in a test-tube, add 1 cm^3 of water. Shake and warm the solution. Allow to cool and add bromine water dropwise, until a precipitate of 2,4,6-tribromophenol is obtained.

What products would you expect to obtain when bromine water is added to (a) ethanol, (b) phenylmethanol?

12. Phenol is nitrated with *dilute* nitric acid (p. 269).

Test for the enol group in phenol

13. Make up a neutral solution of iron(III) chloride by adding ammonia solution to 2 cm^3 of iron(III) chloride solution, until a precipitate just appears. Add the original iron(III) chloride solution, dropwise, until the precipitate just disappears.

Place 5 drops of this solution in two test-tubes. To one add 3 drops of an alcohol and to the other add 1 crystal of phenol.

Reactions of polyhydric alcohols (*choose either ethane-1,2-diol or propane-1,2,3-triol*)

14. To 2 cm^3 of the alcohol in a test-tube, add a small clean pellet of sodium. Note whether there is any effervescence, and then warm the mixture gently. Test any gas evolved with a lighted splint.

15. To 2 cm^3 of the alcohol in a test-tube, add ethanoyl chloride dropwise (taking great **CARE**). Test the gas evolved.

16. To 1 cm^3 of a dilute acidified solution of potassium permanganate, add 5 drops of the alcohol. Warm the mixture gently.

Preparation of propane-1,2,3-triol from a fat

17. Details of the saponification of a fat are given on p. 220.

10.10 Questions

1 How and under what conditions does ethanol react with (a) sodium, (b) phosphorus trichloride, (c) sulphuric acid, (d) chlorine and (e) ethanoic acid.

Suggest a scheme for preparing from ethanol a compound containing **four** atoms of carbon per molecule. (AEB)

2 Name and give the formula of one aliphatic monohydric alcohol and describe how it behaves with (a) phosphorus pentachloride, (b) concentrated sulphuric acid.

Name and give the formula of one aliphatic dihydric alcohol, and write down the formulae of all its possible oxidation products.

Distinguish between a *primary*, *secondary* and *tertiary* alcohol and explain how each behaves on oxidation. (AEB)

ALCOHOLS AND PHENOLS

3 Name the four alcohols represented by the molecular formula C_4H_9OH, and write their structural formulae.
   What is the effect of oxidation upon each of these compounds? Outline an experiment by which, using an acidified dichromate solution as a relatively mild oxidising agent, you could differentiate as far as possible between these four alcohols by recognition of the character of their oxidation products.
   Outline the procedure by which pure ethanol can be obtained industrially from starch. (JMB)

4 Write structural formulae for the isomers corresponding to the molecular formula $C_4H_{10}O$.
   One of these isomers, W, reacts with sodium. A ketone, X, is formed when W is oxidised. Dehydration of W gives a mixture of two hydrocarbons Y and Z, each containing 85·7 per cent of carbon. Explain how these reactions enable W to be identified and specify an appropriate reagent for its oxidation and dehydration respectively. Predict how hydrogen bromide would react with Y and Z. (W)

5 Outline how you would prepare a pure sample of ethanol in the laboratory. What evidence would you cite for the presence of (a) a hydroxyl group (—OH), (b) a methyl group (CH_3—) and (c) a methylene group (—CH_2—) in ethanol? (L)

6 Describe **two** important sources of ethanol and indicate briefly how it is obtained from them.
   Give an account of the behaviour of ethanol with (a) oxidising agents, and (b) dehydrating agents.

7 A compound A, containing C, 60·0; H, 13·3; O, 26·7 per cent, and having molecular weight 60, gave on oxidation B, containing C, 62·1; H, 10·3; O, 27·6 per cent. B did not reduce Fehling's solution.
   When A was heated with concentrated sulphuric acid it gave C. Addition of bromine to C gave D, which on boiling with aqueous sodium carbonate gave E.
   Write down the structural formulae of the compounds A to E. (O and C)

8 Describe briefly **one** method by which benzene can be converted into phenol.
   How would you separate a mixture of benzoic acid and phenol?
   Bromine is added slowly to phenol (0·282 g) dissolved in water until a slight excess of bromine is present. Calculate the *weight* of bromine used. (O and C)

9 Describe the preparation of phenol from benzenesulphonic acid. Compare and contrast the chemical reactivity of the hydroxyl group in ethanol with that in phenol. (W)

10 Outline (a) one process for the manufacture of phenol, (b) a laboratory preparation of phenol from phenylamine.
   How, and under what conditions, does phenol react with (i) sodium carbonate, (ii) 50 per cent nitric acid, (iii) concentrated sulphuric acid, (iv) iodomethane (methyl iodide)? (AEB)

11 (a) A polyhydroxylic compound, $C_4H_8O_4$, was heated with excess ethanoic anhydride. On refluxing 2·87 g of the product with 50·00 cm³ of molar sodium hydroxide solution, the residual alkali required 15·00 cm³ of molar hydrochloric acid for neutralisation. Calculate the number of hydroxyl groups per molecule of the original compound.
   (b) A compound X has the following percentage composition by weight: C = 64·9 per cent, H = 13·5 per cent, O = 21·6 per cent. Oxidation yields a neutral compound Y which does not react with sodium. Further oxidation of Y yields an acid Z. When 8 cm³ of the vapour of Y is sparked with excess oxygen, 32 cm³ of carbon dioxide is formed. Z forms only one silver salt which contains 64·7 per cent by weight of silver.
   Identify X, Y, Z and explain your reasoning. (S(S))

ALCOHOLS AND PHENOLS

12 Outline how phenol is manufactured from petroleum. Explain what happens when phenol reacts with (i) iodomethane, (ii) benzoyl chloride, (iii) bromine water, (iv) nitric acid.

13 Compare and contrast the reactions of the —OH group in phenol and in ethanol.
How, and under what conditions, does phenol react with (a) bromine, (b) benzenediazonium chloride?

14 Describe how phenol may be prepared from benzene.
Outline the simplest methods for effecting the following changes: (a) phenol to phenylamine, (b) phenol to phenyl ethanoate. What action has bromine on phenol?

15 Four isomeric liquids stand side by side on a shelf, each with the label $C_4H_{10}O$, and no other information. *Two* are known to be alcohols, but two have been shown by their infrared spectra to lack an —OH group. Suggest formulae for these compounds and *outline* a scheme by which they may each be identified.
(C Schol.)

16 By what reactions may ethane-1,2-diol be obtained from ethene? Give the structural formula of ethane-1,2-diol and show how this formula may be justified. What substances may be formed from ethane-1,2-diol by oxidation? Give their structural formulae and show how any one of these may be confirmed by an independent method of formation. What are the uses of ethane-1,2-diol?

17 Describe one method by which methanol is manufactured. How does it react with (a) sodium, (b) phosphorus pentachloride, (c) ethanoic acid?
Describe one chemical test which would enable you to distinguish between methanol and ethanol.

18 A compound, A, $C_3H_8O_2$, was ethanoylated; 0·236 g of the ethanoyl derivative was boiled with 50 cm³ of N/10 sodium hydroxide and the resulting solution required 30 cm³ of N/10 sulphuric acid for neutralisation. When A was oxidised it gave B, $C_3H_6O_3$, which could not be ethanoylated, but of which 0·225 g required 25 cm³ of N/10 sodium hydroxide for neutralisation. Assign possible structures to A and B and account for the above results. (C Schol.)

19 An optically active compound W undergoes the following reactions:

$$C_9H_{12}O \xrightarrow[\text{nitric acid}]{\text{dilute}} C_9H_{10}O \xrightarrow{\text{NaOBr}} C_8H_8O \xrightarrow[\text{potassium permanganate}]{\text{alkaline}} C_8H_6O_4$$
$$W \qquad\qquad X \qquad\qquad Y \qquad\qquad Z$$

Z is unchanged on heating. Elucidate these reactions and give structures for W, X, Y and Z.

20 Compare and contrast the properties of phenol and phenylmethanol in their reactions, if any, with sodium carbonate, sodium hydroxide, potassium, bromine, ethanoic acid and an acidified solution of potassium permanganate.

21 How do primary, secondary, and tertiary alcohols differ in their reactions with oxidising agents?
Investigation of the rates of the following reactions of primary, secondary, and tertiary alcohols shows that in (a) the rates are in the order tertiary > secondary > primary, whereas the reverse is true in (b) and (c):

(a) $ROH + HBr = RBr + H_2O$
(b) $2ROH + 2Na = 2RONa + H_2$
(c) $ROH + CH_3COOH = RO.CO.CH_3 + H_2O$

What can you deduce about the esterification reaction (c) from these observations? (O Schol.)

22 Contrast the properties of the hydroxyl groups of ethanol and phenol. Place the following in order of diminishing acidity: C_6H_5OH, C_2H_5OH, CH_3COOH, H_2CO_3, and describe simple tests which would enable you to justify your order.
(O Schol.)

Chapter 11 Ethers

General formula

$$R\text{—}O\text{—}R'$$

11.1 Nomenclature

The two R groups in the structural formula R—O—R' can be the same (the simple ethers) or different (the mixed ethers), and can be either alkyl groups or aromatic groups. According to the I.U.P.A.C. rules, the RO— group is regarded as a substituent of the hydrocarbon R'H; for example, CH_3—O—CH_2—CH_3 is methoxyethane. However, it is common practice to use the name compounded from the two groups R and R' followed by ether, as in Table 11.1.

Table 11.1. Some ethers

NAME	FORMULA	B.P. (°C)
Dimethyl ether	CH_3—O—CH_3	−24
Ethyl methyl ether	CH_3—O—CH_2—CH_3	11
Diethyl ether	CH_3—CH_2—O—CH_2—CH_3	35
Methyl phenyl ether	CH_3—O—C_6H_5	154
Diphenyl ether	C_6H_5—O—C_6H_5	259

11.2 Physical properties of ethers

Dimethyl ether is a colourless gas, and the other lower homologues are colourless liquids with the characteristic 'ether' smell. Their boiling points are much lower than those of the isomeric alcohols, but are about the same as those of the alkanes of similar formula weight (Table 11.2). Molecules of ethers are not associated by hydrogen-bonding in the liquid phase, unlike alcohols.

Table 11.2. The boiling points (°C) of alkanes, alcohols and ethers
(The formula weights are given in brackets)

ALKANE	ALCOHOL	ETHER
Propane (44) −42	Ethanol (46) 78	Dimethyl ether (46) −24
Pentane (72) 36	Butan-1-ol (74) 118	Diethyl ether (74) 35
Heptane (100) 98	Hexan-1-ol (102) 157	Dipropyl ether (102) 91

11.3 Methods of preparation of ethers

Laboratory methods

1. *Simple ethers* can be prepared by the dehydration of an excess of an alcohol with concentrated sulphuric acid at about 140°C (10.5), for example:

$$C_2H_5-OH + H_2SO_4 \rightarrow C_2H_5-O-SO_2-OH + H_2O$$

$$C_2H_5-OH + C_2H_5-O-SO_2-OH \rightarrow C_2H_5-O-C_2H_5 + H_2SO_4$$
$$\text{Diethyl ether}$$

2. *Simple* and *mixed ethers* can be prepared by the reaction between an alkyl halide and the sodium derivative of an alcohol (the alkoxide):

$$R-OH + Na \rightarrow R-O^-Na^+ + \tfrac{1}{2}H_2$$

$$R-O^-Na^+ + R'-X \rightarrow R-O-R' + NaX$$

For example:

$$C_2H_5-O^-Na^+ + CH_3-I \rightarrow C_2H_5-O-CH_3 + NaI$$
$$\text{Ethyl methyl ether}$$

Manufacture

Diethyl ether is obtained as a by-product during the manufacture of ethanol from ethene and concentrated sulphuric acid.

11.4 Chemical properties of ethers

Ethers, like alkanes, are inert towards most inorganic reagents. For example, they are not attacked by sodium or, in the cold, by phosphorus pentachloride, and can therefore be readily distinguished from alcohols. They have three general properties:

1. They are highly inflammable, and mixtures with air are dangerously explosive.

2. They react with a hot, concentrated solution of hydriodic acid to form alkyl iodides, for example:

$$C_2H_5-O-C_2H_5 + 2HI \rightarrow 2C_2H_5-I + H_2O$$

Reaction occurs by the reversible protonation of the ether,

$$C_2H_5-O-C_2H_5 + HI \rightleftharpoons C_2H_5-\overset{+}{\underset{H}{O}}-C_2H_5 \; I^-$$

followed by displacement of a molecule of the alcohol by the nucleophilic iodide ion:

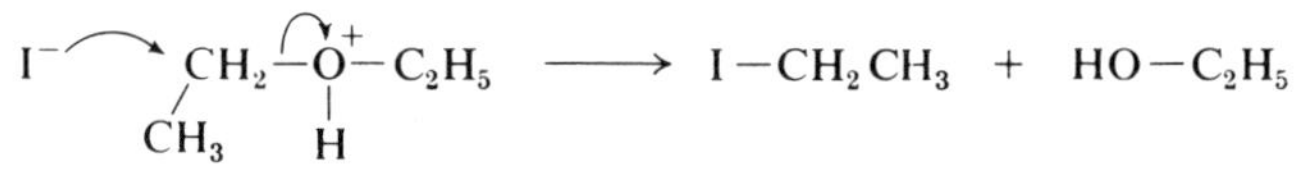

$$I^- \curvearrowright CH_2\underset{CH_3}{-}\overset{+}{\underset{H}{O}}-C_2H_5 \longrightarrow I-CH_2CH_3 + HO-C_2H_5$$

The alcohol reacts with a further molecule of hydrogen iodide.

ETHERS

3. They react with phosphorus pentachloride when heated. No hydrogen chloride is evolved, showing that ethers do not contain a hydroxyl group (10.5):

$$R-O-R' + PCl_5 \rightarrow R-Cl + R'-Cl + POCl_3$$

11.5 Uses of ethers

Diethyl ether is used as an anaesthetic. Inhalation of the vapour depresses the activity of the central nervous system.

Ethers, particularly diethyl ether, are used as solvents for fats, oils and resins.

The ability of ethers to dissolve a wide range of organic compounds, coupled with their resistance to chemical reaction, makes them valuable solvents for organic preparations (e.g. Grignard reagents; 9.8) and for the separation and purification of compounds by extraction.

11.6 Cyclic ethers

Cyclic ethers with five or more members in the ring are, like the corresponding cycloalkanes, relatively strainless and have the properties of their non-cyclic analogues. An example is tetrahydrofuran. However, those with three- and four-membered rings are strained and, like cyclopropane and cyclobutane (5.6), are very reactive towards reagents which are capable of opening the ring. An example is epoxyethane (ethylene oxide).

$$\underset{\text{Epoxyethane}}{\begin{array}{c} H_2C-CH_2 \\ \diagdown \diagup \\ O \end{array}} \qquad \underset{\text{Tetrahydrofuran}}{\begin{array}{c} H_2C-CH_2 \\ H_2C \quad CH_2 \\ \diagdown O \diagup \end{array}}$$

Manufacture of epoxyethane

By passing ethene and oxygen at 250°C over a silver catalyst:

$$CH_2=CH_2 + \tfrac{1}{2}O_2 \longrightarrow \begin{array}{c} H_2C-CH_2 \\ \diagdown \diagup \\ O \end{array}$$

Physical properties of epoxyethane

It is a volatile liquid, b.p. 13°C, which is soluble in water and in organic solvents.

Chemical properties of epoxyethane

1. It is hydrolysed by steam at 200°C under pressure, or by dilute acids at 60°C and atmospheric pressure:

$$\begin{array}{c} H_2C-CH_2 \\ \diagdown \diagup \\ O \end{array} + H_2O \longrightarrow \underset{\text{Ethane-1,2-diol}}{\begin{array}{c} CH_2-CH_2 \\ | \quad\quad | \\ OH \quad OH \end{array}}$$

This is the basis of the manufacture of ethane-1,2-diol.

ETHERS

2. It reacts with the halogen acids, for example:

$$H_2C\underset{O}{-}CH_2 + HCl \longrightarrow \underset{Cl}{CH_2}-\underset{OH}{CH_2}$$
2-Chloroethanol

3. It reacts with Grignard reagents to yield primary alcohols (9.8).

Uses of epoxyethane

1. In the manufacture of ethane-1,2-diol.
2. In the manufacture of the diol ethers which are used as solvents for cellulose ethanoate (p. 286). The diol ethers have an ether and a primary alcohol group:

$$H_2C\underset{O}{-}CH_2 + R-OH \longrightarrow \underset{OR}{CH_2}-\underset{OH}{CH_2}$$

3. In the manufacture of non-ionic detergents (13.10).

11.7 Questions

1 Describe the laboratory preparation of diethyl ether.
   Deduce the probable structural formula of an ether which contains by weight 60 per cent of carbon and 13·3 per cent of hydrogen, and suggest how it could be prepared.

2 Starting with a sample of radioactive potassium cyanide $K\overset{*}{C}N$, suggest reaction schemes enabling you to label each of the carbon atoms in ethyl methyl ether $CH_3OCH_2CH_3$.
   How would you confirm the position of the labelled carbon atom in $CH_3OCH_2\overset{*}{C}H_3$?

3 Describe the preparation of a pure sample of diethyl ether.
   Four unlabelled bottles contain samples of diethyl ether, pentane, propanone, and ethanol. What chemical tests would you use to identify each compound?
   (C(T))

4 Write an equation for the laboratory preparation of diethyl ether from ethanol. Why is an excess of ethanol used in the preparation? What is the main organic impurity in the distillate likely to be and how may it be removed?
   Explain why ether is particularly suitable as a solvent for the extraction of an organic compound from an aqueous solution. What is the main disadvantage in using ether for this purpose?
   Calculate the weight of phenylamine which would be extracted from 100 cm³ of an aqueous solution containing 5·0 g phenylamine by shaking with
   (a) 50 cm³ of ether in one portion,
   (b) two successive 25 cm³ portions of ether.
   Comment on the results.
   (Partition coefficient of phenylamine between ether and water = 5.) (L(X))

5 A is a liquid containing carbon, hydrogen and iodine only. 0·150 g of A in a Victor Meyer's apparatus displaced 25·3 cm³ of air, collected over water at 15°C and 763 mm pressure.

ETHERS

Another compound, B, containing carbon, hydrogen and oxygen only, reacts vigorously with metallic sodium when hydrogen is liberated and a white solid, C, is formed.

When A was heated under reflux with C and the mixture subsequently distilled, a compound D, containing carbon, hydrogen and oxygen only was obtained. D contained C, 64·9 per cent and H, 13·5 per cent and it was not attacked by sodium even on warming.

Identify A and then show that there are two possible compounds for each of B, C and D.

The saturated vapour pressure of water at 15° is 13·0 mm. (L)

6 16 cm³ of a gaseous aliphatic compound A, $C_nH_{3n}O_m$, was mixed with 60 cm³ of oxygen at room temperature and sparked. At the original temperature again, the final gas mixture occupied 44 cm³. After treatment with potassium hydroxide solution the volume of gas remaining was 12 cm³. Deduce the molecular and structural formulae of A, and name it.

Give the name and structural formula of a compound B isomeric with A, and state briefly how A and B react separately with (a) sodium, (b) hydrogen iodide, (c) phosphorus trichloride.

(If there is no reaction in any one case, make this clear.)

Outline a reaction scheme, stating reagents, by which A might be prepared from B. (S)

7 A neutral compound, P, $C_9H_{12}O_2$, fumes when treated with phosphorus pentachloride and, when heated with acidified sodium dichromate solution another neutral compound, Q, is formed. Q produces a characteristic orange-red precipitate with a solution of 2,4-dinitrophenylhydrazine but Q does not react with an ammoniacal solution of silver oxide.

P gives a yellow precipitate with iodine and alkali and if the filtrate from this reaction is acidified, an acid, R, $C_8H_8O_3$, is produced. R, on boiling with hydrogen iodide, gives a further acid, S. A familiar smell of oil of wintergreen is produced if S is warmed with methanol containing a little concentrated sulphuric acid as catalyst.

Deduce the nature of the compounds P, Q, R and S and explain fully all the reactions.

What is the significance of the reaction between S and ethanoic anhydride?

(S(S))

Chapter 12

Aldehydes and ketones

12.1 Introduction

General formula

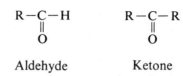

Aldehyde   Ketone

alkanals   *Alkanones*

Both aldehydes and ketones contain the **carbonyl group** ($\mathrm{>C=O}$). This group has characteristic properties which are shown by both classes of compound, so that it is convenient for the two homologous series to be considered together. However, the attachment of a hydrogen atom to the carbonyl group in an aldehyde gives aldehydes certain properties which ketones do not possess and which enable the two classes of compound to be distinguished from one another.

The I.U.P.A.C. nomenclature uses the suffixes **-al** for aldehydes and **-one** for ketones; the main carbon chain is named as usual and, for ketones, the position of the carbonyl group is specified by inserting the number of its carbon atom from the nearer end of the chain. For example:

$$\mathrm{CH_3{-}CH_2{-}CH_2{-}CH_2{-}CHO} \quad \overset{1}{\mathrm{CH_3}}{-}\overset{2}{\mathrm{CH_2}}{-}\overset{3}{\mathrm{CO}}{-}\overset{4}{\mathrm{CH_2}}{-}\overset{5}{\mathrm{CH_2}}{-}\overset{6}{\mathrm{CH_3}}$$
Pentanal   Hexan-3-one

12.2 Nomenclature

The simpler members of the series are often known by their original names. Some examples, with the original names in parentheses, together with their boiling points, are in Table 12.1.

Table 12.1. Some aldehydes and ketones

NAME	STRUCTURAL FORMULA	B.P. (°C)
Methanal (formaldehyde)	H—CHO	−21
Ethanal (acetaldehyde)	$\mathrm{CH_3}$—CHO	21
Propanal (propionaldehyde)	$\mathrm{CH_3}$—$\mathrm{CH_2}$—CHO	49
Benzaldehyde	$\mathrm{C_6H_5}$—CHO	179
Propanone (acetone)	$\mathrm{CH_3}$—CO—$\mathrm{CH_3}$	56
Butanone (ethyl methyl ketone)	$\mathrm{CH_3}$—CO—$\mathrm{CH_2}$—$\mathrm{CH_3}$	80
Pentan-3-one (diethyl ketone)	$\mathrm{CH_3}$—$\mathrm{CH_2}$—CO—$\mathrm{CH_2}$—$\mathrm{CH_3}$	102
Phenylethanone (acetophenone)	$\mathrm{C_6H_5}$—CO—$\mathrm{CH_3}$	202
Diphenylmethanone (benzophenone)	$\mathrm{C_6H_5}$—CO—$\mathrm{C_6H_5}$	306

12.3 Physical properties of aldehydes and ketones

Methanal is a gas, other aldehydes and ketones of relatively low formula weight are liquids and the remainder are solids. Methanal dissolves readily in water; a 40 per cent solution is known as formalin. The liquid aldehydes and ketones of low formula weight are also readily soluble in water (for

12.4 Methods of preparation of aldehydes and ketones

General methods

1. Aldehydes are obtained by the oxidation of primary alcohols, and ketones by the oxidation of secondary alcohols.

A convenient oxidising agent is an acidified solution of sodium dichromate. For example:

$$3CH_3CH_2OH + Na_2Cr_2O_7 + 4H_2SO_4 \rightarrow$$
$$3CH_3CHO + Na_2SO_4 + Cr_2(SO_4)_3 + 7H_2O$$
Ethanal

$$3(CH_3)_2CHOH + Na_2Cr_2O_7 + 4H_2SO_4 \rightarrow$$
$$3CH_3COCH_3 + Na_2SO_4 + Cr_2(SO_4)_3 + 7H_2O$$
Propanone

Sodium dichromate can be regarded as a solution of chromium(VI) oxide:

$$Na_2Cr_2O_7 + H_2O \rightleftharpoons 2CrO_3 + 2NaOH$$

Reaction with the alcohol occurs by formation of an unstable chromium(VI) ester which breaks down to give the carbonyl compound and a chromium(IV) ion:

$$R_2CHOH + CrO_3 \rightleftharpoons R_2CH-O-\underset{\underset{O}{\|}}{\overset{\overset{O}{\|}}{Cr}}-OH$$

$$R_2\overset{H}{\underset{}{C}}-O-\underset{\underset{O}{\|}}{\overset{\overset{O}{\|}}{Cr}}-OH \longrightarrow R_2C=O + \underset{\underset{O}{\|}}{\overset{O^-}{\underset{}{Cr}}}-OH + H^+$$

Three chromium(IV) ions disproportionate to give two chromium(III) ions and one chromium(VI) species:

$$3HCrO_3^- + 9H^+ \rightarrow CrO_3 + 2Cr^{3+} + 6H_2O$$

A difficulty arises in the preparation of aldehydes in this way: the aldehyde is itself readily oxidised (12.5). It is necessary to remove the aldehyde as soon as it is formed in order to prevent this happening, and this can readily be done because aldehydes have lower boiling points than the corresponding alcohols (whose boiling points are relatively high because of their hydrogen-bonded structure; 10.3). Thus, the alcohol is added slowly to the hot oxidising agent so that the aldehyde boils off as fast as it is formed whereas the higher boiling alcohol remains in solution until it is oxidised.

Details of the laboratory preparation of ethanal (p. 176) and propanone (p. 177) are given.

2. Aldehydes and ketones are formed by the hydrolysis of *gem*-dichlorides ($RCHCl_2$ and R_2CCl_2), for example:

$$C_6H_5-CHCl_2 + H_2O \rightarrow C_6H_5-CHO + 2HCl$$
Benzaldehyde

This method is of limited use for aliphatic compounds because of the difficulty of obtaining the dichloro-compounds; in fact, these compounds are usually made from the corresponding aldehyde or ketone with phosphorus pentachloride (12.5). However, the method is particularly useful for aromatic aldehydes because the dichloro-compounds can be obtained by the free-radical chlorination of the corresponding methyl compound (8.4), so that to obtain benzaldehyde, methylbenzene would be the starting material:

$$C_6H_5\text{—}CH_3 + 2Cl_2 \xrightarrow[\text{heat}]{\text{u.v. light}} C_6H_5\text{—}CHCl_2 + 2HCl$$

Specific method for aldehydes

Acid chlorides are reduced to aldehydes by hydrogen on palladium which is supported on barium sulphate:

$$\underset{O}{\underset{\|}{R\text{—}C\text{—}Cl}} + H_2 \longrightarrow \underset{O}{\underset{\|}{R\text{—}C\text{—}H}} + HCl$$

Sulphur and quinoline are added as a poison to prevent the reduction of the aldehyde to the primary alcohol.

Specific method for aromatic ketones

Many aromatic compounds react with acid chlorides in the presence of aluminium trichloride to give aromatic ketones (Friedel-Crafts reaction; 8.3), for example:

$$C_6H_6 + CH_3\text{—}CO\text{—}Cl \xrightarrow{AlCl_3 \text{ as cat.}} C_6H_5\text{—}CO\text{—}CH_3 + HCl$$
Phenylethanone

$$C_6H_6 + C_6H_5\text{—}CO\text{—}Cl \xrightarrow{AlCl_3 \text{ as cat.}} C_6H_5\text{—}CO\text{—}C_6H_5 + HCl$$
Diphenylmethanone

The details of the preparation of phenylethanone are given on p. 177.

Manufacture of ethanal

1. By passing ethyne through dilute sulphuric acid, with mercury(II) sulphate as catalyst, at 60°C:

$$CH\equiv CH + H_2O \xrightarrow{HgSO_4 \text{ as cat.}} CH_3\text{—}CHO$$

2. By the oxidation of ethanol in the gas phase over a silver or a copper catalyst:

$$CH_3CH_2OH + \tfrac{1}{2}O_2 \xrightarrow[500°C]{\text{Ag as cat.}} CH_3CHO + H_2O$$

$$CH_3CH_2OH \xrightarrow[500°C]{\text{Cu as cat.}} CH_3CHO + H_2$$

3. *Wacker process.* By oxidising ethene with palladium(II) chloride in water:

$$CH_2\!\!=\!\!CH_2 + H_2O + PdCl_2 \rightarrow CH_3CHO + 2HCl + Pd$$

By carrying out the reaction in the presence of a copper(II) salt, the palladium which is formed is oxidised back to palladium(II) ion:

$$Pd + 2Cu^{2+} \rightarrow Pd^{2+} + 2Cu^{+}$$

In the presence of air, the copper(I) ion is oxidised back to copper(II) ion:

$$4Cu^{+} + O_2 + 4HCl \rightarrow 4Cu^{2+} + 2H_2O + 4Cl^{-}$$

Therefore, the reaction requires only catalytic amounts of palladium and copper salts, so that the principal cost is that of ethene. Since this is cheaper than ethyne, this method is rapidly superseding the route from ethyne.

Manufacture of methanal

By the oxidation of methanol vapour over heated copper or silver:

$$CH_3OH + \tfrac{1}{2}O_2 \xrightarrow[500°C]{\text{Ag as cat.}} HCHO + H_2O$$

Manufacture of propanone

1. By passing the vapour of propan-2-ol over copper at 500°C. The alcohol is obtained from propene (10.4).

$$(CH_3)_2CHOH \xrightarrow[500°C]{\text{Cu as cat.}} (CH_3)_2CO + H_2$$

2. By oxidation of propan-2-ol in the liquid phase (20.4).

3. From the Cumene process for the manufacture of phenol (10.8).

12.5 Chemical properties of aldehydes and ketones

The reactions can be divided into three types:
(a) Reactions of the carbonyl group.
(b) Reactions of the alkyl group(s) adjacent to the carbonyl group.
(c) Oxidation reactions. Reactions of this type constitute the principal difference between aldehydes and ketones.

Many of the reactions which follow are shown by all aldehydes and ketones, but some members of each series show exceptional behaviour which is mentioned below and also in a section on anomalous properties (p. 174).

(a) Reactions of the carbonyl group

These can be subdivided into (i) **addition reactions** and (ii) **condensation reactions**.

(i) Addition reactions

Both aldehydes and ketones undergo addition reactions. The following are examples:

1. When treated with an alkali–metal cyanide, such as KCN, in the presence of alkali, they form **2-hydroxynitriles** (**cyanohydrins**). The reaction corresponds to the addition of hydrogen cyanide to the double bond, for example:

ALDEHYDES AND KETONES

$$CH_3CHO + HCN \rightarrow CH_3-\underset{\underset{CN}{|}}{\overset{\overset{OH}{|}}{C}}-H$$

2-Hydroxypropanonitrile

The reaction occurs in two stages: the cyanide ion adds to the carbonyl group, and then the new anion takes up a proton from the solvent (usually water):

$$CH_3-\overset{\overset{\displaystyle O}{\|}}{\underset{\underset{\underset{N}{\|||}}{\underset{C}{|}}}{C}}-H \longrightarrow CH_3-\underset{\underset{\underset{N}{\|||}}{\underset{C}{|}}}{\overset{\overset{O^-}{|}}{C}}-H \xrightarrow{H_2O} CH_3-\underset{\underset{\underset{N}{\|||}}{\underset{C}{|}}}{\overset{\overset{OH}{|}}{C}}-H$$

Thus, the essential reagent is cyanide ion, which is a nucleophilic reagent. However, a solution of potassium cyanide in water contains only a small proportion of cyanide ion, owing to hydrolysis:

$$CN^- + H_2O \rightleftharpoons HCN + OH^-$$

Addition of the alkali is necessary to increase the concentration of cyanide ion by displacing the above equilibrium to the left-hand side.

The mechanism of formation of cyanohydrins is typical of all nucleophilic additions to aldehydes and ketones: that is, reaction occurs by addition of a nucleophile to form an anion (Fig. 12.1) followed by uptake of a proton.

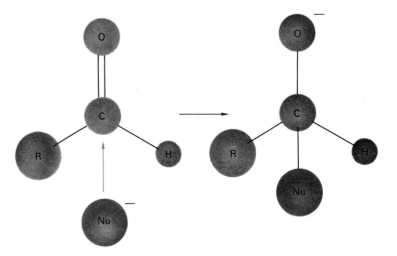

FIG. 12.1. *The reaction of a nucleophile with an aldehyde*

An understanding of the mechanism enables us to understand why the double bond in an aldehyde or ketone is reactive towards nucleophilic reagents whereas that in an alkene is not. In the anion formed by addition to a carbonyl group, the negative charge resides on oxygen:

$$CH_3-\underset{\underset{CN}{|}}{\overset{\overset{O^-}{|}}{C}}-H$$

However, if a nucleophile adds to an alkene, the negative charge resides on carbon; since carbon is much less strongly electron-attracting than oxygen, this species is less stable and less readily formed.

Aromatic aldehydes, such as benzaldehyde, react with potassium cyanide in a different manner from other aldehydes or ketones (p. 175).

2. Aldehydes and ketones react with sodium hydrogen sulphite to form addition compounds, for example:

$$(CH_3)_2C=O \;+\; NaHSO_3 \;\longrightarrow\; (CH_3)_2C\begin{subarray}{l}\diagup OH\\ \diagdown SO_2-O^-\;Na^+\end{subarray}$$

In this reaction, the hydrogen sulphite anion acts as a nucleophilic reagent:

$$\underset{OH}{\overset{O}{\underset{\|}{O^-\text{–}S:}}} \quad \underset{CH_3}{\overset{CH_3}{\diagdown}}C=O \;\longrightarrow\; \underset{HO\;\;CH_3}{\overset{O\;\;CH_3}{O=S-C-O^-}}$$

Ketones only undergo this reaction if at least one of the two groups attached to the carbonyl group is methyl. The probable reason is that methyl is the smallest group, and when two larger groups are attached, their size hinders the approach of the nucleophilic reagent (the hydrogen sulphite ion) to the carbonyl group.

3. Aldehydes and ketones are reduced to **alcohols** by treatment with sodium borohydride in water. Aldehydes give primary alcohols and ketones give secondary alcohols, for example:

$$4CH_3\text{—}CH=O + NaBH_4 \rightarrow (CH_3\text{—}CH_2\text{—}O)_4B^-Na^+$$

$$(CH_3\text{—}CH_2\text{—}O)_4B^-Na^+ + 3H_2O \rightarrow$$
$$4CH_3\text{—}CH_2\text{—}OH + NaH_2BO_3$$

The reducing agent is the borohydride anion, BH_4^-. It acts as a nucleophile by transferring a hydride ion to the carbonyl group:

$$H_3\bar{B}\text{–}H \quad \underset{CH_3}{\diagup}CH=O \;\longrightarrow\; BH_3 + CH_3\text{–}CH_2\text{–}O^-$$

The species BH_3 immediately reacts with the alkoxide ion,

$$CH_3\text{—}CH_2\text{—}O^- + BH_3 \rightarrow CH_3\text{—}CH_2\text{—}O\text{—}\bar{B}H_3$$

and the resulting anion then reduces another carbonyl group. Further reactions of this type occur until all four hydrogens of the BH_4^- anion have been replaced.

Lithium aluminium hydride, $LiAlH_4$, can be used in place of sodium borohydride. It is a much more powerful reducing agent and reacts violently with water, so that it is necessary to use an inert solvent such as ether. The product of the reaction is a salt of the alcohol, $(R_2CHO)_4Al^-Li^+$; excess of lithium aluminium hydride is destroyed by adding ethyl ethanoate (which acts by being reduced to ethanol), and the alcohol is then liberated from the salt by addition of dilute sulphuric acid:

$$4R_2CO + LiAlH_4 \rightarrow (R_2CHO)_4Al^-Li^+$$

$$2(R_2CHO)_4Al^-Li^+ + 4H_2SO_4 \rightarrow 8R_2CHOH + Al_2(SO_4)_3 + Li_2SO_4$$

The aldehyde or ketone reacts with the aluminohydride anion, AlH_4^-, in a similar way to the borohydride anion (p. 169).

4. Aldehydes and ketones are also reduced with sodium amalgam and water, with sodium and ethanol or with zinc and ethanoic acid, for example:

$$C_2H_5CH{=}O + 2Na + 2C_2H_5OH \rightarrow C_2H_5CH_2OH + 2C_2H_5O^-Na^+$$

$$(CH_3)_2C{=}O + Zn + 2CH_3CO_2H \rightarrow (CH_3)_2CHOH + (CH_3CO_2)_2Zn$$

These reactions do not occur by nucleophilic attack but by the transfer of electrons from the electropositive metal to the carbonyl group and the uptake of two protons from the water, ethanol, or acid; a simple representation is:

$$\diagdown_{\diagup}\!\!C{=}O + 2e + 2H^+ \longrightarrow \diagdown_{\diagup}\!\!CH{-}OH$$

Sodium provides one electron, so that two atoms of sodium per molecule of carbonyl compound are required, whereas one atom of zinc, which can provide two electrons (giving Zn^{2+}), reduces one molecule of carbonyl compound.

5. Aldehydes and ketones react with phosphorus pentachloride to give **gem-dichloro compounds**, for example:

$$CH_3{-}CH{=}O + PCl_5 \rightarrow CH_3{-}CHCl_2 + POCl_3$$
1,1-Dichloroethane

$$CH_3{-}CO{-}CH_3 + PCl_5 \rightarrow CH_3{-}CCl_2{-}CH_3 + POCl_3$$
2,2-Dichloropropane

(ii) Condensation reactions

Aldehydes and ketones react with compounds which contain the $-NH_2$ group with the elimination of water.

Hydroxylamine forms **oximes**, for example:

$$CH_3CH{=}O + NH_2OH \rightarrow CH_3CH{=}N{-}OH + H_2O$$
Ethanal oxime

$$(CH_3)_2C{=}O + NH_2OH \rightarrow (CH_3)_2C{=}N{-}OH + H_2O$$
Propanone oxime

Hydrazine forms **hydrazones** which, since they still contain an $-NH_2$ group, can react with more of the carbonyl compound to give **azines**, for example:

$$CH_3CH{=}O + NH_2NH_2 \rightarrow CH_3CH{=}N{-}NH_2 + H_2O$$

$$CH_3CH{=}N{-}NH_2 + CH_3CH{=}O \rightarrow CH_3CH{=}N{-}N{=}CHCH_3 + H_2O$$

Phenylhydrazine forms **phenylhydrazones**, for example:

$$C_6H_5CH=O + H_2N-NH-C_6H_5 \rightarrow C_6H_5CH=N-NH-C_6H_5 + H_2O$$
Phenylhydrazine   Benzaldehyde phenylhydrazone

2,4-Dinitrophenylhydrazine forms **2,4-dinitrophenylhydrazones**, for example:

$$\underset{CH_3}{\overset{C_6H_5}{>}}C=O + H_2N-NH-\underset{O_2N}{C_6H_3}-NO_2 \rightarrow \underset{CH_3}{\overset{C_6H_5}{>}}C=N-NH-\underset{O_2N}{C_6H_3}-NO_2 + H_2O$$

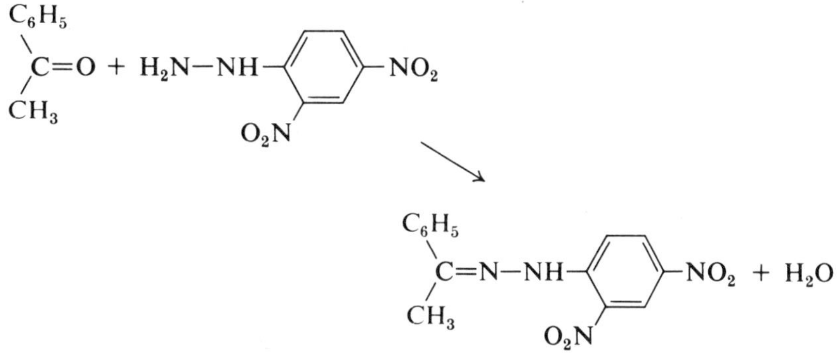

Phenylethanone 2,4-dinitrophenylhydrazone

Primary amines form **imines** (also called Schiff bases), for example:

$$C_6H_5CH=O + H_2N-C_6H_5 \rightarrow C_6H_5CH=N-C_6H_5 + H_2O$$

However, many of the derivatives with amines are unstable.

The reactions with hydroxylamine, phenylhydrazine and 2,4-dinitrophenylhydrazine are used for the characterisation of aldehydes and ketones because the products are mostly crystalline solids and the melting points of the derivatives from closely similar aldehydes or ketones are usually sufficiently different to enable the carbonyl compound to be recognised.

The reactions all occur by nucleophilic addition to the carbonyl group followed by the movement of a proton from one atom to another and then the elimination of water, for example:

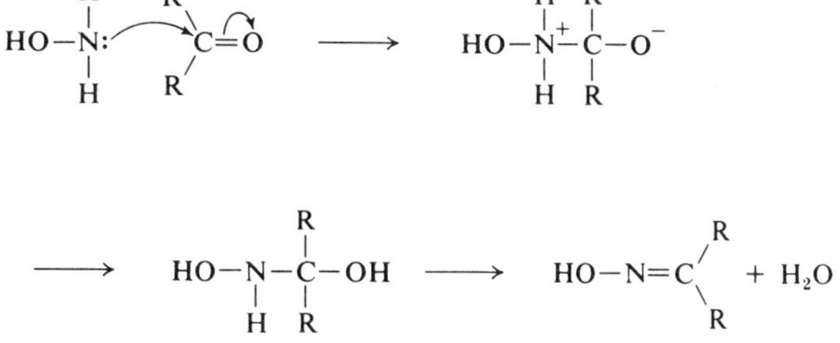

$$HO-\underset{H}{\overset{H}{N}}: \quad \underset{R}{\overset{R}{>}}C=O \longrightarrow HO-\underset{H}{\overset{H}{\overset{|}{N^+}}}-\underset{R}{\overset{R}{\underset{|}{C}}}-O^-$$

$$\longrightarrow HO-\underset{H}{\overset{R}{\underset{|}{N}}}-\underset{R}{\overset{R}{\underset{|}{C}}}-OH \longrightarrow HO-N=C\underset{R}{\overset{R}{<}} + H_2O$$

(b) Reactions of the alkyl group(s)

1. Aldehydes and ketones which possess at least one hydrogen atom on the carbon atom adjacent to the carbonyl group (i.e. >CH—CO—, etc.)

undergo condensation reactions in the presence of a base. For example, ethanal and dilute alkali give 3-hydroxybutanal:

$$2CH_3CHO \xrightarrow{NaOH} CH_3-\underset{\underset{OH}{|}}{CH}-CH_2-CH=O$$
$$\text{3-Hydroxybutanal}$$

If the solution is warmed, water is eliminated:

$$CH_3-\underset{\underset{OH}{|}}{CH}-CH_2-CH=O \rightarrow CH_3-CH=CH-CH=O + H_2O$$
$$\text{But-2-enal}$$

Propanone gives 4-hydroxy-4-methylpentan-2-one:

$$2(CH_3)_2CO \xrightarrow{NaOH} (CH_3)_2\underset{\underset{OH}{|}}{C}-CH_2-\underset{\underset{O}{\|}}{C}-CH_3$$

With concentrated alkali, ethanal forms a resin which precipitates from the solution. It is a polymeric compound of structure

$$CH_3(CH=CH)_xCH=O$$

where x is large and varies with the conditions of the reaction.

However, those aldehydes which do not contain at least one hydrogen atom on the carbon next to the carbonyl group (e.g. methanal and benzaldehyde) do not undergo condensation reactions with alkali (see above) but instead, with concentrated alkali, undergo the **Cannizzaro reaction** in which one half of the quantity of the aldehyde is oxidised and the other half is reduced, for example:

$$2C_6H_5-CH=O + NaOH \rightarrow C_6H_5-CO_2^-Na^+ + C_6H_5-CH_2OH$$
$$\text{Sodium benzoate} \quad \text{Phenylmethanol}$$

Reaction occurs by addition of hydroxide ion to the carbonyl group of one molecule of the aldehyde, to give an intermediate which transfers a hydride ion to a second molecule of the aldehyde. Reaction is completed by transference of a proton:

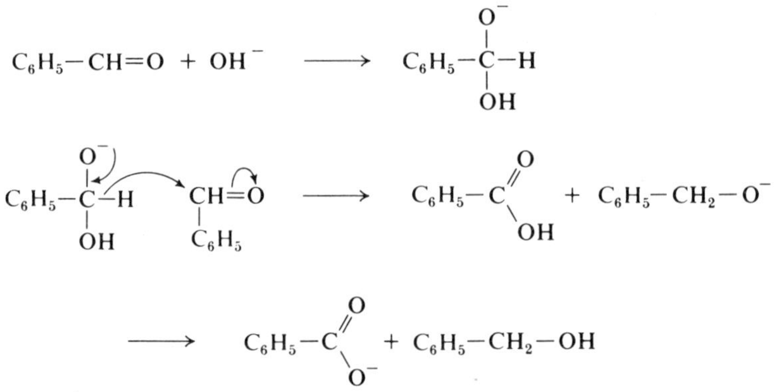

$$C_6H_5-CH=O + OH^- \longrightarrow C_6H_5-\underset{\underset{OH}{|}}{\overset{\overset{O^-}{|}}{C}}-H$$

$$C_6H_5-\underset{\underset{OH}{|}}{\overset{\overset{O^-}{|}}{C}}-H \quad \underset{\underset{C_6H_5}{|}}{CH=O} \longrightarrow C_6H_5-C\underset{OH}{\overset{\overset{O}{\|}}{\diagdown}} + C_6H_5-CH_2-O^-$$

$$\longrightarrow C_6H_5-C\underset{O^-}{\overset{\overset{O}{\|}}{\diagdown}} + C_6H_5-CH_2-OH$$

2. Aldehydes and ketones which possess at least one hydrogen atom on the carbon atom adjacent to the carbonyl group react with the halogens in the presence of dilute acid or alkali; one or more of these hydrogen atoms is replaced by halogen atoms. In the acid-catalysed reaction, it is possible to isolate the monohalogenated product, since introduction of a second halogen atom occurs more slowly than that of the first, for example:

$$CH_3\text{—}CH_2\text{—}CO\text{—}CH_2\text{—}CH_3 + I_2 \xrightarrow{H^+} CH_3\text{—}CHI\text{—}CO\text{—}CH_2\text{—}CH_3 + HI$$

However, in the base-catalysed reaction, the reaction rate increases as each halogen atom is introduced, so that it is possible to isolate only the product in which all the appropriate hydrogen atoms have been replaced, for example:

$$CH_3\text{—}CH_2\text{—}CO\text{—}CH_2\text{—}CH_3 + 4I_2 \xrightarrow{OH^-} CH_3\text{—}CI_2\text{—}CO\text{—}CI_2\text{—}CH_3 + 4HI$$

If the carbonyl compound contains the group —CO—CH$_3$ (i.e. ethanal or a methyl ketone), reaction in the presence of alkali leads first to the replacement of all three hydrogen atoms in the methyl group by halogen atoms, for example,

$$CH_3\text{—}CH_2\text{—}CO\text{—}CH_3 + 3Cl_2 + 3NaOH \rightarrow CH_3\text{—}CH_2\text{—}CO\text{—}CCl_3 + 3H_2O + 3NaCl$$

and then, with excess of alkali, to breakage of a C—C bond:

$$CH_3\text{—}CH_2\text{—}CO\text{—}CCl_3 + NaOH \rightarrow CH_3\text{—}CH_2\text{—}CO_2^- Na^+ + CHCl_3$$
$$\text{Trichloromethane}$$

Similarly, bromine gives tribromomethane, CHBr$_3$, and iodine gives tri-iodomethane, CHI$_3$. Tri-iodomethane is a yellow crystalline solid which is easily recognised, and its formation from a carbonyl compound indicates that this must have been either ethanal or a methyl ketone, since no others contain the group —CO—CH$_3$. This reaction is known as the **iodoform test**, after the original name for tri-iodomethane. Since alcohols which contain the group —CH(OH)—CH$_3$ (e.g. ethanol and propan-2-ol) are oxidised by iodine to give the group —CO—CH$_3$, these also give a positive iodoform test (p. 144).

(c) Oxidation reactions

The principal difference between an aldehyde and a ketone is that the —CH=O group of an aldehyde is readily oxidised to the carboxylic acid group, —CO$_2$H, whereas ketones are difficult to oxidise in solution.

Oxidation of aldehydes

Aldehydes are oxidised to carboxylic acids by sodium (or potassium) dichromate in acidic solution:

$$R\text{—}\underset{O}{\overset{\|}{C}}\text{—}H \xrightarrow{Na_2Cr_2O_7/H_2SO_4} R\text{—}\underset{O}{\overset{\|}{C}}\text{—}OH$$

Likewise, they are oxidised by potassium permanganate.

The ready oxidation of aldehydes compared with ketones enables the two types of carbonyl compound to be simply distinguished. Three reactions for this purpose (details for two of which are on p. 178) are:

1. A solution of silver nitrate in an excess of a solution of ammonia, which contains the complex ion, $Ag(NH_3)_2^+$, is reduced by an aldehyde to silver, which deposits on the wall of a test-tube as an easily recognised mirror:

$$RCHO + 2Ag(NH_3)_2^+ + H_2O \rightarrow RCO_2H + 2Ag + 2NH_4^+$$

2. Fehling's solution, which is made by mixing a solution of copper(II) 2,3-dihydroxybutanoic sulphate with an alkaline solution of a salt of (tartaric) acid and which contains a deep-blue complexed copper(II) ion, is reduced by an aldehyde to copper(I) oxide, which deposits as a red precipitate.

However, some aldehydes, such as benzaldehyde, do not react in this way, so that a negative result must be interpreted with care.

3. Fuchsin is a pink dye which forms a colourless complex when treated with sulphur dioxide. The addition of an aldehyde to this colourless solution restores the pink colour of the dye (Schiff's test).

Oxidation of ketones

Ketones are oxidised by strong oxidising agents, such as alkaline potassium permanganate and hot nitric acid. The bond between the carbonyl group and the adjacent carbon atom is broken, for example:

$$CH_3-CH_2-CO-CH_2-CH_3 \xrightarrow{HNO_3} CH_3-CH_2-CO_2H + CH_3-CO_2H$$

The acids formed contain *fewer* carbon atoms than the ketone. The acid formed on oxidation of an aldehyde contains the *same* number of carbon atoms.

Specific reactions of aldehydes and ketones

Methanal, HCHO, differs from other aldehydes in the following ways:

1. When an aqueous solution of methanal is gently evaporated, a linear polymer, polymethanal, is formed. Polymethanal has the structure

$$HO(CH_2-O)_xCH_2OH$$

When it is distilled from an acidified solution, methanal forms the cyclic trimer:

$$3HCHO \longrightarrow \begin{array}{c} \text{cyclic trimer of methanal} \end{array}$$

2. It does not undergo a condensation reaction of the type ethanal

ALDEHYDES AND KETONES

undergoes when treated with alkali, but with concentrated alkali it undergoes the Cannizzaro reaction:

$$2HCHO + NaOH \rightarrow H-CO_2^-Na^+ + CH_3OH$$

This behaviour of methanal is typical of that of aldehydes which do not have a hydrogen atom attached to the carbon atom next to the carbonyl group.

Ethanal, CH_3CHO, forms cyclic polymers with acids:

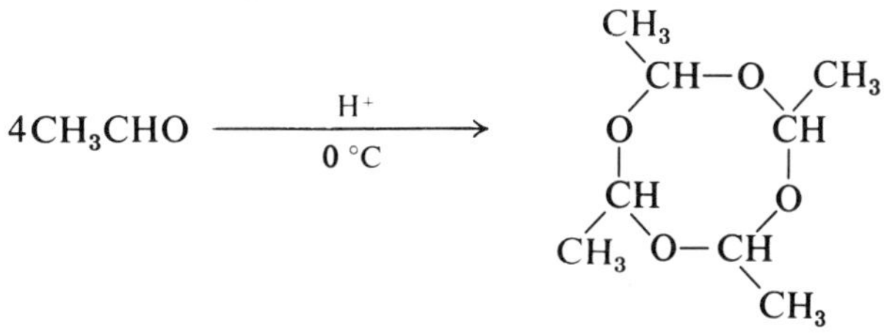

$$4CH_3CHO \xrightarrow[0\ °C]{H^+} \text{Ethanal tetramer}$$

$$3CH_3CHO \xrightarrow[\text{room temp.}]{H^+} \text{Ethanal trimer}$$

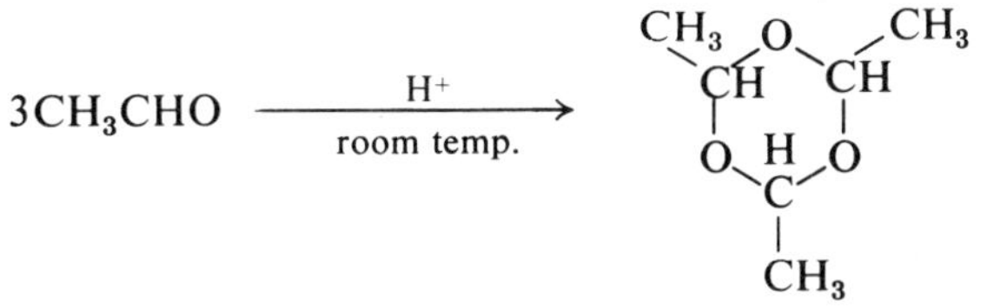

Benzaldehyde, C_6H_5CHO, does not reduce Fehling's solution or undergo condensations with alkali, but it undergoes the Cannizzaro reaction. In these respects it is typical of aldehydes which do not have a C—H bond adjacent to carbonyl.

Benzaldehyde is peculiar in not giving a cyanohydrin with potassium cyanide and alkali. Instead, it undergoes a condensation reaction to form 2-hydroxy-1,2-diphenylethanone, in which cyanide ion acts as a catalyst:

$$2C_6H_5CHO \xrightarrow{CN^-} C_6H_5-\underset{O}{\overset{\|}{C}}-\underset{OH}{\overset{|}{CH}}-C_6H_5$$

2-Hydroxy-1,2-diphenylethanone

12.6 Uses of aldehydes and ketones

Methanal

1. In the manufacture of thermosetting plastics, in particular Bakelite, carbamide-methanal resins and polyoxymethylene (p. 316).
2. In solution (formalin) it is used as a disinfectant and to preserve animal specimens.

Ethanal

In the manufacture of ethanoic acid (p. 184), which is required for the manu-

ALDEHYDES AND KETONES

facture of ethenyl ethanoate (p. 315) and thence poly(ethenyl ethanoate) (p. 315), and of ethanoic anhydride (p. 213) and thence cellulose ethanoate (p. 289).

Propanone

1. In the manufacture of Perspex (p. 316).
2. In the manufacture of ethenone, used to make ethanoic anhydride (14.4).
3. As a solvent for plastics, varnishes and greases.

12.7 Practical work

Small-scale preparation of ethanal

To 6 cm³ of water in a flask, add 2 cm³ of *concentrated* sulphuric acid, and set up the apparatus as shown in Figure 12.2.

Make up a solution containing 5 g of sodium dichromate in 5 cm³ of water, add 4 cm³ of ethanol and pour the solution into the dropping funnel.

Boil the acid in the flask and then turn out the bunsen flame. Add the mixture containing ethanol slowly so that the dilute acid remains near its boiling point.

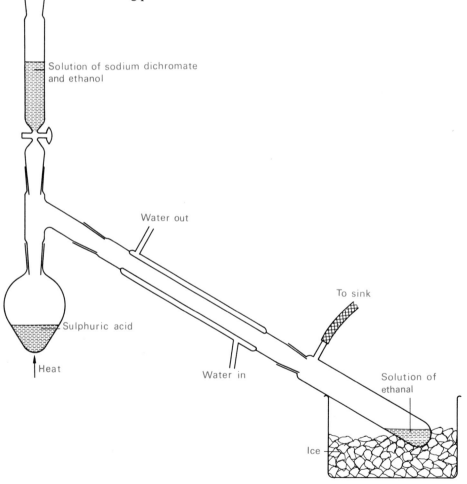

FIG. 12.2. *Preparation of ethanal*

Collect the distillate and redistil (Fig. 2.2), using a hot-water bath. Collect the fraction boiling between 20 and 23°C.

ALDEHYDES AND KETONES

Small-scale preparation of propanone

Repeat the experiment above, using 4 cm³ of propan-2-ol in place of 4 cm³ of ethanol. Collect the distillate and redistil it, collecting the fraction boiling between 54 and 57°C.

Small-scale preparation of phenylethanone. An example of the Friedel-Crafts reaction

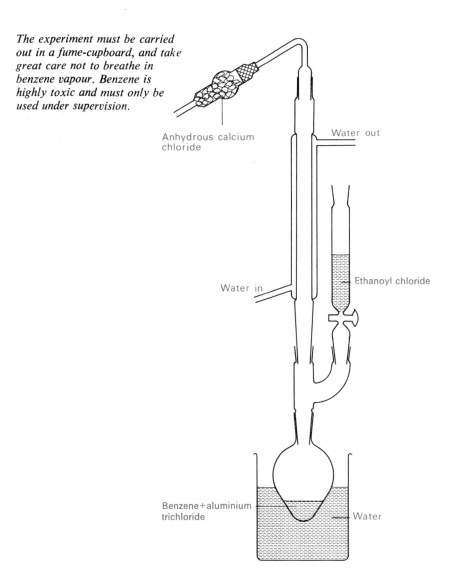

FIG. 12.3. *Preparation of phenylethanone*

The experiment must be carried out in a fume-cupboard, and take great care not to breathe in benzene vapour. Benzene is highly toxic and must only be used under supervision.

Set up the apparatus (Fig. 12.3) with 6 g of anhydrous aluminium trichloride and 4 cm³ of dry benzene in the flask. Add 4 cm³ of ethanoyl chloride dropwise, while shaking the mixture. When the acid chloride has been added, heat the water-bath to about 60°C and keep it at this temperature for about 30 minutes.

Pour the mixture into a beaker containing 30 cm³ of water, with stirring. Transfer to a separating funnel, add 10 cm³ of benzene and shake. When two layers have separated, run off the lower (aqueous) layer and then add 10 cm^3 of dilute sodium hydroxide to the organic layer. Shake, and then run off the lower layer again. Run the upper layer into a conical flask, add two or three pieces of *anhydrous* calcium chloride, cork the flask and shake it gently.

Distil the mixture, using a water-bath (cf. Fig. 9.6), to remove benzene. Then, using a bunsen burner, distil off phenylethanone, collecting the fraction boiling between 195 and 205°C.

Reactions of methanal, ethanal, benzaldehyde, propanone, pentan-3-one and phenylethanone

(a) Reactions of the carbonyl group

Addition reactions

1. *Preparation of ethanal hydrogen sulphite*. Make up 10 cm^3 of a saturated solution of sodium metabisulphite, in a flask, and pass sulphur dioxide through it for about 3 minutes. Add 1 cm^3 of ethanal, shake, stopper the flask and leave for 1–2 hours. Crystals of the addition product, ethanal hydrogen sulphite, are formed.

The experiment can be repeated with other aldehydes and ketones, for example with propanone (1 cm^3) or benzaldehyde (1 cm^3), instead of ethanal.

Condensation reactions

2. To 5 drops of one of the aldehydes or ketones in a test-tube, add methanol until the compound just dissolves. Add 5 cm^3 of the solution of 2,4-dinitrophenylhydrazine (see Appendix II, p. 350, for the preparation of this solution). Cork the test-tube and shake the mixture. Allow it to stand. If a precipitate is not formed within 5 minutes, add dilute sulphuric acid dropwise.

Filter the precipitate using a small Buchner funnel and flask and wash it with a minimum amount of methanol. Transfer the precipitate to a filter paper and squeeze it between two papers to dry it. Recrystallise the solid from the minimum quantity of a (1:1) mixture of ethanol and water. Filter, dry the crystals and obtain the melting point (p. 354).

(b) Reactions of the alkyl group

3. To 3 drops of ethanal in a test-tube, cautiously add concentrated sodium hydroxide solution drop by drop. A brown resin is formed with a characteristic smell.

4. To 5 drops of propanone, add 10 drops of iodine solution and 10 drops of 2M hydrochloric acid. Warm the mixture gently.

Repeat the experiment with pentan-3-one.

5. *The iodoform reaction*. In four separate test-tubes, add 5 drops of (i) ethanal, (ii) propanone, (iii) benzaldehyde and (iv) phenylethanone, followed by 10 drops of iodine solution in each (Appendix II). Add dilute sodium hydroxide solution dropwise until the brown colour just disappears.

Observe what happens (a) in the cold, (b) when the mixtures are gently warmed.

(c) Oxidation reactions

6. *Silver mirror test*. Clean 3 test-tubes by washing them thoroughly with distilled water and then with propanone and drying them.

ALDEHYDES AND KETONES

To 3 cm³ of a solution of silver nitrate, add 1 drop of a dilute solution of sodium hydroxide. Add dilute ammonia solution dropwise until the brown precipitate of silver oxide just redissolves. Divide this solution into three test-tubes.

To one of the test-tubes, add 3 drops of ethanal, to the second 3 drops of propanone and to the third 3 drops of benzaldehyde.

Warm the test-tubes in a beaker of boiling water for about 5 minutes.

7. *Fehling's test.* Make up some Fehling's solution by adding solution II (an alkaline solution of sodium potassium 2,3-dihydroxybutanoate) to 5 cm³ of solution I (a solution of copper(II) sulphate) until a deep blue solution is formed (Appendix II). Divide the solution equally between 3 test-tubes.

To one, add 3 drops of ethanal, to the second, 3 drops of propanone and to the other, 3 drops of phenylethanone. Boil the mixtures for a few minutes.

8. *Reaction with potassium permanganate.* To 10 cm³ of dilute sulphuric acid in a test-tube, add 2 cm³ of a 1 per cent solution of potassium permanganate and divide the solution between 5 test-tubes. Add 3 drops of the following to the test-tubes: (i) methanal, (ii) ethanal, (iii) propanone, (iv) benzaldehyde, (v) phenylethanone. At first shake the mixtures without warming, then warm them very gently.

In particular, contrast the behaviour of (a) ethanal and propanone, (b) propanone and phenylethanone.

(d) Specific reactions of some aldehydes and ketones

9. Pour 1–2 cm³ of methanal solution (formalin) on a watch-glass, and place it on a beaker containing water that is boiling gently.

10. Pour 1–2 cm³ of ethanal into a test-tube and place the tube in a beaker containing an ice-salt freezing mixture. Add two drops of concentrated sulphuric acid and stir the mixture gently. Observe whether there is a rise in temperature (why?) and whether a new compound is formed.

11. *The Cannizzaro Reaction.* To a cool solution of potassium hydroxide (5 g in 5 cm³ of water) in a boiling tube, add about 5 cm³ of benzaldehyde. Cork the boiling tube, shake and allow it to stand overnight. Add about 20 cm³ of water to dissolve the potassium benzoate, and extract the aqueous layer with ether. To the aqueous portion, add dilute hydrochloric acid to precipitate benzoic acid. Filter and recrystallise from hot water (m.p. 121°C). Dry the ethereal extract over solid potassium carbonate. Fractionate the dry extract and collect phenylmethanol (b.p. 204–207°C). Benzaldehyde does not form a resin (cf. reaction 3 with ethanal).

(e) To identify an aldehyde or ketone

12. You are given a compound which is either an aldehyde or ketone.
    (i) Test the compound with Fehling's solution (reaction 7) to see whether it is either an aldehyde or ketone.
    (ii) Prepare a solid derivative of the compound, the 2,4-dinitrophenylhydrazone (reaction 2), and determine its melting point. A list of melting points of 2,4-dinitrophenylhydrazones is given in Appendix V (p. 354).

12.8 Questions

1. Give the structural formulae of the compounds obtained when propanone reacts with
   (a) hydroxylamine,
   (b) sodium hydrogen sulphite,
   (c) concentrated sulphuric acid,
   (d) lithium aluminium hydride.

2. To a solution of propanone (y g) in water was added an excess of bromine water followed by sodium hydroxide solution. The tribromomethane, $CHBr_3$, obtained weighed 4·365 g. Calculate the value of y. (O and C)

3. In terms of the functional groups found in aldehydes and ketones, explain why these two classes of compound show similarities and differences in chemical behaviour. Illustrate your answer by reference to the reactions of ethanal and propanone with (i) sodium hydrogen sulphite, (ii) phosphorus pentachloride, (iii) ammoniacal silver nitrate, (iv) sulphuric acid, (v) Fehling's solution (AEB)

4. What do you understand by (a) *homologous series*, (b) *isomerism*?
   Give the full structural formulae for (i) **all** compounds having the molecular formula $C_4H_{10}O$, (ii) **two** compounds having the molecular formula C_3H_6O. Name **four** of the compounds in (i) and describe two experimental methods by which you could distinguish between the two compounds in (ii). (C(N, T))

5. Compare the reactions of ethanal, benzaldehyde and propanone with (a) hydrogen cyanide, HCN, (b) aqueous sodium hydroxide, (c) concentrated hydrochloric acid, (d) ammoniacal silver nitrate, (e) phenylhydrazine, $C_6H_5NHNH_2$. (O)

6. Outline the preparation of propanone starting from ethanol.
   Write the structural formulae of all compounds with the general formula $C_4H_{10}O$. Indicate which of these compounds give the iodoform reaction.
   How is the iodoform test performed? (C(N))

7. (a) A compound contains 60 per cent carbon, 13·3 per cent hydrogen and 26·7 per cent oxygen and has a vapour density of 30. Derive the molecular formula of the compound and write down all the isomers corresponding to this formula.
   (b) 10 cm³ of a gaseous hydrocarbon were mixed with 60 cm³ of oxygen. After sparking and cooling, 50 cm³ of gas remained and this when shaken with potassium hydroxide contracted to a volume of 30 cm³. The residual gas was oxygen and all measurements were made at the same conditions of temperature and pressure. Calculate the formula of the hydrocarbon. (AEB)

8. Analysis of a compound X (M.W. = 72), containing carbon, hydrogen and oxygen only, gave C = 66·7 per cent and H = 11·1 per cent. The compound neither reacted with phenylhydrazine nor gave the iodoform reaction. Reaction with trioxygen and subsequent hydrolysis gave methanal as one of the products. A mole of X absorbed a mole of hydrogen when catalytically hydrogenated to give a product Y which could be oxidized with potassium permanganate to a product Z (M.W. = 72) which did not react with Fehling's solution.
   Identify the compounds X, Y, Z, giving your reasoning. How would you confirm the structure of Z? (W)

9. Describe analytical and synthetic methods you would employ to establish that the product of a reaction possessed the structural formula:

$$\underset{\underset{CO \cdot CH_3}{}}{\overset{\overset{Cl}{|}}{C_6H_4}}$$

(W(S))

10 Propanone reacts with iodine in aqueous solution according to the following reaction, which is catalysed by H^+:

$$I_2 + CH_3-\underset{\underset{O}{\|}}{C}-CH_3 \rightarrow CH_2I-\underset{\underset{O}{\|}}{C}-CH_3 + HI$$

In three experiments the rate of the reaction was studied in aqueous acidic solutions using concentrations of propanone much larger than that of the iodine. The results are given below:

	Initial concentrations		[I_2] in *millimoles* per dm^3 at time t in minutes					
	[Propanone]	[H^+]	0	5	9	12	14	15
Exp. 1	1·00 M	0·100 M	2·50	1·65	0·97	0·46	0·12	0
Exp. 2	1·00 M	0·150 M	3·20	1·93	0·90	0·14	0	0
Exp. 3	2·00 M	0·100 M	6·00	4·30	2·94	1·92	1·24	0·90

Plot the concentrations of iodine against time.

How does the reaction rate depend on the concentrations of the various species? Can you express all the data with one rate expression containing only one rate constant?

Given that

$$CH_2=\underset{\underset{OH}{|}}{C}-CH_3$$

is an intermediate in the reaction, devise a mechanism which will explain the data.

(O Schol.)

Chapter 13 — Carboxylic acids

13.1 Introduction

General formula

$$R-CO_2H$$

Carboxylic acids contain the **carboxyl group**

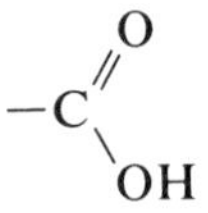

$$-C\underset{OH}{\overset{O}{\lVert}}$$

which is a combination of the **carb**onyl group, $>C=O$, and the hydr**oxyl** group, $-OH$. It will be seen that the properties of each group separately are modified when they are combined.

There are compounds with one carboxylic acid group (monocarboxylic acids), two (dicarboxylic acids) and more than two.

13.2 Nomenclature of monocarboxylic acids

Monocarboxylic acids are named, according to the I.U.P.A.C. system, by replacing the final **e** of the corresponding hydrocarbon by **oic acid**, for example:

$$CH_3-CH_2-CH_2-CH_2-CO_2H \qquad CH_3-\underset{\underset{CH_3}{|}}{CH}-CH_2-CO_2H$$

Pentanoic acid 3-Methylbutanoic acid

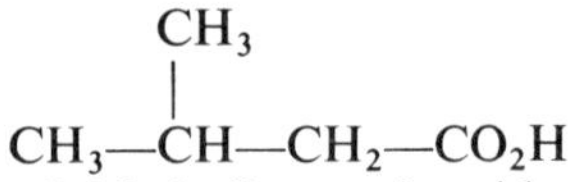

The two lowest members, methanoic and ethanoic acid, are often known by their original names: formic acid and acetic acid.

Table 13.1. **Some monocarboxylic acids**

NAME	FORMULA	M.P. (°C)	B.P. (°C)
Methanoic acid	$H-CO_2H$	8	101
Ethanoic acid	CH_3-CO_2H	17	118
Propanoic acid	$CH_3CH_2-CO_2H$	−22	141
Butanoic acid	$CH_3CH_2CH_2-CO_2H$	−5	163
2-Methyl-propanoic acid	$(CH_3)_2CH-CO_2H$	−47	154
Benzoic acid	$C_6H_5-CO_2H$	121	

13.3 Physical properties of monocarboxylic acids

The lowest members are liquids with pungent odours, Ethanoic acid smells of vinegar, and the higher acids smell of rancid butter, which is partly butanoic acid. Methanoic acid and ethanoic acid are miscible with water, but as the formula weight increases the solubility decreases.

The formula weight of lower members of the series as determined by, for example, the depression of freezing point of a solvent such as benzene is about twice that expected for the molecular formula $R-CO_2H$. This is because carboxylic acids exist as dimers: pairs of molecules are linked by two hydrogen bonds, for example:

$$CH_3-C\underset{O-H\cdots O}{\overset{O\cdots H-O}{\lessgtr}}C-CH_3$$

A further consequence of this is that the boiling points of carboxylic acids are much higher than those of alkanes of similar formula weight, since more energy is required in order to break the hydrogen bonds in the vaporisation of carboxylic acids; for example, ethanoic acid (formula weight, 60) boils at 118°C whereas butane (formula weight, 58) boils at −0·5°C.

13.4 Methods of preparation of monocarboxylic acids

General methods

1. By the oxidation of primary alcohols and aldehydes with acidified sodium dichromate solution (10.5, 12.5):

$$R-CH_2OH \xrightarrow{Na_2Cr_2O_7} R-CHO \xrightarrow{Na_2Cr_2O_7} R-CO_2H$$

2. By the hydrolysis, with dilute mineral acid or alkali, of acid nitriles (14.6) and acid amides (14.5):

$$R-CN + H_2O \xrightarrow{H^+ \text{ or } OH^-} R-CONH_2$$

$$R-CONH_2 + H_2O \xrightarrow{H^+ \text{ or } OH^-} R-CO_2^- \; NH_4^+$$

$$R-CO_2^- \; NH_4^+ \begin{array}{c} \xrightarrow{NaOH} R-CO_2^- \; Na^+ + H_2O + NH_3 \\ \xrightarrow{HCl} R-CO_2H + NH_4Cl \end{array}$$

3. By the reaction of carbon dioxide with a Grignard reagent followed by hydrolysis (9.8):

$$R-MgX + CO_2 \rightarrow R-CO-O-MgX \xrightarrow{H_2O} R-CO_2H + Mg(OH)X$$

Details of the preparation of benzoic acid from phenylmagnesium bromide are given on p. 196.

Methanoic acid

By heating a solution of ethanedioic acid in propane-1,2,3-triol at 150°C:

$$\begin{array}{c} CO_2H \\ | \\ CO_2H \end{array} \rightarrow H-CO_2H + CO_2$$

Benzoic acid

By the oxidation of methylbenzene with hot, alkaline potassium permanganate solution, followed by acidification:

$$C_6H_5-CH_3 \xrightarrow[NaOH]{KMnO_4} C_6H_5-CO_2^- \; Na^+$$

$$C_6H_5-CO_2^- \; Na^+ + HCl \longrightarrow C_6H_5-CO_2H + NaCl$$
$$\text{Benzoic acid}$$

Practical details are described on p. 108.

Manufacture of monocarboxylic acids

Methanoic acid is made by heating powdered sodium hydroxide with carbon monoxide at 150°C under a pressure of about 8 atmospheres:

$$NaOH + CO \rightarrow H-CO_2^- \; Na^+$$
$$\text{Sodium methanoate}$$

and then adding dilute sulphuric acid to a slurry of crushed sodium methanoate and heating the mixture to distil the methanoic acid:

$$H-CO_2^- \; Na^+ + H_2SO_4 \rightarrow H-CO_2H + NaHSO_4$$

Ethanoic acid is mainly manufactured from ethanal and from butane or the naphtha fraction obtained from the distillation of petroleum which contains C_4-C_{10} alkanes (19.4):

1. *From ethanal*:

$$CH_3-CHO + \tfrac{1}{2}O_2 \xrightarrow[60-70°C]{(CH_3CO_2)_2Mn \text{ as cat.}} CH_3-CO_2H$$

2. *From butane or naphtha*, by oxidation in air at 200°C in the presence of cobalt(II) ethanoate as catalyst.

The main by-products are methanoic acid (from butane) and methanoic acid and propanoic acid (from naphtha).

Benzoic acid is made by passing air under pressure into methylbenzene at 150°C in the presence of an organic cobalt salt as a catalyst:

$$2C_6H_5-CHO + O_2 \rightarrow 2C_6H_5-CO_2H$$

13.5 Chemical properties of monocarboxylic acids

1. They are weak acids, dissociating to a small extent (1–2 per cent) in water:

$$R-CO_2H \rightleftharpoons R-CO_2^- + H^+$$

The dissociation constants, K_a, of some typical members of the series are

	K_a at 25°C
H—CO$_2$H	1.7×10^{-4}
CH$_3$—CO$_2$H	1.7×10^{-5}
CH$_3$CH$_2$—CO$_2$H	1.3×10^{-5}
C$_6$H$_5$—CO$_2$H	6.3×10^{-5}
C$_6$H$_5$CH$_2$—CO$_2$H	4.9×10^{-5}

Thus, the hydroxyl group is far more acidic than in an alcohol, its properties being modified in this respect by the carbonyl group, for a reason described in Section 4.9. They are also stronger acids than phenols, but weaker than sulphonic acids.

2. They react with bases to form salts, for example:

$$R—CO_2H + NaOH \rightarrow R—CO_2^- \, Na^+ + H_2O$$

They react with carbonates and hydrogen carbonates to liberate carbon dioxide:

$$R—CO_2H + NaHCO_3 \rightarrow R—CO_2^- \, Na^+ + H_2O + CO_2$$

The last reaction enables them to be distinguished from simple phenols, for although phenols are acidic enough to turn blue litmus red and to form salts with sodium hydroxide, they are weaker acids than carbonic acid and thus do not liberate carbon dioxide from sodium hydrogen carbonate (10.8).

The reaction with sodium hydrogen carbonate also enables carboxylic acids to be separated from simple phenols. For example, if a mixture of benzoic acid and phenol is partitioned between a solution of sodium hydrogen carbonate and ether, the acid dissolves in the aqueous layer (with liberation of carbon dioxide) and the phenol mainly dissolves in the ether. The two solutions are separated; the ether is evaporated to leave phenol, and hydrochloric acid is added to the aqueous solution to precipitate benzoic acid:

$$C_6H_5—CO_2^- \, Na^+ + HCl \rightarrow C_6H_5—CO_2H + NaCl$$

3. Acids react with alcohols, in the presence of an acid catalyst, to form esters:

$$R—CO_2H + R'—OH \underset{}{\overset{H^+}{\rightleftharpoons}} R—CO—O—R' + H_2O$$

The reaction is discussed on p. 205.

4. They react with phosphorus halides to form acid halides:

$$R—CO_2H + PCl_5 \longrightarrow R—CO—Cl + POCl_3 + HCl$$

$$R—CO_2H + PBr_5 \xrightarrow{\text{Red P/Br}_2} R—CO—Br + POBr_3 + HBr$$

Sulphur dichloride oxide (thionyl chloride) can also be used:

$$R—CO_2H + SOCl_2 \rightarrow R—CO—Cl + SO_2 + HCl$$

5. They are reduced to primary alcohols by lithium aluminium hydride:

$$R-CO_2H \xrightarrow{LiAlH_4} R-CH_2OH$$

However, unlike aldehydes and ketones (p. 169), they are not reduced by sodium borohydride.

6. A C—H bond adjacent to the carboxyl group is converted into a C—Cl bond when chlorine gas is passed into the hot acid in the presence of ultraviolet light, for example:

$$(CH_3)_2\underset{H}{C}-CO_2H + Cl_2 \xrightarrow[\text{heat}]{\text{u.v. light}} (CH_3)_2\underset{Cl}{C}-CO_2H + HCl$$

When there is more than one such C—H bond, further reaction can occur, for example:

$$CH_3-CO_2H + Cl_2 \xrightarrow[\text{heat}]{\text{u.v. light}} \underset{\text{Chloroethanoic acid}}{CH_2Cl-CO_2H} + HCl$$

$$CH_2Cl-CO_2H + Cl_2 \xrightarrow[\text{heat}]{\text{u.v. light}} \underset{\text{Dichloroethanoic acid}}{CHCl_2-CO_2H} + HCl$$

$$CHCl_2-CO_2H + Cl_2 \xrightarrow[\text{heat}]{\text{u.v. light}} \underset{\text{Trichloroethanoic acid}}{CCl_3-CO_2H} + HCl$$

7. Although carboxylic acids contain the carbonyl group, C=O, they do not undergo the addition reactions with, e.g., sodium hydrogen sulphite or the condensation reactions with, e.g., hydroxylamine which are characteristic of this group in aldehydes and ketones.

This is because an orbital on the oxygen atom of the hydroxyl group which contains two unshared electrons interacts with the p orbital of the adjacent carbon atom (Fig. 13.1). This provides extra delocalisation in the molecule, as compared with that in an aldehyde or ketone, which would be lost by addition of a reagent to the carbonyl group, so that a carboxylic acid is more resistant to addition than an aldehyde or ketone.

Methanoic acid

Methanoic acid, $H-CO_2H$ differs in the following respects from the other monocarboxylic acids:

1. It is dehydrated by concentrated sulphuric acid:

$$H-CO_2H \xrightarrow{H_2SO_4} CO + H_2O$$

This is the basis of a laboratory preparation of carbon monoxide.

2. It reduces Fehling's solution and gives a silver mirror when warmed with a solution of silver nitrate and ammonia. These are characteristics of

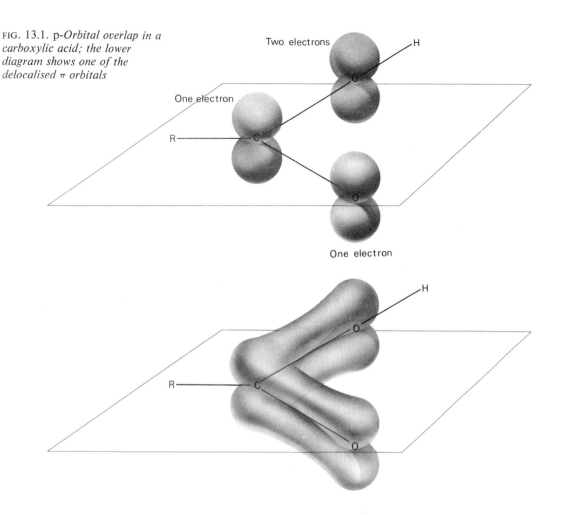

FIG. 13.1. p-*Orbital overlap in a carboxylic acid; the lower diagram shows one of the delocalised π orbitals*

aldehydes (12.5), and it is because methanoic acid contains the readily oxidised aldehydic group (HO—CH=O) that it undergoes these reactions, that is:

$$\text{H—C—OH} \xrightarrow{\text{'O'}} (\text{HO—C—OH}) \rightarrow CO_2 + H_2O$$
$$\qquad \|\qquad\qquad\qquad\quad\|$$
$$\qquad O\qquad\qquad\qquad\quad O$$

3. It does not form acid halides.

13.6 Uses of monocarboxylic acids

Ethanoic acid is used in the manufacture of ethenyl ethanoate, required for the production of poly(ethenyl ethanoate) (21.2), and ethanoic anhydride (14.4), required for making cellulose ethanoate (18.3) and other ethanoate esters.

Long-chain monocarboxylic acids are used in the manufacture of soaps and detergents (p. 194).

13.7 Dicarboxylic acids

Some examples of dicarboxylic acids, together with their original names, are in Table 13.2.

Table 13.2. Some dicarboxylic acids

NAME	FORMULA	M.P. (°C)
Ethanedioic acid (oxalic acid)	HO_2C-CO_2H	189
Propanedioic acid (malonic acid)	$HO_2C-CH_2-CO_2H$	136 (decomposes)
Butanedioic acid (succinic acid)	$HO_2C-CH_2-CH_2-CO_2H$	185
Hexanedioic acid (adipic acid)	$HO_2C-(CH_2)_4-CO_2H$	149
Benzene-1,2-dicarboxylic acid (phthalic acid)	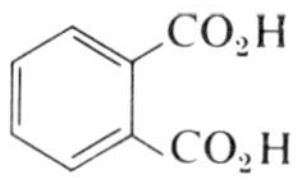 benzene ring with two ortho CO_2H groups	200 (decomposes)
Benzene-1,4-dicarboxylic acid (terephthalic acid)	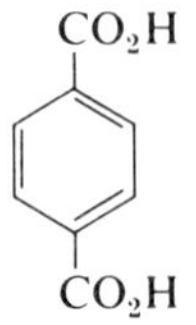 benzene ring with two para CO_2H groups	300 (sublimes)

They are white crystalline solids. The lower members are soluble in water and ethanol but are insoluble in ether.

Laboratory preparations of dicarboxylic acids

Ethanedioic acid is made by heating sodium methanoate at 400°C to give disodium ethanedioate,

$$2H-CO_2^- Na^+ \rightarrow Na^+ {}^-O_2C-CO_2^- Na^+ + H_2$$

dissolving this salt in water and adding calcium hydroxide to precipitate calcium ethanedioate,

$$Na^+ {}^-O_2C-CO_2^- Na^+ + Ca(OH)_2$$

$$\longrightarrow Ca^{2+} \begin{matrix} {}^-O_2C \\ | \\ {}^-O_2C \end{matrix} + 2NaOH$$

and then adding to the dried calcium salt the exact quantity of dilute sulphuric acid needed to liberate the acid:

$$Ca^{2+} \begin{matrix} {}^-O_2C \\ | \\ {}^-O_2C \end{matrix} + H_2SO_4 \longrightarrow HO_2C-CO_2H + CaSO_4$$

Calcium sulphate precipitates, and ethanedioic acid is crystallised from the filtrate as the hydrate, $(CO_2H)_2 \cdot 2H_2O$.

Propanedioic acid is made by the reaction of sodium cyanide in water with the sodium salt of chloroethanoic acid, followed by hydrolysis of the nitrile with concentrated hydrochloric acid:

$$Cl-CH_2-CO_2^- \; Na^+ + NaCN \rightarrow N\equiv C-CH_2-CO_2^- \; Na^+ + NaCl$$
$$N\equiv C-CH_2-CO_2^- \; Na^+ + 2H_2O + 2HCl \rightarrow$$
$$HO_2C-CH_2-CO_2H + NH_4Cl + NaCl$$

Butanedioic acid is prepared from 1,2-dibromoethane:

$$Br-CH_2-CH_2-Br + 2NaCN \rightarrow N\equiv C-CH_2-CH_2-C\equiv N + 2NaBr$$
$$N\equiv C-CH_2-CH_2-C\equiv N + 4H_2O + 2HCl \rightarrow$$
$$HO_2C-CH_2-CH_2-CO_2H + 2NH_4Cl$$

Manufacture of dicarboxylic acids

Ethanedioic acid is made from sodium methanoate as described above. Hexanedioic acid, required for the production of nylon-6.6 (21.3), is manufactured from cyclohexane (20.6), and benzene-1,4-dicarboxylic acid, required for the production of Terylene, is made from 1,4-dimethylbenzene (20.5).

Chemical properties of dicarboxylic acids

In most respects dicarboxylic acids resemble monocarboxylic acids. They react with bases to form two series of salts, for example:

$$HO_2C-CO_2H + NaOH \rightarrow HO_2C-CO_2^- \; Na^+ + H_2O$$
Sodium hydrogen
ethanedioate

$$HO_2C-CO_2^- \; Na^+ + NaOH \rightarrow Na^+ \; {}^-O_2C-CO_2^- \; Na^+ + H_2O$$
Disodium ethanedioate

and with phosphorus halides and sulphur dichloride oxide to form acid halides, for example:

$$HO_2C-CO_2H + 2PCl_5 \rightarrow Cl-CO-CO-Cl + 2POCl_3 + 2HCl$$
Ethanedioyl chloride

The lower members of the series have certain special properties:

1. Ethanedioic acid and propanedioic acid are decarboxylated on being heated strongly:

$$HO_2C-CO_2H \xrightarrow{150°C} H-CO_2H + CO_2$$

$$HO_2C-CH_2-CO_2H \xrightarrow{150°C} CH_3-CO_2H + CO_2$$

2. Ethanedioic acid is a reducing agent, being readily oxidised to carbon dioxide and water, for example with a *warm* solution of acidified potassium permanganate:

$$HO_2C-CO_2H \xrightarrow{\text{'O'}} 2CO_2 + H_2O$$

3. Ethanedioic acid is dehydrated by concentrated sulphuric acid:

$$HO_2C-CO_2H \xrightarrow{\text{conc. } H_2SO_4} CO + CO_2 + H_2O$$

CARBOXYLIC ACIDS

This is a convenient method for the preparation of carbon monoxide in the laboratory.

4. Propanedioic acid is dehydrated to form tricarbon dioxide when heated with phosphorus pentoxide:

$$HO_2C-CH_2-CO_2H \xrightarrow{P_4O_{10}} O=C=C=C=O + 2H_2O$$
Tricarbon dioxide

5. Butanedioic acid forms a cyclic anhydride when heated with ethanoic anhydride:

$$HO_2C-CH_2-CH_2-CO_2H + CH_3-CO-O-CO-CH_3$$

$$\longrightarrow \begin{array}{c} H_2C-CH_2 \\ \diagup \quad \diagdown \\ O=C \quad\quad C=O \\ \diagdown \quad \diagup \\ O \end{array} + 2CH_3CO_2H$$

Butanedioic anhydride

13.8 Substituted carboxylic acids

Chloroethanoic acid

Chloroethanoic acid is prepared by passing chlorine into hot ethanoic acid in the presence of ultraviolet light:

$$CH_3-CO_2H + Cl_2 \xrightarrow[\text{heat}]{\text{u.v. light}} CH_2Cl-CO_2H + HCl$$

Further chlorination can occur, giving dichloroethanoic acid and then trichloroethanoic acid (p. 186), and in order to optimise the yield of the monochloro compound the reaction is stopped when there has been the appropriate increase in weight.

Chloroethanoic acid is a deliquescent solid (m.p. 61°C) which is soluble in water. It is a stronger acid that ethanoic acid (4.8).

The chlorine atom behaves as in alkyl halides. Thus, it is readily displaced, as chloride ion, by nucleophilic reagents, for example:

1. With dilute sodium hydroxide, the sodium salt of hydroxyethanoic acid is formed:

$$Cl-CH_2-CO_2H + 2NaOH \rightarrow HO-CH_2-CO_2^- \; Na^+ + NaCl + H_2O$$
Sodium hydroxyethanoate

2. With sodium cyanide, the sodium salt of cyanoethanoic acid is formed:

$$Cl-CH_2-CO_2H + 2NaCN \rightarrow$$
$$N\equiv C-CH_2-CO_2^- \; Na^+ + NaCl + HCN$$
Sodium cyanoethanoate

3. With ammonia, aminoethanoic acid (glycine) is formed:

$$Cl-CH_2-CO_2H + 2NH_3 \rightarrow H_2N-CH_2-CO_2H + NH_4Cl$$
Glycine

Hydroxyethanoic acid

Hydroxyethanoic acid is prepared by the hydrolysis of chloroethanoic acid with sodium hydroxide, followed by acidification:

$$Cl-CH_2-CO_2H + 2NaOH \rightarrow HO-CH_2-CO_2^- Na^+ + NaCl + H_2O$$

$$HO-CH_2-CO_2^- Na^+ + HCl \rightarrow HO-CH_2-CO_2H + NaCl$$
Hydroxyethanoic acid

It is a white solid, readily soluble in water, and is a stronger acid than ethanoic acid though weaker than chloroethanoic acid.

It exhibits the properties of both a monocarboxylic acid and a primary alcohol:

1. As an acid, it can be converted into an ester and an acid halide, for example:

$$HO-CH_2-CO_2H + CH_3-OH \underset{}{\overset{H^+}{\rightleftharpoons}} HO-CH_2-CO-O-CH_3 + H_2O$$
Methyl hydroxyethanoate

2. As a primary alcohol, it can be oxidised to an aldehyde by, for example, acidified sodium dichromate solution; further oxidation occurs readily:

$$HO-CH_2-CO_2H \xrightarrow{Na_2Cr_2O_7} O=CH-CO_2H \xrightarrow{Na_2Cr_2O_7}$$
Oxoethanoic acid

$$HO_2C-CO_2H \xrightarrow{Na_2Cr_2O_7} 2CO_2 + H_2O$$
Ethanedioic acid

In addition, it undergoes two reactions which neither a carboxylic acid nor a primary alcohol can undergo:

3. On being heated, it forms a cyclic ester (an example of a **lactide**):

$$2HO-CH_2-CO_2H \longrightarrow \begin{array}{c}\text{cyclic ester}\end{array} + 2H_2O$$

4. When warmed with concentrated sulphuric acid, it forms methanal and carbon monoxide:

$$HO-CH_2-CO_2H \xrightarrow{conc. H_2SO_4} H-CH=O + CO + H_2O$$

2-Hydroxypropanoic acid

2-Hydroxypropanoic acid (lactic acid) is prepared from ethanal or from propanoic acid:

$$CH_3-CH=O \xrightarrow[(KCN+HCl)]{HCN} CH_3-\underset{OH}{\overset{|}{CH}}-CN \xrightarrow[Boil]{HCl} CH_3-\underset{OH}{\overset{|}{CH}}-CO_2H$$

$$CH_3-CH_2-CO_2H \xrightarrow{Br_2} CH_3-\underset{Br}{\overset{|}{CH}}-CO_2H \xrightarrow[\text{(ii) HCl}]{\text{(i) NaOH}} CH_3-\underset{OH}{\overset{|}{CH}}-CO_2H$$

2-Hydroxypropanoic acid exists as two optically active isomers (p. 230), each of which has m.p. 26°C. When prepared in the laboratory as above, it is obtained as a racemic mixture of the two isomers which is optically inactive and has m.p. 18°C; the individual isomers can be obtained by resolution of this mixture (p. 237).

The chemical properties of 2-hydroxypropanoic acid are similar to those of hydroxyethanoic acid:

1. It has approximately the same acid dissociation constant as hydroxyethanoic acid.
2. As an acid, it forms esters and acid halides.
3. As an alcohol, the secondary alcohol group is oxidised to a keto group. To prevent oxidation of the product, 2-oxopropanoic acid, a weak oxidising agent such as silver oxide must be used:

$$CH_3-\underset{OH}{CH}-CO_2H + Ag_2O \rightarrow CH_3-\underset{O}{\overset{\parallel}{C}}-CO_2H + H_2O + 2Ag$$
$$\text{2-Oxopropanoic acid}$$

4. As a 2-hydroxy-acid, it forms a cyclic ester (lactide) on being heated:

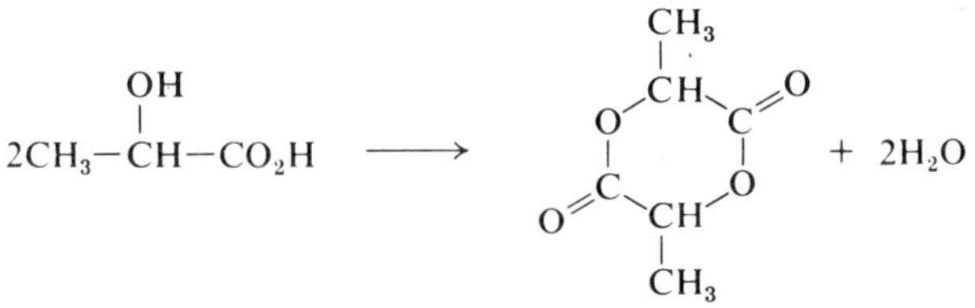

$$2CH_3-\underset{OH}{CH}-CO_2H \longrightarrow \text{[cyclic lactide]} + 2H_2O$$

5. As a 2-hydroxy-acid, it yields carbon monoxide when warmed with concentrated sulphuric acid:

$$CH_3-\underset{OH}{CH}-CO_2H \xrightarrow{\text{conc. } H_2SO_4} CH_3-CH=O + CO + H_2O$$

13.9 Salts of carboxylic acids

Carboxylic acids react with inorganic bases to form salts, for example:

$$CH_3-CO_2H + NaOH \rightarrow CH_3-CO_2^- Na^+ + H_2O$$

Physical properties of salts

Sodium and potassium salts are white, crystalline solids which are soluble in water but practically insoluble in ether and most other organic solvents. The calcium salts are much less soluble in water, and the silver salts are practically insoluble. All the salts have relatively high melting points (200°–400°C).

The solutions in water have a pH above 7 as the result of hydrolysis:

$$R-CO_2^- Na^+ + H_2O \rightleftharpoons R-CO_2H + Na^+ + OH^-$$

Chemical properties of salts

1. When the *anhydrous* sodium salt is heated with soda-lime, an **alkane** is formed:

$$R-CO_2^- Na^+ + NaOH \rightarrow R-H + Na_2CO_3$$

This is an example of **decarboxylation**, the term used when the elements of carbon dioxide are removed from a molecule.

A well-known example of this reaction is the formation of methane from sodium ethanoate:

$$CH_3CO_2^- \, Na^+ + NaOH \xrightarrow{\text{soda-lime}} CH_4 + Na_2CO_3$$

These reactions also occur with the acid itself, the sodium salt of the acid being formed during the reaction, for example when benzoic acid is heated with soda-lime:

$$C_6H_5CO_2H + NaOH \rightarrow C_6H_5CO_2^- \, Na^+ + H_2O$$

$$C_6H_5CO_2^- \, Na^+ + NaOH \rightarrow C_6H_6 + Na_2CO_3$$

2. *Kolbe's reaction.* When a strong aqueous solution of a salt is electrolysed, an **alkane** is formed:

$$2R\text{—}CO_2^- \, Na^+ + 2H_2O \rightarrow R\text{—}R + 2CO_2 + 2NaOH + H_2$$

At the *anode*, the carboxylate ion releases one electron to give a radical, R—CO—O·. This fragments with formation of an alkyl radical and carbon dioxide, and two alkyl radicals combine to form the alkane:

$$R\text{—}CO_2^- \rightarrow R\text{—}CO\text{—}O\cdot + e$$
$$R\text{—}CO\text{—}O\cdot \rightarrow R\cdot + CO_2$$
$$2R\cdot \rightarrow R\text{—}R$$

At the *cathode*, there is competition between sodium ions and hydrogen ions (from water) for discharge. Although the concentration of hydrogen ions is very low, the discharge potential for hydrogen favours the formation of hydrogen gas:

$$H^+ + e \rightarrow H; \quad 2H \rightarrow H_2$$

3. When the *anhydrous* sodium salt is heated with an acid chloride, an **acid anhydride** is formed:

$$R\text{—}CO_2^- \, Na^+ + R\text{—}CO\text{—}Cl \rightarrow R\text{—}CO\text{—}O\text{—}CO\text{—}R + NaCl$$

4. The sodium salts of monocarboxylic acids are stable when heated alone, except for sodium methanoate which decomposes at 400°C, yielding **disodium ethanedioate** and hydrogen:

$$2H\text{—}CO_2^- \, Na^+ \rightarrow Na^+ \, {}^-O_2C\text{—}CO_2^- \, Na^+ + H_2$$

5. The silver salts of carboxylic acids react on heating with alkyl halides to give **esters**:

$$R\text{—}CO_2^- \, Ag^+ + R'\text{—}Cl \rightarrow R\text{—}CO\text{—}O\text{—}R' + AgCl$$

6. When the calcium salt of a carboxylic acid is heated strongly, a **ketone**

CARBOXYLIC ACIDS

is formed, but the yield is usually very low.

$$(R-CO_2^-)_2Ca^{2+} \rightarrow R-CO-R + CaCO_3$$

13.10 Soaps and detergents

Soaps are obtained from natural fats, known as **glycerides** because they are esters formed by the trihydric alcohol, propane-1,2,3-triol, which was formerly known as glycerol, with long-chain carboxylic acids. The glycerides are hydrolysed by heating with caustic soda (**saponification**) to form soaps—the sodium salts of the acids—and propane-1,2,3-triol:

$$\begin{array}{c} CH_2-O-CO-R \\ | \\ CH-O-CO-R \\ | \\ CH_2-O-CO-R \end{array} + 3NaOH \longrightarrow \begin{array}{c} CH_2-OH \\ | \\ CH-OH \\ | \\ CH_2-OH \end{array} + 3RCO_2^- Na^+$$

A glyceride   Propane-1,2,3-triol   A soap

Sodium chloride is added to precipitate the soap, and this is then processed into bars or soap powder. Glycerides contain saturated carboxylic acids which have an even number of carbon atoms, generally within the range 12–20, for example, octadecanoic acid (stearic acid), $CH_3-(CH_2)_{16}-CO_2H$.

Soaps act by lowering the surface tension between water and an oil or other insoluble material. They do so by virtue of containing both a hydrophilic ('water-loving') group ($-CO_2^-$) and a hydrophobic ('water-hating') group (the alkyl chain); molecules of water tend to congregate near the former and molecules of the water-insoluble material congregate around the latter.

One disadvantage of soaps is that they form insoluble calcium salts with the calcium ions in hard water and in the clays which are present in dirt; a good deal of the soap is wasted in this way. This problem is avoided by the use of synthetic detergents in which a sulphonate group, $-SO_2-O^-$, or sulphate group, $-O-SO_2-O^-$, replaces the carboxylate group as the hydrophilic component, since the corresponding calcium salts are more soluble in water than the calcium salts of carboxylic acids.

Until about 1965, the commonest detergents contained alkylbenzene-sulphonates made from a polymer of propene by a Friedel-Crafts reaction:

$$CH_3-\underset{\underset{CH_3}{|}}{CH}-CH_2-\underset{\underset{CH_3}{|}}{CH}-CH_2-\underset{\underset{CH_3}{|}}{CH}-CH_2-CH=CH_2 + C_6H_6$$

$$\xrightarrow{AlCl_3 \text{ as cat.}} CH_3-(\underset{\underset{CH_3}{|}}{CH}-CH_2)_3-\underset{\underset{CH_3}{|}}{CH}-C_6H_5$$

followed by sulphonation (p. 103) and neutralisation of the sulphonic acid with sodium hydroxide:

$$CH_3-(\underset{\underset{CH_3}{|}}{CH}-CH_2)_3-\underset{\underset{CH_3}{|}}{CH}-C_6H_5 \xrightarrow[(2)\ NaOH]{(1)\ H_2SO_4} CH_3-(\underset{\underset{CH_3}{|}}{CH}-CH_2)_3-\underset{\underset{CH_3}{|}}{CH}-C_6H_4-SO_2-O^-\ Na^+$$

However, these detergents suffer from the disadvantage that they are not degraded by bacteria in sewage plants, which meant that many rivers suffered from the foam and there was also the danger that detergents could be 'recycled' into the drinking-water supplies. The failure of bacteria to attack the materials results from the presence of the large number of branches in their alkyl groups, and the use of such detergents has been abandoned in the United Kingdom. Reduction in the number of branches increases their capacity for biodegradation and so most detergents now contain linear or only singly branched alkyl groups. There are three important types; the first two are **anionic** detergents while the third is **non-ionic**.

All three types are obtained from non-branched alkenes, which themselves are derived partly from the cracking of waxes (19.8), partly from the polymerisation of ethene with a Ziegler catalyst (20.4) and partly from alkanes by free-radical chlorination followed by catalytic dehydrochlorination:

$$R-CH_2-CH_3 + Cl_2 \rightarrow R-CHCl-CH_3 \xrightarrow[300°C]{\text{Silica gel as cat.}} R-CH=CH_2$$

1. *Alkylbenzenesulphonates*, which are made from non-branched alkenes ($C_{10}-C_{14}$):

$$R-CH=CH_2 + C_6H_6 \xrightarrow{\text{HF as cat.}} R-CH(CH_3)-C_6H_5$$

The alkylbenzene is sulphonated and neutralised. These are used in powder detergents (e.g. Omo, Surf, Tide), and are slowly biodegradable. The formation of an alkylbenzenesulphonate is described on p. 198.

2. *Alkyl sulphates*, which are made from straight-chain alcohols ($C_{10}-C_{14}$) derived from alkenes by the OXO process (p. 308) or with a Ziegler catalyst (p. 303):

$$R-CH_2OH \xrightarrow{H_2SO_4} R-CH_2-O-SO_2OH \xrightarrow{NaOH} R-CH_2-O-SO_2-O^- Na^+$$

3. *Ethoxylates* (which are non-ionic detergents) are made from long-chain alcohols and epoxyethane, for example:

$$CH_3-(CH_2)_{10}-CH_2-OH + 8\, H_2C\overset{\displaystyle O}{-}CH_2$$

$$\longrightarrow CH_3-(CH_2)_{10}-CH_2-(OCH_2CH_2)_8-OH$$
An ethoxylate

These do not contain an ionic group as their hydrophilic component, but hydrophilic properties are conferred on them by the presence of a number of oxygen atoms in one part of the molecule which are capable of forming hydrogen-bonds with molecules of water.

A packet of detergent contains about 20 per cent of active detergent and an equal amount of sodium sulphate, which increases the bulk of the powder. A further 30 per cent is made up of inorganic phosphates, which are added to remove soluble, complex calcium salts by reaction with the

CARBOXYLIC ACIDS

calcium ions in the dirt. Other ingredients include sodium peroxoborate (about 5 per cent), which is a bleaching agent, and fluorescers, which are organic compounds which absorb ultraviolet light and re-emit the energy in the blue part of the visible spectrum, thereby making 'yellow' clothes appear white.

The inorganic phosphates which enter lakes and rivers *via* sewage are nutrients for algae and are responsible for the proliferation of these plants, as a green surface sludge, in seas and lakes in various parts of the world. It is likely that legislation will be introduced to remove, or at least reduce, the phosphate content of detergents so as to eliminate this source of pollution.

13.11 Practical work

Preparation of a carboxylic acid from an ester

The example given is the preparation of benzoic acid from ethyl benzoate.

Reflux a mixture of 5 cm^3 of ethyl benzoate and 25 cm^3 of sodium hydroxide solution (made by dissolving 10 g of sodium hydroxide in 25 cm^3 of water) for 30 minutes (cf. Fig. 2.1).

Distil the mixture and collect the first 2 cm^3 of distillate (cf. Fig. 2.2). Carry out some of the tests for ethanol (for example, No. 3, p. 155, and No. 7, p. 155).

Pour the residue from the flask into a beaker and add M sulphuric acid until the solution is acid to litmus. Filter the crystals, using a Buchner funnel (cf. Fig. 2.8). Dissolve the crystals in a minimum of boiling water. Allow the solution to cool and filter the crystals, dry them between pads of filter paper and determine their melting point.

Small-scale preparation of phenylmagnesium bromide and benzoic acid

1. Preparation of phenylmagnesium bromide solution

All apparatus and materials used in the preparation of a Grignard reagent must be thoroughly dry since traces of moisture not only react with the reagent but also inhibit its formation. Make sure that the apparatus is clean and that it has been dried in an oven before the experiment.

Diethyl ether must stand over clean sodium wire or pellets and the bromobenzene over anhydrous calcium chloride in carefully stoppered tubes or flasks, if possible, overnight before the experiment is to be carried out.

Place 0·5 g of magnesium turnings in a flask and add one or two crystals of iodine, followed by 10 cm^3 of dry ether. Stand the flask in a cold water-bath, add 2 cm^3 of dry bromobenzene, then fit a dry reflux condenser which is attached to a calcium chloride drying tube at its open end.

Raise the temperature of the water-bath to 40–45°C, turn out the bunsen and allow the contents of the flask to reflux for 20–25 minutes. The disappearance of the colour of the iodine and the formation of a cloudiness in the reaction mixture are indications that the reaction is proceeding satisfactorily. Remove the warm water-bath and replace it with a freezing mixture of ice and salt.

2. Preparation of benzoic acid

Arrange for the carbon dioxide, generated in the Kipp's apparatus, to be washed in water to remove traces of hydrochloric acid and to be dried by passing through *concentrated* sulphuric acid. A more satisfactory method is to place a few pieces of solid carbon dioxide in a conical flask and lead the carbon dioxide straight into the flask containing the solution, which should be immersed in ice.

Pass a gentle stream of dry carbon dioxide through the solution of phenylmagnesium bromide for 5–10 minutes. Decant the contents of the flask into a small beaker and place the latter in the freezing mixture. Dilute 3 cm^3 of *concentrated* hydrochloric acid by adding 3 cm^3 of water. Introduce the acid slowly into the beaker, with stirring, in order to liberate the benzoic acid.

Remove the beaker from the freezing mixture, add 15 cm^3 of ether and stir. Decant the liquid into a separating funnel and return the lower aqueous layer to the beaker. Repeat the ether extraction twice, using 5 cm^3 of ether each time.

Combine the ether extracts and shake with 10 cm^3 of 2M sodium hydroxide solution in a separating funnel. Remove the stopper from the funnel occasionally to release the pressure. Most of the benzoic acid enters the lower aqueous layer as the sodium salt. Transfer the aqueous layer to a beaker. (If a precipitate of magnesium hydroxide should appear, remove it by filtration through a Buchner funnel.) Acidify the filtrate with 2M hydrochloric acid (testing the solution with litmus paper). A white precipitate of benzoic acid is obtained. Precipitation is hastened by cooling in the freezing mixture and scratching with a glass rod. Filter off the precipitate; wash *in situ* with distilled water. Dry the solid and take its melting point. If time, purify by recrystallisation from hot water. Dry the crystals in the oven at about 100°C and redetermine the melting point.

Reduction of a carboxylic acid to an alcohol with lithium aluminium hydride

Lithium aluminium hydride is a specific reagent; for example, it does not reduce unsaturated carbon–carbon bonds (as in alkenes and alkynes).

The acid chosen for this reaction is 2-chlorobenzoic acid as the product, 2-chlorophenylmethanol, is a solid.

Lithium aluminium hydride is a dangerous chemical. It reacts violently with water and only dry solvents and dry apparatus must be used. Experiments should be done in a fume cupboard with a satisfactory exhaust for the hydrogen formed during reactions to be led away. There should be no naked flames nearby.

Place 30 cm^3 of diethyl ether, previously dried over sodium, in a round-bottom flask, which has been carefully dried, and fit a reflux condenser. Weigh out 0·5 g of lithium aluminium hydride on a watch glass and add it, in *very small* quantities, to the ether. If there is any effervescence, it means that the ether is not dry.

Add, in very small quantities at a time, 1·5 g of 2-chlorobenzoic acid to the solution of lithium aluminium hydride in ether. After all the acid has been added, reflux the mixture, using a beaker of hot water, for about 30 minutes.

Then add, dropwise, 5 cm^3 of ethyl ethanoate which will be reduced by the excess of lithium aluminium hydride and thus decompose it. Allow the mixture to reflux for a further 10 minutes.

Cool the mixture and then add 20 cm^3 of M sulphuric acid. 2-Chlorophenylmethanol is liberated from the complex aluminium compound formed and dissolves in the ether layer.

Place the mixture in a separating funnel and run the ether layer into a conical flask. Add some anhydrous sodium sulphate, fit a stopper to the flask and swirl the mixture for about 5 minutes.

Decant the solution into a flask and distil off the ether using a beaker of hot water.

Detach the flask and place it in a beaker of ice. Filter off the crystals of 2-chlorophenylmethanol and dry them between filter papers. Take the melting point of the crystals, which, before further purification, will be about 70°C.

If time permits, recrystallise the product from a dilute aqueous solution of ethanol.

Preparation of detergents

1. To 5 g of dodecanol in a flask, add 2 cm^3 of chlorosulphonic acid dropwise (**CARE**), with stirring. Keep the mixture below 35°C by immersing the flask in a large beaker of cold water. Stir the mixture for a further 10 minutes and then divide into two parts.

$$C_{12}H_{25}-OH + Cl-SO_2-OH \rightarrow C_{12}H_{25}-O-SO_2-OH + HCl$$
$$\text{Dodecyl sulphate}$$

(a) To one part, add a dilute solution of sodium hydroxide, until the mixture is neutral. Transfer the solution to an evaporating basin and evaporate it until the solid detergent is formed. The detergent can be recrystallised from ethanol.

$$C_{12}H_{25}-O-SO_2-OH + NaOH \rightarrow C_{12}H_{25}-O-SO_2-O^- Na^+ + H_2O$$

(b) To the second part, add a solution of tris-(2-hydroxyethyl)amine (2 cm^3 in 20 cm^3 of water):

$$C_{12}H_{25}-O-SO_2-OH + N(CH_2CH_2OH)_3 \rightarrow$$
$$C_{12}H_{25}-O-SO_2-O^- \overset{+}{H}N(CH_2CH_2OH)_3$$

Finally, make the mixture neutral to litmus by adding dropwise a solution of sodium hydrogen carbonate.

(c) Examine the lathering properties of the two detergents in both hard and soft water.

2. *Great care must be taken when using oleum. Wear safety glasses and carry out the experiment in a fume cupboard.*

Place 11·5 cm^3 of an alkylbenzene in a flask fitted with a thermometer, and cool the hydrocarbon to about 5°C in an ice-bath. Add 5 cm^3 of oleum (sulphuric acid containing about 20 per cent free sulphur trioxide) using a dropping-pipette, shaking between each addition. The temperature rises slowly but must not be allowed to rise above 56°C. The temperature should be about 55 ± 1°C at the end of the addition.

Replace the ice-bath with a beaker of hot water to keep the mixture at 55 ± 1°C for a further 30 minutes. Cool the mixture.

Place 4 g of crushed ice in a beaker and then surround it with a larger beaker containing an ice-water mixture. Stir in a solution of 1·5 g of sodium hydroxide dissolved in 6 cm^3 of water.

Transfer the reaction mixture containing the alkylbenzenesulphonic acid to a tap-funnel and add it to the alkali solution, with stirring, making sure that the temperature of the mixture does not rise above 50°C. Add the sulphonic acid until the pH of the mixture in the beaker is between 6·5 and 7·5 (using narrow-range Universal Indicator papers). If the pH becomes too low, add 2M sodium hydroxide solution to adjust it to 6·5–7·5.

A solid will precipitate out, and the mixture is then heated in an evaporating basin over a beaker of boiling water until most of the liquid is removed.

Although it is difficult to purify the detergent any further, its detergent properties can be studied by transferring a small amount of the mixture in the evaporating basin to a test-tube and dissolving it in water.

The detergent may be discoloured owing to the formation of 'hot-spots' on adding oleum to the hydrocarbon.

Reactions of methanoic acid and sodium methanoate

1. To some solid sodium hydrogen carbonate in a test-tube, add some methanoic acid.

2. Make up neutral iron(III) chloride solution by adding dilute ammonia solution to 2 cm^3 of iron(III) chloride solution until a precipitate appears. Add the original iron(III) chloride solution until the precipitate just disappears.

Place 5 drops of methanoic acid in a test-tube. Add dilute ammonia solution until just alkaline and boil off excess of ammonia. To this neutral solution, add 5 drops of neutral iron(III) chloride solution.

3. To 1 cm^3 of methanoic acid in a test-tube, add 1 cm^3 of *concentrated* sulphuric acid and warm. Test the gas evolved with (a) a lighted splint, (b) lime-water.

4. Place sodium methanoate crystals in a test-tube to a depth of about 1 cm. Heat vigorously and test the gas evolved with a lighted splint.

Cool the residue and add 1 cm^3 of concentrated sulphuric acid. Test the gas evolved with (a) a lighted splint, (b) lime-water.

The reducing properties of methanoic acid are illustrated by the reactions 5–7.

5. To 0·5 cm^3 of methanoic acid in a test-tube, add 1 cm^3 of dilute sulphuric acid and warm gently. Add a 1 per cent solution of potassium permanganate drop by drop, and note whether the permanganate is decolorised.

6. To 1 cm^3 of ammoniacal silver nitrate solution (p. 178), add 2 or 3 drops of methanoic acid. A white precipitate of silver methanoate is formed. If the test-tube is immersed in hot water, some deposit of silver is formed.

7. Add methanoic acid dropwise to 1 cm^3 of a solution of mercury(II) chloride. A white precipitate of mercury(I) chloride is slowly formed. On warming, the precipitate darkens owing to reduction to mercury.

Reactions of ethanoic acid and sodium ethanoate

8. Repeat experiment 1, using ethanoic acid.

9. Repeat experiment 2, using ethanoic acid.

10. Repeat experiment 5, using ethanoic acid.

11. Mix together some *anhydrous* sodium ethanoate and soda-lime in a test-tube. Heat the mixture strongly. Test the gas evolved with a lighted splint.

12. Warm a mixture of 5 drops of ethanol, 5 drops of ethanoic acid

and 1 drop of *concentrated* sulphuric acid. Note the characteristic odour of the product.

Reactions of benzoic acid

13. Repeat experiment 1 using a hot solution of benzoic acid, and compare the result with that obtained when a solution of phenol is added to sodium hydrogen carbonate.

14. Repeat experiment 5 using a hot solution of benzoic acid.

15. Repeat experiment 11 using benzoic acid in place of sodium ethanoate.

16. Repeat experiment 12 using benzoic acid in place of ethanoic acid.

Reactions of ethanedioic acid and disodium ethanedioate

17. Repeat experiment 1 using a solution of ethanedioic acid.

18. Repeat experiment 3 using solid ethanedioic acid.

19. Repeat experiment 4 using solid disodium ethanedioate.

20. Repeat experiment 5 using solid ethanedioic acid. Shake until ethanedioic acid has dissolved. Note whether any reaction occurs (a) in the cold, (b) when the mixture is warmed.

21. Repeat experiment 11 using disodium ethanedioate instead of sodium ethanoate.

13.12 Further reading

Detergents. Elaine Moore (reprinted 1970). Unilever Educational Booklet. Revised Ordinary Series No. 1.

Theory of Detergency. R. J. Taylor (1969). Unilever Educational Booklet. Advanced Series No. 7.

13.13 Films

Outline of Detergency. Unilever Film Library*.
Detergents up to date. Petroleum Films Bureau*.
* On free loan.

13.14 Questions

1 Outline by means of balanced equations and essential reaction conditions (a) **two** general methods for the synthesis of aliphatic carboxylic acids from alkyl iodides, (b) **one** method for the synthesis of benzoic acid from benzene.

Give **two** reactions of methanoic acid which are not shown by other aliphatic carboxylic acids.

Describe with practical details how you would detect the presence of a hydroxyl group in benzoic acid. (AEB)

2 Describe, with equations, a chemical test you would employ to distinguish between the following compounds:
(a) Methanoic acid and ethanoic acid
(b) Methanoic acid and ethanedioic acid
(c) Ethanoic acid and ethanedioic acid
(d) Phenol and benzoic acid.

3 Outline practical reaction schemes to obtain as many compounds as possible from ethanoic acid. Give the names and formulae of the products and intermediates, and indicate the reagents which are used.

CARBOXYLIC ACIDS

4 Starting with ethanoic acid, by what reactions can the following be prepared: (a) ethanoic anhydride, (b) ethanamide, (c) methane, (d) chloroethanoic acid, (e) ethane?

5 Give three general reactions by which an aliphatic carboxylic acid may be prepared.
  Describe how you would carry out a practical test to show that alcohols and acids both contain hydroxyl groups.
  State briefly how, starting from ethanoic acid, you would prepare (a) methane, (b) ethanoic anhydride.

6 Describe in outline two methods by which benzoic acid could be made in the laboratory from benzene, giving the equations for the reactions and conditions required.
  Describe the reactions by which the following could be obtained from benzoic acid: (a) benzoyl chloride, (b) benzoic anhydride, (c) benzamide.

7 What reactions are characteristic of carboxylic acids? To what extent may methanoic acid, ethanoic acid and ethanedioic acid be regarded as typical of this class of compounds?

8 Name five organic substances which can be obtained *directly* from salts of ethanoic acid. State what other reagents, if any, would be required. Give the conditions and equations for the reactions.

9 Starting with carbon monoxide, outline in each case **one** method by which (a) methanoic acid and (b) sodium ethanedioate are obtained. What products are obtained when methanoic acid is treated with mercury(II) chloride solution?

10 The molecular weight of a weak, monobasic, organic acid A was calculated from (a) the osmotic pressure of its aqueous solution, (b) the depression of the freezing point of benzene observed when A was dissolved in this solvent. The two methods gave different values for the molecular weight of A. Suggest an explanation for this difference.

11 Most of the reactions of carboxylic acids can be classified as belonging to one of four types: (a) reactions involving cleavage of the O—H bond; (b) reactions at the carbonyl carbon; (c) reactions at the 2-carbon atom; (d) decarboxylation. Discuss the chemistry of carboxylic acids under these headings. (W(S))

12 Compare: (a) the properties of the CO group in ethanal, propanone and ethanoic acid; (b) the properties of the hydroxyl group in phenol and ethanol. What explanations have been suggested for these differences? How do you account for the fact that phenylamine is a weaker base than ethylamine? (O(S))

13 A substance A ($C_4H_6O_2$) rapidly decolourised a solution of potassium permanganate in the cold. On treatment with trioxygen and hydrolysis of the products, A gave a neutral substance B (C_2H_4O) and an acidic substance C ($C_2H_2O_3$), both of which gave silver mirrors with ammoniacal silver solutions, and orange precipitates with 2,4-dinitrophenylhydrazine sulphate in aqueous methanol (DNPH). B, on oxidation, gave an acid D, whose calcium salt, on heating, gave E, C_3H_6O, which gave an orange precipitate with DNPH, but did not react with the ammoniacal silver solution. Oxidation of C gave an acid F, which decolourised acidified potassium permanganate on warming, and which, on heating with concentrated sulphuric acid gave off some gas. This gas was passed through lime water, which turned cloudy, and the effluent gas was found to burn in air.
  Deduce the structures of A, B, C, D, E and F, and outline the course of the above reactions. (L)

14 The following table gives the values of the dissociation constants of ethanoic acid and some of its related acids:

Acid	Dissociation constant (K_a)
Chloroethanoic	1.4×10^{-3}
Ethanoic	1.86×10^{-5}
Phenylethanoic	5.2×10^{-5}
Aminoethanoic	1.67×10^{-10}

(a) Discuss the theoretical reasons for the differences between the values for CH_3CO_2H, CH_2ClCO_2H and $CH_2NH_2CO_2H$.

(b) Calculate the pH of:

(i) 0·1M phenylethanoic acid;

(ii) a mixture of equal volumes of 0·2M phenylethanoic acid and 0·2M sodium phenylethanoate;

(iii) a mixture of equal volumes of 0·2M phenylethanoic acid and 0·2M ethanoic acid.

(c) Which of the solutions in (b) would change least in pH on dilution ten times? Explain your answer. (L(XS))

15 On combustion 0·1575 g of a hydroxy-monocarboxylic acid gave 0·2310 g of carbon dioxide and 0·0945 g of water. The vapour density of its ethyl ester is 59.
Calculate (a) the percentage composition, (b) the empirical formula, (c) the molecular weight, (d) molecular formula, of the acid.
Give the names and structural formula of all the hydroxy-acids having this molecular formula and account for their existence. (AEB)

16 Outline how pure ethanedioic acid crystals can be made, starting from carbon monoxide. How, and under what conditions, does ethanedioic acid react with

(a) potassium permanganate,

(b) methanol,

(c) sulphuric acid? (C(N))

17 Outline a synthesis of butanedioic acid,

$$\begin{array}{c} CH_2.CO.OH \\ | \\ CH_2.CO.OH \end{array}$$

from ethene indicating the reagents and conditions for each reaction you mention.

By means of equations and brief notes on reaction conditions show how the following compounds could be prepared from butanedioic acid:

(a) butanedioic anhydride,

(b) 2-hydroxybutanedioic acid, $\begin{array}{c} CH(OH).CO.OH \\ | \\ CH_2.CO.OH \end{array}$

(c) 4-phenyl-4-oxobutanoic acid, $C_6H_5.CO.CH_2.CH_2.CO.OH$,

(d) a mixture of *cis*- and *trans*-butenedioic acids.

How could you convert *trans*-butenedioic acid into its *cis*-isomer? (JMB(S))

18 A substance, A, of molecular formula $C_3H_4OCl_2$ reacted with cold water to give a compound, B, $C_3H_5O_2Cl$. A on treatment with ethanol gave a liquid C, $C_5H_9O_2Cl$. When A was boiled with water, a compound D, $C_3H_6O_3$ was obtained. D was optically active and could be ethanoylated.
Deduce the nature of the compounds A, B, C and D, and account for the above reactions. (C Entrance)

19 Treatment of an aromatic compound A, C_8H_{10}, with ethanoyl chloride in the presence of aluminium chloride gives B, $C_{10}H_{12}O$. On being warmed with iodine and sodium hydroxide, B forms the sodium salt of C, $C_9H_{10}O_2$. Both B and C are converted to D, $C_9H_6O_6$, by vigorous oxidation with chromic acid. When

CARBOXYLIC ACIDS

heated, D readily forms E, $C_9H_4O_5$.
Deduce structures for the compounds A to E and elucidate the above reactions.
(O Schol.)

20 Three isomeric organic acids have the following composition:

H = 2·09 per cent; C = 43·98 per cent; O = 16·75 per cent; Cl = 37·17 per cent.

When heated with soda-lime, all three isomers are converted to the same dichlorobenzene.

Give the structural formulae of the dichlorobenzene and each of the three acids from which it is obtained.
(O Schol.)

21 A solid A, m.p. 32·5° and formula $C_5H_8O_3$, can be isolated from the mixture formed when starch is boiled with hydrochloric acid. The following reactions can be carried out with A:

(i) $A \xrightarrow{Na_2CO_3} B\ (C_5H_7O_3Na) + CO_2$

(ii) $A \xrightarrow[HCl]{CH_3OH} C_6H_{10}O_3$

(iii) $A \xrightarrow{hydroxylamine} C_5H_9O_3N$

(iv) $A \xrightarrow{NaOH + I_2}$ A yellow precipitate $C + (C_4H_4O_4Na_2)$

(v) $A \xrightarrow[with\ HNO_3]{oxidation}$ Ethanoic acid + an acid $D\ (C_3H_4O_4)$

(vi) $A \xrightarrow[Na/Hg\ and\ water]{reduction\ with} F\ (C_5H_9O_3Na)$.

Elucidate the structure of A, giving your reasoning.

Acidification of F gives a neutral substance $C_5H_8O_2$, instead of the expected acid, and electrolysis of B affords octane-2,7-dione. Can you write formulae to explain these reactions?
(O Schol.)

Chapter 14

Derivatives of carboxylic acids

14.1 Introduction

The hydroxyl group, —OH, in a carboxylic acid can be replaced by other functional groups, so that there is a series of compounds which contain the **acyl** group, RCO—, which form a parallel series to the derivatives of the alkanes:

Functional group	Acyl derivatives		Alkyl derivatives	
—OH	RCO—OH	Acid	R—OH	Alcohol
—OR′	RCO—OR′	Ester	R—OR′	Ether
—X (—F, —Cl, —Br, —I)	RCO—X	Acid halide	R—X	Alkyl halide
—OCOR′	RCO—OCOR′	Acid anhydride	R—OCOR′	Ester
—NH$_2$	RCO—NH$_2$	Acid amide	R—NH$_2$	Amine

In addition, carboxylic acids can be converted into **acid nitriles**, R—CN.

14.2 Esters

General formula

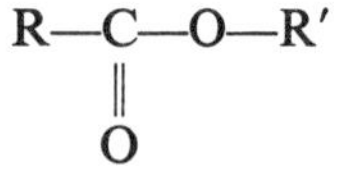

$$R-\underset{\underset{O}{\|}}{C}-O-R'$$

Physical properties

Esters are neutral liquids with pleasant, fruity smells. They are usually insoluble in water but are soluble in organic solvents.

Their melting points and boiling points are below those of the corresponding acids because ester molecules, unlike acid molecules, are not associated by hydrogen-bonding.

Table 14.1. Some esters

NAME	FORMULA	B.P. (°C)
Methyl methanoate	H—CO$_2$—CH$_3$	32
Ethyl methanoate	H—CO$_2$—C$_2$H$_5$	53
Methyl ethanoate	CH$_3$—CO$_2$—CH$_3$	56
Ethyl ethanoate	CH$_3$—CO$_2$—C$_2$H$_5$	77
Methyl propanoate	C$_2$H$_5$—CO$_2$—CH$_3$	79
Ethyl propanoate	C$_2$H$_5$—CO$_2$—C$_2$H$_5$	98
Methyl benzoate	C$_6$H$_5$—CO$_2$—CH$_3$	200

Preparation

1. By the reaction between an acid and an alcohol in the presence of a small amount of a strong acid such as sulphuric acid as catalyst (**esterification**), for example:

$$CH_3-CO_2H + C_2H_5-OH \underset{}{\overset{H^+}{\rightleftharpoons}} CH_3-CO-O-C_2H_5 + H_2O$$
Ethyl ethanoate

The equilibrium constants for esterification are usually close to 1. In order to obtain a good yield of an ester from a given amount of an acid, it is necessary to use an excess of the alcohol. Suppose the equilibrium constant in the above reaction is 1·0. Then, if 1 mol of both ethanol and ethanoic acid are used, and if the total volume is V cm^3 and the amounts of ester and water at equilibrium are x mol,

$$\frac{x^2/V^2}{(1-x)^2/V^2} = 1$$

Thus, $x = 0.5$ mol, so that the yield of ester, based on ethanoic acid or ethanol, is 50 per cent. However, if 10 mol of ethanol are used per mol of ethanoic acid, then

$$\frac{x^2/V^2}{(1-x)(10-x)/V^2} = 1$$

so that $x \simeq 0.9$ mol; that is, the yield of ethyl ethanoate based on ethanoic acid is about 90 per cent.

2. By the reaction between an alcohol and either an acid chloride (14.3) or an acid anhydride (14.4), for example:

$$CH_3-CO-Cl + C_2H_5-OH \rightarrow CH_3-CO-O-C_2H_5 + HCl$$
Ethanoyl chloride

$$CH_3-CO-O-CO-CH_3 + CH_3-OH \rightarrow$$
Ethanoic anhydride
$$CH_3-CO-O-CH_3 + CH_3CO_2H$$

3. By the reaction between the silver salt of an acid and an alkyl halide (13.9):

$$R-CO_2^- \, Ag^+ + R'-X \rightarrow R-CO-O-R' + AgX$$

Chemical properties

1. Esters are hydrolysed by heating with mineral acids or alkalis. The catalysed reaction is reversible, and is the exact opposite of esterification, for example:

$$CH_3-CO-O-C_2H_5 + H_2O \underset{}{\overset{H^+}{\rightleftharpoons}} CH_3-CO-OH + C_2H_5-OH$$

Consequently, because the equilibrium constant is usually close to 1, some ester always remains at the end of the reaction.

The alkali-catalysed reaction is also reversible:

$$R-CO-O-R' + H_2O \underset{}{\overset{OH^-}{\rightleftharpoons}} R-CO-OH + R'-OH$$

However, in this case the carboxylic acid formed reacts with hydroxide ion to give the acid anion:

$$R\text{—}CO_2H + OH^- \rightleftharpoons R\text{—}CO_2^- + H_2O$$

Equilibrium in this step lies almost completely to the right-hand side. Consequently, as the acid is formed in the first reaction it is removed by the second, so that eventually practically all the ester is converted into its hydrolysis products. This makes the alkali-catalysed reaction more efficient than the acid-catalysed one, and it is normally the process chosen. It is often referred to as **saponification**; naturally occurring esters are converted into soaps in this way (13.10).

In the hydrolysis of an ester, either the bond between the carbonyl group and the oxygen atom, or the bond between the alkyl group and the oxygen atom, might be broken.

$$R\text{—}CO\!\mid\!O\text{—}R' \quad \longrightarrow \quad R\text{—}CO \quad + \quad O\text{—}R'$$
$$HO\!\mid\!H \qquad\qquad\qquad\qquad HO \qquad\quad H$$

or

$$R\text{—}CO\text{—}O\!\mid\!R' \quad \longrightarrow \quad R\text{—}CO\text{—}O \quad + \quad R'$$
$$H\!\mid\!OH \qquad\qquad\qquad\qquad H \qquad\quad OH$$

A study of the hydrolysis of esters which are labelled with an ^{18}O isotope has shown that it is the former bond that is broken. Thus, when the ester $R\text{—}C^{16}O\text{—}^{18}O\text{—}R'$ is hydrolysed, the ^{18}O isotope is found in the resulting alcohol and not in the carboxylic acid:

$$R\text{—}C\!\mid\!^{18}O\text{—}R' + H_2^{16}O \longrightarrow R\text{—}C\text{—}^{16}OH + R'\text{—}^{18}OH$$
$$\overset{\|}{^{16}O} \qquad\qquad\qquad\qquad \overset{\|}{^{16}O}$$

The mechanism of alkali-catalysed ester hydrolysis is as follows. The nucleophilic hydroxide ion adds to the carbonyl group of the ester to give an intermediate like that formed in addition to an aldehyde or ketone (12.5):

$$R\text{—}\underset{\underset{O}{\|}}{C}\text{—}OR' + OH^- \longrightarrow R\text{—}\underset{\underset{O^-}{|}}{\overset{\overset{OH}{|}}{C}}\text{—}OR'$$

This intermediate then fragments into the acid and an alkoxide ion:

$$R\text{—}\underset{\underset{O^-}{|}}{\overset{\overset{OH}{|}}{C}}\text{—}OR' \longrightarrow R\text{—}\underset{\underset{O}{\|}}{\overset{\overset{OH}{|}}{C}} + R'\text{—}O^-$$

The alkoxide ion reacts with the solvent, for example water, to give the alcohol, and the carboxylic acid dissociates:

$$R'\text{—}O^- + H_2O \rightleftharpoons R'\text{—}OH + OH^-$$
$$R\text{—}CO_2H + OH^- \rightleftharpoons R\text{—}CO_2^- + H_2O$$

2. Esters, like acids, can be reduced with lithium aluminium hydride to form alcohols:

$$R-CO-OR' \xrightarrow{LiAlH_4} R-CH_2-OH + R'-OH$$

e.g.

$$CH_3-CO-OCH_3 \xrightarrow{LiAlH_4} CH_3CH_2OH + CH_3OH$$

Esters cannot be reduced with sodium borohydride (p. 169).

Unlike acids, esters can be reduced with sodium in ethanol to form alcohols:

$$R-CO-OR' \xrightarrow{Na/C_2H_5OH} R-CH_2-OH + R'-OH$$

3. Esters react with ammonia, in either concentrated aqueous or alcoholic solution, to form acid amides:

$$R-CO-OR' + NH_3 \rightarrow R-CO-NH_2 + R'-OH$$

e.g.

$$CH_3-CO-OC_2H_5 + NH_3 \rightarrow CH_3-CO-NH_2 + C_2H_5OH$$
$$\text{Ethanamide}$$

However, the reaction is much slower than with acid chlorides or anhydrides, and amides are therefore more easily made from these acid derivatives.

Uses

Esters are responsible for the smell and flavour of many fruits and flowers. Hence, artificial flavouring essences are prepared from esters. Ethyl methanoate is used in raspberry essence and 3-methylbutyl ethanoate in pear essence. Esters are also used in artificial scents.

Esters are extensively used as solvents for drugs, antibiotics, etc.

Waxes are esters of higher carboxylic acids and higher alcohols. For example, a constituent of beeswax is $C_{15}H_{31}CO_2C_{31}H_{63}$.

Fats and oils are esters of higher carboxylic acids and propane-1,2,3-triol. These esters are known as **glycerides**, and some are used to make soaps (13.10).

Diethyl propanedioate

Diethyl propanedioate (often called malonic ester) is made from the sodium salt of chloroethanoic acid by nucleophilic displacement of chloride by cyanide ion (by heating with potassium cyanide) followed by heating with ethanol in the presence of sulphuric acid:

$$Cl-CH_2-CO_2^- Na^+ \xrightarrow{CN^-} N\equiv C-CH_2-CO_2^- Na^+ \xrightarrow[H_2SO_4]{2C_2H_5OH}$$
$$C_2H_5O_2C-CH_2-CO_2C_2H_5$$
$$\text{Diethyl propanedioate}$$

Diethyl propanedioate (a liquid, b.p. 198°C) is a valuable reagent in organic synthesis. Its usefulness stems from the fact that it forms a sodium salt when treated with sodium ethoxide in ethanol,

$$CH_2(CO_2C_2H_5)_2 + C_2H_5O^- Na^+ \rightleftharpoons Na^+ \bar{C}H(CO_2C_2H_5)_2 + C_2H_5OH$$

the anion of which is a nucleophilic reagent. For example, its reaction with an alkyl halide gives an alkylpropanedioic ester which, after hydrolysis and decarboxylation, gives a carboxylic acid:

$$Na^+ \bar{C}H(CO_2C_2H_5)_2 + R-X \rightarrow R-CH(CO_2C_2H_5)_2 + NaX$$

$$R-CH(CO_2C_2H_5)_2 \xrightarrow{2NaOH} R-CH(CO_2^- Na^+)_2 \xrightarrow{2HCl}$$

$$R-CH(CO_2H)_2 \xrightarrow{heat} R-CH_2-CO_2H + CO_2$$

Ethyl 3-oxobutanoate

Ethyl 3-oxobutanoate (often called acetoacetic ester) is made by refluxing ethyl ethanoate in the presence of a little ethanol over sodium wire. Sodium ethoxide is formed,

$$Na + C_2H_5OH \rightarrow C_2H_5O^- Na^+ + \tfrac{1}{2}H_2$$

and catalyses the reaction:

$$2CH_3-CO_2C_2H_5 \rightarrow CH_3-CO-CH_2-CO_2C_2H_5 + C_2H_5OH$$

Ethyl 3-oxobutanoate is a colourless liquid (b.p. 181°C) which exists as a mixture of two tautomers (15.2). Like diethyl propanedioate, it is useful in organic synthesis as a result of its forming a sodium salt,

$$CH_3COCH_2CO_2C_2H_5 + C_2H_5O^- Na^+ \rightleftharpoons Na^+$$
$$CH_3CO\bar{C}HCO_2C_2H_5 + C_2H_5OH$$

whose anion is a nucleophilic reagent. For example, its reaction with an alkyl halide gives a derivative which, after hydrolysis with aqueous acid, yields a ketone and, after hydrolysis with concentrated alkali, yields a carboxylic acid:

$$Na^+CH_3CO\bar{C}HCO_2C_2H_5 + R-X \longrightarrow CH_3CO\overset{R}{\underset{|}{C}}HCO_2C_2H_5 + NaX$$

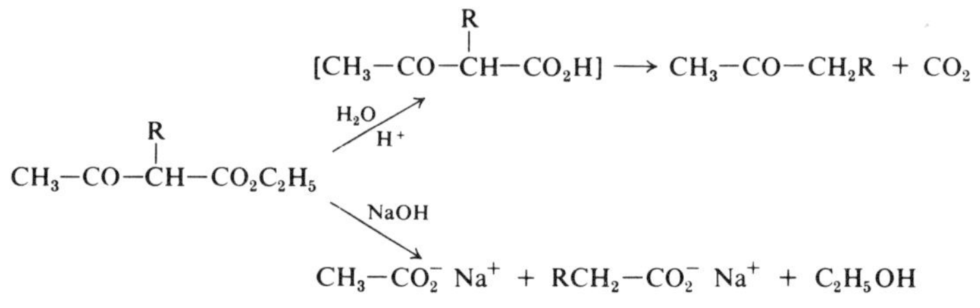

$$[CH_3-CO-\overset{R}{\underset{|}{C}}H-CO_2H] \rightarrow CH_3-CO-CH_2R + CO_2$$

$$CH_3-CO-\overset{R}{\underset{|}{C}}H-CO_2C_2H_5 \xrightarrow{H_2O/H^+} \nearrow$$

$$\xrightarrow{NaOH} CH_3-CO_2^- Na^+ + RCH_2-CO_2^- Na^+ + C_2H_5OH$$

14.3 Acid halides

General formula

$$R-\underset{\underset{O}{\|}}{C}-X \quad (X \text{ is } F, Cl, Br \text{ or } I)$$

Physical properties

The lower acid halides are mobile, colourless liquids with pungent odours. They fume in moist air owing to their ready hydrolysis to the corresponding halogen acid. Some typical members are in Table 14.2.

Table 14.2. Some acid halides

NAME	FORMULA	B.P. (°C)
Ethanoyl chloride	$CH_3-CO-Cl$	52
Ethanoyl bromide	$CH_3-CO-Br$	77
Ethanoyl iodide	CH_3-CO-I	108
Propanoyl chloride	$CH_3CH_2-CO-Cl$	80
Benzoyl chloride	$C_6H_5-CO-Cl$	197

(Methanoyl chloride has not been isolated.)

Preparation

Acid chlorides are prepared by the reaction between a carboxylic acid and phosphorus trichloride, phosphorus pentachloride, or sulphur dichloride oxide:

$$3R-CO-OH + PCl_3 \rightarrow 3R-CO-Cl + H_3PO_3$$

$$R-CO-OH + PCl_5 \rightarrow R-CO-Cl + POCl_3 + HCl$$

$$R-CO-OH + SOCl_2 \rightarrow R-CO-Cl + SO_2 + HCl$$

The choice of reagent is governed by the boiling points of the products. If the acid halide has a very low boiling point (e.g. ethanoyl chloride), PCl_3 is used and the halide is easily separated by distillation from phosphorous acid (which decomposes at 200°C). If the acid halide has a very high boiling point (e.g. benzoyl chloride), PCl_5 can be used; fractional distillation gives first $POCl_3$ (b.p. 107°C) and then the acid chloride. If it has an intermediate boiling point (e.g. propanoyl chloride), $SOCl_2$ is suitable; the gaseous SO_2 and HCl pass off first.

Acid bromides and iodides are made from carboxylic acids and phosphorus tribromide and phosphorus tri-iodide respectively. These phosphorus halides are prepared *in situ* (p. 113).

Chemical properties

The principal reactions of acid halides are with water, ammonia and amines, in which the overall reaction can be described by the equation:

$$R-CO-X + H-Y \rightarrow R-CO-Y + H-X$$

The process is known as **acylation**. **Ethanoylation** is the name reserved for the

DERIVATIVES OF CARBOXYLIC ACIDS

substitution of the **ethanoyl** group, CH_3CO-, for example by ethanoyl chloride or ethanoic anhydride. The introduction of the **benzoyl** group, C_6H_5CO-, by, for example, benzoyl chloride, is known as **benzoylation**.

Benzoylation generally takes place more slowly than ethanoylation.

1. Aliphatic acid halides are readily hydrolysed:

$$R-CO-X + H_2O \rightarrow R-CO-OH + HX$$

e.g.

$$CH_3-CO-Cl + H_2O \rightarrow CH_3-CO-OH + HCl$$

Thus, when the stopper is removed from a bottle of ethanoyl chloride, white fumes are seen, owing to the interaction of hydrogen chloride with the moist air.

Reaction occurs by addition of water to the carbonyl group:

$$R-\underset{\underset{O}{\|}}{C}-X + H_2O \longrightarrow R-\underset{\underset{O^-}{|}}{\overset{\overset{H\overset{+}{O}H}{|}}{C}}-X$$

followed by loss of a proton and the halide ion, X^-:

$$R-\underset{\underset{O^-}{|}}{\overset{\overset{H\overset{+}{O}H}{|}}{C}}-X \longrightarrow R-\underset{\underset{O}{\|}}{\overset{\overset{OH}{|}}{C}} + H^+ + X^-$$

Hydrolysis is far easier than with esters, which are unaffected by water alone although they react with hydroxide ion. The reason is that a halogen substituent $-X$ is more strongly electron-attracting than an alkoxide substituent $-OR'$; the carbon atom of the carbonyl group is therefore more electron-deficient in an acid halide than in an ester and reacts with water, H_2O, a weak nucleophilic reagent, whereas for an ester a much stronger nucleophilic reagent, the hydroxide ion, OH^-, is necessary.

2. Acid halides react with alcohols and phenols to form esters, for example:

$$CH_3-CO-Cl + CH_3-OH \rightarrow CH_3-CO-O-CH_3 + HCl$$
Methyl ethanoate

$$CH_3-CO-Cl + C_6H_5-OH \rightarrow CH_3-CO-O-C_6H_5 + HCl$$
Phenyl ethanoate

$$C_6H_5-CO-Cl + C_6H_5-OH \rightarrow C_6H_5-CO-O-C_6H_5 + HCl$$
Phenyl benzoate

The mechanisms of these reactions are similar to those between acid halides and water, with the oxygen atom of the alcohol, $R-O-H$, or the phenol, C_6H_5-O-H, acting as the nucleophile.

3. Acid halides react with ammonia to form acid amides, for example:

$$CH_3-CO-Cl + NH_3 \rightarrow CH_3-CO-NH_2 + HCl$$
$$\text{Ethanamide}$$

$$NH_3 + HCl \rightarrow \overset{+}{N}H_4\ Cl^-$$

Acid halides react with primary amines in a similar way:

$$CH_3-CO-Cl + CH_3-NH_2 \rightarrow CH_3-CO-NH-CH_3 + HCl$$
$$\textit{N}\text{-Methylethanamide}$$

$$CH_3-NH_2 + HCl \rightarrow CH_3-\overset{+}{N}H_3\ Cl^-$$

The reactions have mechanisms similar to that in hydrolysis, with the nitrogen atom acting as the nucleophile.

4. Acid halides react with *anhydrous* sodium salts of carboxylic acids to form acid anhydrides (14.4), for example:

$$CH_3-CO-Cl + CH_3-CO_2^-\ Na^+ \rightarrow$$
$$CH_3-CO-O-CO-CH_3 + NaCl$$
$$\text{Ethanoic anhydride}$$

The carboxylate ion acts as the nucleophile and the mechanism of the reaction is as described above.

The reactions of acid halides with nucleophilic reagents, such as reactions 1–4, can be compared with those of aldehydes and ketones with nucleophilic reagents on the one hand and with those of primary alkyl halides with nucleophilic reagents on the other.

Like aldehydes and ketones, acid halides react by addition of a nucleophile (symbolised as Y^-):

$$\underset{O}{\overset{\|}{R-C-R'}} \xrightarrow{Y^-} R-\underset{O^-}{\overset{Y}{\underset{|}{\overset{|}{C}}}}-R'$$

$$\underset{O}{\overset{\|}{R-C-X}} \xrightarrow{Y^-} R-\underset{O^-}{\overset{Y}{\underset{|}{\overset{|}{C}}}}-X$$

In the case of the acid halide, the halide ion then breaks off:

$$R-\underset{O^-}{\overset{Y}{\underset{|}{\overset{|}{C}}}}-X \longrightarrow R-\underset{O}{\overset{Y}{\underset{\|}{\overset{|}{C}}}} + X^-$$

and the overall process is a *substitution reaction*. In contrast, in the case of aldehydes and ketones, the group —H or —alkyl does not form a stable anion, so that instead of one of these groups breaking off, a proton is transferred to the intermediate adduct from the solvent:

$$R-\underset{O^-}{\overset{Y}{\underset{|}{\overset{|}{C}}}}-R' + H^+ \longrightarrow R-\underset{OH}{\overset{Y}{\underset{|}{\overset{|}{C}}}}-R'$$

and the overall process is an *addition reaction*.

Primary alkyl halides, like acid halides, undergo substitution with nucleophiles, but in this case an intermediate adduct cannot be formed since carbon cannot form 5 bonds; instead, the approach of the nucleophile to the alkyl group is concerted with the departure of halide ion:

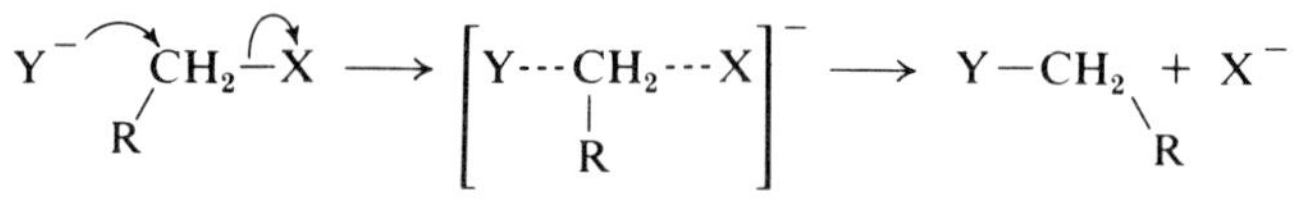

$$Y^- \; CH_2-X \longrightarrow [Y \cdots CH_2 \cdots X]^- \longrightarrow Y-CH_2 + X^-$$
(with R substituent)

5. Acid halides are reduced by lithium aluminium hydride to alcohols:

$$R-\underset{O}{\underset{\|}{C}}-X \xrightarrow{LiAlH_4} R-CH_2-OH + HX$$

This reduction occurs by way of the aldehyde, RCHO, and in another method of reduction, due to Rosenmund, it is possible to obtain the aldehyde as the product. The method is to pass hydrogen into a solution of the acid chloride in the presence of palladium as catalyst suspended on barium sulphate; a mixture of sulphur and quinoline is added as a 'poison' to prevent further reduction:

$$R-\underset{O}{\underset{\|}{C}}-Cl + H_2 \xrightarrow[\text{'poison'}]{\text{Pd as cat.}} R-\underset{O}{\underset{\|}{C}}-H + HCl$$

Uses

Acid halides (normally the chlorides) are used mainly as acylating agents (i.e. to introduce the group RCO—) for the preparation of esters and acid amides.

14.4 Acid anhydrides

General formula

$$R-\underset{O}{\underset{\|}{C}}-O-\underset{O}{\underset{\|}{C}}-R'$$

Physical properties

The lower aliphatic members are mobile, colourless liquids with pungent smells. The simplest aromatic member, benzoic anhydride, is a white solid, m.p. 42°C. Some typical acid anhydrides are in Table 14.3.

Table 14.3. Some acid anhydrides

NAME	FORMULA	B.P. (°C)
Ethanoic anhydride	$(CH_3CO)_2O$	136
Propanoic anhydride	$(CH_3CH_2CO)_2O$	168
Benzoic anhydride	$(C_6H_5CO)_2O$	360

Preparation

By the reaction between an acid chloride and the anhydrous sodium salt of a carboxylic acid, for example:

$$CH_3-CO-Cl + CH_3-CO_2^- Na^+ \rightarrow$$
$$CH_3-CO-O-CO-CH_3 + NaCl$$
$$\text{Ethanoic anhydride}$$

'Mixed' acid anhydrides can be formed, using an acid chloride and a sodium salt with different groups, R and R':

$$R-CO-Cl + R'-CO_2^- Na^+ \rightarrow R-CO-O-CO-R' + NaCl$$

Manufacture of ethanoic anhydride

Ethanoic anhydride is manufactured by passing ethenone through ethanoic acid:

$$CH_2=C=O + CH_3CO_2H \rightarrow CH_3-CO-O-CO-CH_3$$

Ethenone is manufactured by passing propanone vapour over an alloy of nickel and chromium at 700°C:

$$CH_3-CO-CH_3 \rightarrow CH_2=C=O + CH_4$$

Chemical properties

Acid anhydrides react with nucleophilic reagents in the same way as acid halides except that, because the group R'CO—O— in an anhydride RCO—OCOR' is less strongly electron-attracting than the halogen X in an acid halide RCO—X, reaction is slower. Typical examples are:

1. With water, to give the corresponding acid, for example:

$$(CH_3CO)_2O + H_2O \rightarrow 2CH_3CO_2H$$

2. With alcohols or phenols, to give esters, for example:

$$(CH_3CO)_2O + C_2H_5OH \rightarrow CH_3CO_2C_2H_5 + CH_3CO_2H$$

3. With concentrated aqueous ammonia, to give acid amides, for example:

$$(CH_3CO)_2O + NH_3 \rightarrow CH_3CONH_2 + CH_3CO_2H$$
$$\text{Ethanamide}$$

Uses of ethanoic anhydride

The principal use of ethanoic anhydride is in the manufacture of two important plastics, cellulose ethanoate (p. 286) and poly(ethenyl ethanoate) (p. 315). In the laboratory, it is used to make esters (ethanoates) (p. 205) and acid amides (p. 247).

14.5 Acid amides

General formula

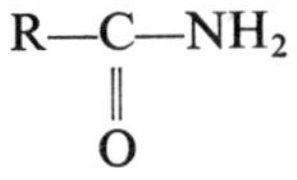

$$R-\underset{\underset{O}{\|}}{C}-NH_2$$

Physical properties

Table 14.4. Some acid amides

NAME	FORMULA	M.P. (°C)	B.P. (°C)
Methanamide	$H-CO-NH_2$	2	193
Ethanamide	$CH_3-CO-NH_2$	82	222
Propanamide	$CH_3CH_2-CO-NH_2$	79	213

With the exception of methanamide, amides are white crystalline solids. The lower members are soluble in water; all amides are soluble in organic solvents. As ordinarily prepared, ethanamide has the characteristic smell of mice, owing to the presence of the methyl derivative $CH_3-CO-NH-CH_3$; pure ethanamide has no smell.

The high melting points and boiling points of amides are due to the formation of hydrogen-bonds between the oxygen atom of one molecule and the amino-hydrogen atom of another:

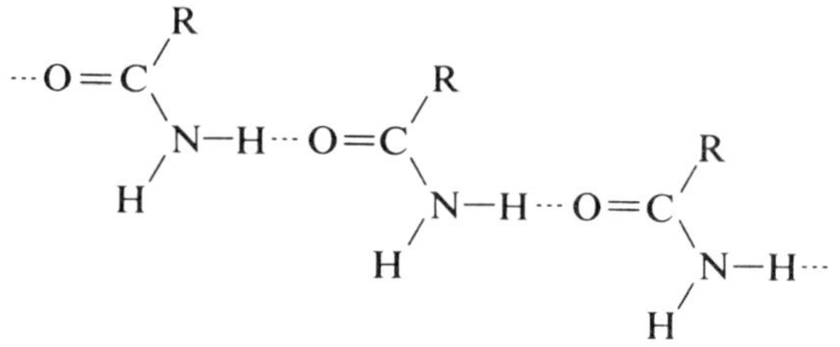

$$\cdots O=C\begin{matrix}R\\ \diagdown\\ N-H\cdots O=C\begin{matrix}R\\ \diagdown\\ N-H\cdots O=C\begin{matrix}R\\ \diagdown\\ N-H\cdots\\ \diagup\\ H\end{matrix}\\ \diagup\\ H\end{matrix}\\ \diagup\\ H\end{matrix}$$

Preparation

1. By the dehydration of the ammonium salt of a carboxylic acid. For example, ethanamide is generally prepared by refluxing a solution of ammonium carbonate in an excess of ethanoic acid for about 4 hours; ammonium ethanoate is first formed and then dehydrated:

$$CH_3CO_2H \xrightarrow{(NH_4)_2CO_3} CH_3CO_2^- NH_4^+ \rightarrow CH_3CONH_2 + H_2O$$

2. By the action of ammonia on esters (14.2), acid halides (14.3) or acid anhydrides (14.4).

Chemical properties

1. Amides are much weaker bases than amines, even though both contain the group $-NH_2$; thus, amides are neutral to litmus and do not dissolve in hydrochloric acid to form salts.

This is because the orbital on the nitrogen atom which contains the unshared pair of electrons interacts with the p orbital of the adjacent carbon atom (Fig. 14.1), giving increased delocalisation in an amide as compared with an amine.

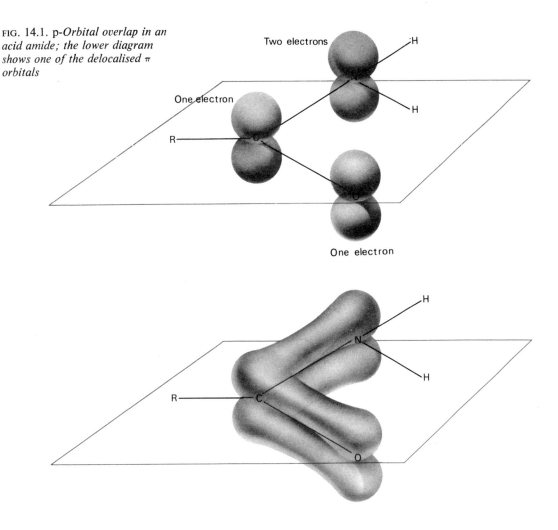

FIG. 14.1. p-*Orbital overlap in an acid amide; the lower diagram shows one of the delocalised π orbitals*

Therefore, relative to an amine, an amide resists reaction with a proton since this process requires the use of the unshared pair of electrons on nitrogen in the formation of the new N—H bond and so results in the loss of the extra bonding.

2. Amides are hydrolysed by heating with a mineral acid (usually hydrochloric acid), for example:

$$\text{CH}_3\text{—CO—NH}_2 \xrightarrow[\text{H}^+]{\text{H}_2\text{O}} \text{CH}_3\text{—CO}_2\text{H} + \text{NH}_4^+$$

They are also hydrolysed by heating with an alkali such as caustic soda, for example:

$$\text{CH}_3\text{—CO—NH}_2 \xrightarrow[\text{OH}^-]{\text{H}_2\text{O}} \text{CH}_3\text{—CO}_2^- + \text{NH}_3$$

The alkali-catalysed reaction occurs by the same mechanism as the hydrolysis of esters and acid halides. The hydroxide ion adds to the carbonyl group of the amide to give an adduct from which amide ion is eliminated; this reacts with water to give ammonia, while the carboxylic acid liberated forms its sodium salt:

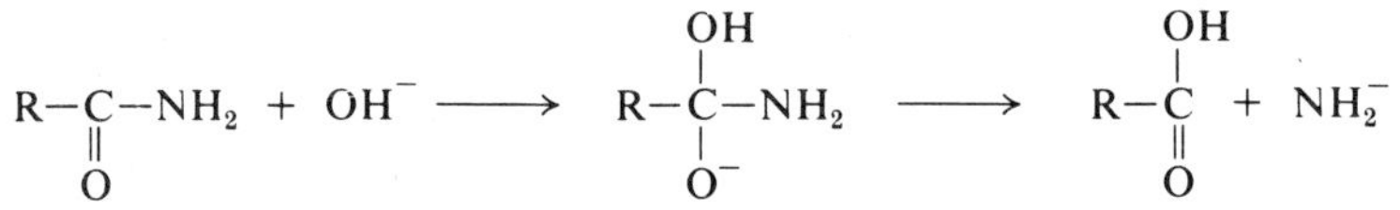

$$R-\underset{\underset{O}{\|}}{C}-NH_2 + OH^- \longrightarrow R-\underset{\underset{O^-}{|}}{\overset{\overset{OH}{|}}{C}}-NH_2 \longrightarrow R-\underset{\underset{O}{\|}}{\overset{\overset{OH}{|}}{C}} + NH_2^-$$

$$NH_2^- + H_2O \longrightarrow NH_3 + OH^-$$

$$R-CO_2H + OH^- \longrightarrow R-CO_2^- + H_2O$$

The group —NH_2 is less strongly electron-attracting than —OR', so that amides are less easily hydrolysed by alkali than esters, RCO_2R', and much less easily than acid chlorides.

The fact that heating an amide with caustic soda liberates ammonia while an amine does not react enables the two types of compound to be readily distinguished. Amides can also be distinguished from ammonium salts in this way, for the latter liberate ammonia in the cold,

$$R-CO_2^- NH_4^+ + NaOH \rightarrow R-CO_2^- Na^+ + NH_3 + H_2O$$

whereas the former do so only on heating.

3. When amides are distilled over phosphorus pentoxide, acid nitriles are formed, for example:

$$CH_3-CO-NH_2 \xrightarrow{P_4O_{10}} CH_3-C\equiv N + H_2O$$
$$\text{Ethanonitrile}$$

4. Like amines, amides react with nitrous acid to liberate nitrogen, for example:

$$CH_3-CO-NH_2 + HNO_2 \rightarrow CH_3-CO_2H + N_2 + H_2O$$

5. Amides react with bromine in a solution of sodium hydroxide to give amines (**Hofmann reaction**), for example:

$$CH_3-CO-NH_2 \xrightarrow{Br_2/NaOH} CH_3-NH_2 + CO_2$$
$$\text{Methylamine}$$

Carbamide

Although carbamic acid, H_2N-CO_2H, has never been isolated, its acid amide, carbamide (urea) ($H_2N-CO-NH_2$), is stable. It can be made by heating ammonium cyanate to dryness (1.1) or by the action of ammonia on carbonyl chloride:

$$Cl-CO-Cl + 4NH_3 \rightarrow H_2N-CO-NH_2 + 2NH_4Cl$$

Physical properties of carbamide

Carbamide is a white, crystalline solid (m.p. 133°C) which is soluble in water and alcohol but insoluble in most organic solvents.

Manufacture of carbamide

By heating excess of ammonia with carbon dioxide at 200°C and 200 atmospheres pressure:

$$CO_2 + 2NH_3 \rightarrow H_2N-CO_2^- \; NH_4^+ \rightarrow H_2N-CO-NH_2 + H_2O$$

Chemical properties of carbamide

1. Carbamide is a monoacidic base; it is a stronger base than simple acid amides like ethanamide but a weaker base than amines. For example, it forms an insoluble nitrate, $H_2N-CO-NH_3^+ \; NO_3^-$.

2. Carbamide is hydrolysed when heated with a solution of alkali:

$$H_2N-CO-NH_2 + 2NaOH \rightarrow Na_2CO_3 + 2NH_3$$

3. Carbamide reacts with nitrous acid to form nitrogen:

$$H_2N-CO-NH_2 + 2HNO_2 \rightarrow CO_2 + 3H_2O + 2N_2$$

4. When carbamide is heated above its melting point, a compound known as **biuret** is formed:

$$2H_2N-CO-NH_2 \rightarrow H_2N-CO-NH-CO-NH_2 + NH_3$$
$$\text{Biuret}$$

If an alkaline solution of biuret is treated with a drop of copper(II) sulphate solution, a violet colour appears. This is known as the **biuret test**, and is a test for all compounds containing the **peptide** linkage, —CO—NH—, such as proteins (18.2).

Uses of carbamide

1. In the manufacture of carbamide-methanal plastics (p. 317).
2. In the manufacture of the barbiturates, a group of drugs with sedative properties of which the parent is barbituric acid; for example:

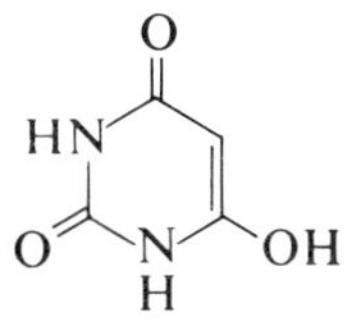

Barbituric acid

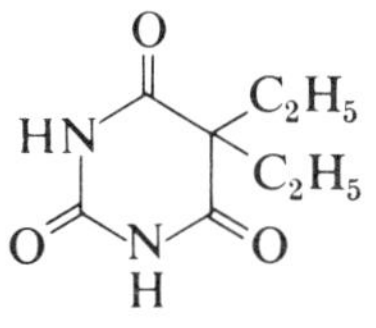

Veronal

3. As a fertiliser. This is a convenient way of providing a source of nitrogen in the soil for utilisation by plants.

14.6 Acid nitriles

The compounds are also known as cyanides.

General formula

$$R-C\equiv N$$

Physical properties

The lowest members (except for hydrogen cyanide) are colourless liquids with pleasant smells. They are fairly soluble in water and very soluble in organic compounds.

Some typical members are in Table 14.5.

Table 14.5. Some acid nitriles

NAME	FORMULA	B.P. (°C)
Methanonitrile (Hydrogen cyanide)	H—CN	26
Ethanonitrile (Methyl cyanide)	CH_3—CN	81
Propanonitrile (Ethyl cyanide)	C_2H_5—CN	97
Benzonitrile (Phenyl cyanide)	C_6H_5—CN	190

Preparation

1. By refluxing an alcoholic solution of an alkyl halide and potassium cyanide, for example:

$$CH_3-I + KCN \rightarrow CH_3-CN + KI$$
$$\text{Ethanonitrile}$$

Reaction takes place by nucleophilic displacement of halide ion by cyanide ion ($S_N 2$ reaction; 9.3).

2. By dehydration of the ammonium salt of a carboxylic acid or an amide with phosphorus pentoxide, for example:

$$CH_3-CO_2^- \; NH_4^+ \xrightarrow{P_4O_{10}} CH_3-CO-NH_2 \xrightarrow{P_4O_{10}} CH_3-C\equiv N$$

Chemical properties

1. Nitriles are hydrolysed, *via* the amide, by refluxing with either mineral acid:

$$CH_3-C\equiv N \xrightarrow[H^+]{H_2O} CH_3-CO-NH_2 \xrightarrow[H^+]{H_2O} CH_3-CO_2H + NH_4^+$$

or alkali:

$$CH_3-C\equiv N \xrightarrow[OH^-]{H_2O} CH_3-CO-NH_2 \xrightarrow[OH^-]{H_2O} CH_3-CO_2^- + NH_3$$

2. Nitriles are reduced to primary amines by sodium and an alcohol, and by lithium aluminium hydride, for example:

$$CH_3-C\equiv N \rightarrow CH_3-CH_2-NH_2$$
$$\text{Ethylamine}$$

$$C_6H_5-C\equiv N \rightarrow C_6H_5-CH_2-NH_2$$
$$\text{(Phenylmethyl)amine}$$

14.7 Isocyano-compounds

Isocyano-compounds (isonitriles) have the structure

$$R-\overset{+}{N}\equiv\overset{-}{C}$$

and are isomers of the nitriles, $R-C\equiv N$. The lower members are unpleasant-smelling liquids.

The compounds are made by refluxing an alcoholic solution of an alkyl halide and silver cyanide:

$$R-X + AgCN \rightarrow R-\overset{+}{N}\equiv\overset{-}{C} + AgX$$

or by heating a primary amine and trichloromethane with an ethanolic solution of alkali (the carbylamine reaction; 9.4):

$$R-NH_2 + CHCl_3 + 3KOH \rightarrow R-\overset{+}{N}\equiv\overset{-}{C} + 3KCl + 3H_2O$$

They are stable to alkali (unlike nitriles) but are readily hydrolysed by mineral acid to primary amines:

$$R-\overset{+}{N}\equiv\overset{-}{C} + 2H_2O \xrightarrow{H^+} R-NH_2 + HCO_2H$$

They can thus be distinguished from their isomers, the nitriles, which are hydrolysed to the ammonium salt of the corresponding carboxylic acid.

14.8 Practical work

Small-scale preparation of an ester

In this example, ethyl ethanoate is prepared from ethanoic acid and ethanol.

Set up the apparatus (Fig. 14.2), with 5 cm³ of ethanol and 5 cm³ of *concentrated* sulphuric acid in the flask (the acid must be added slowly, with gentle shaking), and a mixture of 5 cm³ of ethanol and ethanoic acid in the dropping funnel.

FIG. 14.2. *Preparation of ethyl ethanoate*

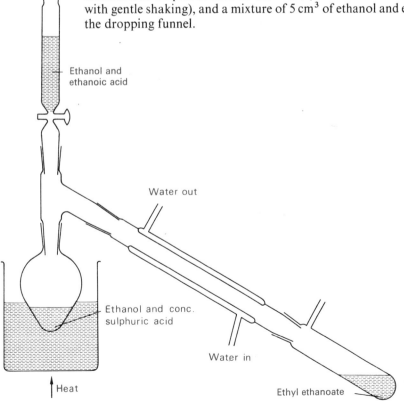

Heat the oil-bath to 140°C and add the mixture from the funnel at the same rate as the ester distils over.

When distillation stops, transfer the distillate to a separating funnel and add 5 cm^3 of a 30 per cent solution of sodium carbonate. Shake, removing the stopper from time to time to relieve the pressure due to carbon dioxide.

Remove the lower (aqueous) layer and add a solution of 5 g of calcium chloride in 5 cm^3 of water to the separating funnel and shake the mixture (to remove excess of ethanol). Remove the lower layer again.

Pour the ester into a test-tube and add 2 or 3 pieces of anhydrous calcium chloride. Stopper the tube and shake it. Decant the clear liquid into a flask and distil it (cf. Fig. 2.2), collecting the fraction boiling between 75 and 79°C.

Preparation of phenyl benzoate

Details of the preparation of phenyl benzoate are given on p. 155.

Reactions of esters

1. *Saponification of an ester*. Details of the saponification of an ester, ethyl benzoate, are given on p. 196.

2. *Saponification of a fat*. Boil for 20 minutes under reflux about 1 g of lard or olive oil (or any fat), 1 g of potassium hydroxide and 10 cm^3 of ethanol, until no more oil is observed when a few drops of the mixture are added to water. Distil the reaction mixture to remove the ethanol, and dissolve the residue in 15 cm^3 of hot water.

(a) To 5 cm^3 of the solution of the residue, add slowly a saturated solution of sodium chloride. Filter the precipitate and test the soap for its lathering properties.

(b) To another 5 cm^3 of the solution, add 5 cm^3 of water. Shake well.

Reactions of ethanoyl chloride and benzoyl chloride

Take great care when handling these compounds.

1. Carry out the following reactions with both ethanoyl chloride and benzoyl chloride and compare their reactivities:
    (a) To 5 drops of the acid chloride in a test-tube, add 5 drops of water. Test the gas evolved.
    (b) To 5 drops of ethanol, add 2 drops of the acid chloride. Warm the mixture *if necessary*. Dilute the solution with 5 drops of water and note the smell of the organic product.

2. (a) To 5 drops of phenylamine, add 2 drops of ethanoyl chloride (**CARE**). A vigorous reaction takes place, and a white precipitate of *N*-phenylethanamide is formed. *N*-Phenylethanamide can be recrystallised from hot water. M.p. 114°C.

(b) To 3 drops of phenylamine, add 5 cm^3 of 10 per cent sodium hydroxide solution. Add 5 drops of benzoyl chloride. Shake. Filter the residue of *N*-phenylbenzamide and wash with water. It is recrystallised from hot ethanol. M.p. 163°C.

Reactions of ethanoic anhydride

1. To 5 drops of ethanoic anhydride in a test-tube, add 5 drops of water.

Shake the mixture and note the rate of reaction compared with that between ethanoyl chloride and water.

2. *Preparation of* N-*phenylethanamide and* N-*phenylbenzamide.* For details, see pp. 258 and 257 respectively.

Preparation of cellulose ethanoate (p. 289).

Reactions of ethanamide

1. To about 0·1 g of ethanamide in a test-tube, add 2 cm^3 of dilute sodium hydroxide solution. Boil, and test the vapour evolved with moist red litmus paper.

2. To about 0·1 g of ethanamide in a test-tube, add 3 drops of bromine (**CARE**). Add 1–2 cm^3 of dilute sodium hydroxide solution; cork the test-tube and shake for about a minute.

Remove the cork, add one pellet of sodium hydroxide and boil the solution gently. Test the gas by (a) its smell, (b) its action on moist red litmus paper.

Reactions of ethanonitrile

1. To 5 drops of ethanonitrile in a test-tube, add 5 drops of dilute sodium hydroxide solution. Warm gently and test the gas evolved with moist red litmus paper.

2. To 5 drops of ethanonitrile in a test-tube, add about 0·1 g of zinc dust followed by 10 drops of *concentrated* hydrochloric acid. When the effervescence due to the evolution of hydrogen has more or less stopped, add sodium hydroxide solution until the mixture is alkaline. Warm the mixture and test the gas evolved with moist red litmus paper.

Reactions of carbamide

1. Carbamide is a monoacidic base. To 1 cm^3 of hot water in a test-tube, add some crystals of carbamide until the solution is saturated. Decant this solution into a clean test-tube and add *concentrated* nitric acid dropwise. A white precipitate of the nitrate is formed.

2. *Hydrolysis of carbamide.* (a) To 0·1 g of carbamide in a test-tube, add 1 cm^3 of 2M sodium hydroxide solution and boil the mixture. Test the vapour evolved with moist red litmus paper and note the smell of the gas evolved.

This reaction shows that the amide group, —CONH$_2$, is present.

(b) Action of the enzyme, urease. Experiment 3(a) in Section 18.4.

3. Dissolve a few crystals of sodium nitrite in water in a test-tube and cool the solution in a beaker of ice. Add 1 cm^3 of dilute hydrochloric acid. (There may be some effervescence due to reaction between sodium nitrite and the acid.) Add a few crystals of carbamide. Observe whether any gas is evolved.

4. Heat about 0·1 g of carbamide in a test-tube until it melts. Test gases evolved with moist red litmus paper.

Continue to heat the residue gently for a further 3 minutes. Cool it and add 10 drops of water followed by 2 drops of a dilute solution of copper(II) sulphate followed by dilute sodium hydroxide solution until the mixture is alkaline.

A violet coloration confirms that the residue contains a compound which

DERIVATIVES OF CARBOXYLIC ACIDS

has a peptide link, —CO—NH—. This compound is called biuret and the test is named after it, the **biuret test**.

14.9 Questions

1 Name and give the structural formulae of the aliphatic acids and esters which have an empirical formula C_2H_4O and a molecular weight of 88.
   Outline the chemical tests that you would apply to enable you to distinguish between each isomer. (L)

2 Give **three** general methods for the preparation of esters. State which of them you would select for the preparation of phenyl benzoate and outline the procedure you would adopt.
   How would you prepare a pure water-free specimen of ethanol from ethyl benzoate? Briefly indicate the necessary conditions for each step.
   Give a brief account of the constitution of naturally occurring fats, and show what useful products may be derived from them. (JMB)

3 Describe the preparation from ethanol of pure samples of ethanoic acid and of ethyl ethanoate.
   How, and under what conditions, does ethanoic acid react with (a) thionyl chloride, (b) methylamine, and (c) soda lime? (C(T))

4 0·6 g of ethanoic acid were mixed with an equimolecular amount of ethanol and the mixture was sealed in a small glass tube which was immersed in a bath of boiling water. When equilibrium had been reached the glass tube was removed from the water bath and broken open under the surface of about 30 cm³ of water in a conical flask. The resulting mixture was titrated with 0·1 M sodium hydroxide using phenolphthalein as indicator and required 33·30 cm³ of the alkali.
   Calculate the equilibrium constant for the esterification of ethanol by ethanoic acid.
   Describe in outline how you would modify the above reaction conditions for the preparation of a pure sample of ethyl ethanoate. (AEB)

5 What do you understand by the equilibrium constant of a reversible reaction? What do you need to know if you are to predict correctly the qualitative effect of temperature on this constant? State and explain the principle involved.
   When one gram-mole each of ethanoic acid and ethanol are mixed together at 20°C, two-thirds of a gram-mole of ethyl ethanoate is formed. How many gram-moles of ethyl ethanoate would be present eventually if one gram-mole of water was added to the above mixture? (O)

6 Give the structural formulae of the functional groups characteristic of (a) carboxylic acids, (b) acid chlorides, (c) acid anhydrides, (d) primary amines, (e) amides.
   Suggest a scheme whereby ethanoic acid might be obtained from methane as starting material. Indicate by equations how ethanoic acid might be converted into the corresponding acid chloride, acid anhydride, acid amide and ethyl ester respectively. (AEB)

7 How would you prepare ethanoyl chloride from ethanoic acid?
   Name the products and write the equations for the reactions of ethanoyl chloride with: (a) sodium hydroxide, (b) ammonia, (c) ethanol, (d) phenol, (e) anhydrous sodium ethanoate. (C(N, T))

8 Describe, with essential practical details, the preparation of ethanamide from ammonium ethanoate.
   Describe the reactions which occur between ethanamide and (a) phosphorus pentoxide, (b) nitrous acid, and (c) bromine and sodium hydroxide solution.

9 Starting with ethanoic acid as the only organic compound, outline the preparation of a sample of ethanoic anhydride.
   Ethanoic anhydride (y g) was shaken with water and the resulting solution was

found to neutralise 40 cm³ of molar sodium hydroxide solution. The same weight (y g) of ethanoic anhydride was added to an excess of cold phenylamine. Calculate the weight of N-phenylethanamide formed. (O and C)

10 For each of the following pairs of substances describe **one** simple chemical test which would serve to distinguish between them:

(a) ethanoyl chloride and ethanoic anhydride;

(b) ethanal and propanone;

(c) methanoic acid and ethanoic acid;

(d) ethanamide and phenylamine.

In each case state the conditions under which the reaction occurs and give the equation for it. (L)

11 Compare the hydrolysis of (a) ethyl benzoate, (b) benzoyl chloride, (c) benzamide. Describe how you would isolate a pure specimen of the common product of hydrolysis and explain how it may be reconverted into (a), (b) and (c). (W)

12 By means of equations, supplemented by brief notes on relative speeds of reaction, indicate what reactions occur between ammonia and the following compounds: (a) chloromethane, (b) ethanoyl chloride, (c) chlorobenzene, (d) ethanal, (e) ethanoic acid, (f) methyl propanoate.

By what *simple* chemical means could you quickly distinguish between the product formed from (b) and that formed from (e)? (JMB)

13 In this problem the molecular formula of some of the compounds is given in brackets.

(i) A substance X (C_4H_8ONBr) yielded ammonia when boiled under reflux with an aqueous solution of sodium hydroxide.

(ii) A portion of the resulting solution was acidified, treated with chlorine, and then shaken with trichloromethane. Two layers were formed, the lower of which was orange in colour.

(iii) The remainder of the solution from (i) was evaporated to a solid substance which on acidification and distillation yielded an acidic substance Y ($C_4H_8O_3$).

(iv) On oxidation Y yielded Z ($C_4H_6O_3$). Z gave a precipitate with 2,4-dinitrophenylhydrazine, but not with Fehling's solution.

Suggest a possible structure for X, Y and Z, and describe what happens in the reactions outlined above. (L(Nuffield)(S))

14 On analysis a compound X of molecular weight 59 was found to contain 40·67 per cent carbon, 8·5 per cent hydrogen, 23·72 per cent nitrogen, the remainder being oxygen. Derive the formula of X, write a structural formula and name the compound.

Stating essential conditions of reaction, describe how the compound you name for X reacts with (a) phosphorus pentoxide, (b) sodium hydroxide, (c) nitrous acid, (d) bromine and potassium hydroxide. (AEB)

15 The reaction between ethanoic acid and ethanol is catalysed by hydrogen ions. Describe how you would attempt to prove this.

16 The percentage composition of an aliphatic compound was found to be C = 20·0, H = 6·7, O = 26·7, N = 46·6. A solution of 0·25 g of the compound in 20 g of water froze at −0·39°C. What was the compound and what would be the action of heat on it?

(Molecular depression constant for water is 18·6° per 100 g.)

17 How, and under what conditions, does ethanamide react with (i) dilute hydrochloric acid, (ii) bromine and sodium hydroxide, (iii) phosphorus pentoxide, (iv) nitrous acid (acidified sodium nitrite solution)?

Give balanced equations and essential conditions for the reactions by which ethylammonium chloride (ethylamine hydrochloride) could be prepared from ethanamide.

What volume of 0·50M (0·50N) hydrochloric acid would be required to react completely with the gas evolved when 1·00 g of ethanamide is boiled with an excess of sodium hydroxide? (C(T))

18 What general methods are available for the preparation of (a) ethers, (b) acid anhydrides, (c) esters?

Compare the structures of diethyl ether, ethyl ethanoate, and ethanoic anhydride, and their reactions and methods of preparation, so as to bring out the similarities in structure and the effect of the modifications in structure.

19 What is the importance in organic chemistry of derivatives of hydrogen cyanide?

20 A compound has the structural formula $CH_3COCH_2CH=CHCH_2CN$. How would you expect this compound to react with (a) sodium and ethanol, (b) bromine, (c) sodium hydroxide, and (d) phosphorus pentachloride?

21 A colourless liquid, A, contains 58·54 per cent of carbon, 7·32 per cent of hydrogen and 34·14 per cent of nitrogen. When boiled with hydrochloric acid, A produced a compound B. When a pure sample of B was fused with soda-lime, a colourless inflammable gas was formed. When A was treated with dilute sulphuric acid and zinc, a colourless liquid, C, was formed, which reacted with iron(III) chloride to give a brown precipitate.

Identify A, B, and C and explain, giving equations, the reactions which occurred above.

Describe two reactions by which A could be made.

22 An organic acid P, on treatment with phosphorus pentachloride yielded a substance Q, which contained 78·0 per cent of chlorine and had a molecular weight of 182. Q when heated with dilute sodium carbonate solution yielded a substance R, which when crystallised and heated with soda lime produced a volatile liquid S. The product when heated with phenylamine and sodium hydroxide yielded a substance possessing a highly offensive smell. Derive a formula for P and explain the course of the reactions described. (L(X,S))

23 2·000 g of a neutral aliphatic compound X, containing carbon, hydrogen and oxygen only, gave on combustion 3·617 g of carbon dioxide and 1·233 g of water. If the vapour density of X is 73, calculate its molecular formula.

When X was refluxed with aqueous sodium hydroxide and then distilled, the distillate contained only one organic compound and this gave a positive response to the iodoform test. The solution left in the flask was evaporated to give a solid, which reacted with hot concentrated sulphuric acid to give a mixture of two gases, one of which turned lime-water milky whilst the other burnt with a blue flame.

Identify X, give its structural formula and explain fully the reactions which have been used to find out what it is. (O)

24 Describe briefly the preparation of ethanonitrile starting from (a) methanol, and (b) ethanol.

Ethanonitrile (x g) was boiled under reflux with sodium hydroxide solution and the ammonia which was evolved was passed into molar hydrochloric acid solution (50 cm³). The excess of acid required 26 cm³ of molar sodium hydroxide solution for neutralisation. Calculate x. (O and C)

25 Outline **two** methods of preparing carbamide. What is the historical importance of its first synthesis?

How does carbamide react with (a) nitrous acid solution (sodium nitrite and hydrochloric acid), (b) concentrated nitric acid, (c) sodium hydroxide solution, (d) a dilute alkaline solution of bromine? (O and C)

26 When 0·357 g of a colourless liquid, A, which was almost insoluble in water, was burnt, 0·616 g of carbon dioxide and 0·189 g of water were obtained. When

A was warmed with an excess of aqueous sodium hydroxide and the resulting solid obtained on evaporation was strongly heated, a gas *B* was evolved. When 10 cm³ of *B* was exploded with 30 cm³ of oxygen (an excess) and the resulting gas, after cooling, was shaken with caustic potash solution there was a diminution in the volume of the gas by 10 cm³.

When *A* was warmed with aqueous ammonia, under suitable conditions a crystalline solid, *C*, was obtained, which contained 23·75 per cent of nitrogen.

Identify *A*, *B* and *C* and explain the reactions that take place.

What would happen if *A* was heated with calcium hydroxide in equimolecular proportions? (L)

27 A compound *A* has the molecular formula $C_6H_{10}O_4$. It is a sweet-smelling liquid, sparingly soluble in water.

   (a) When *A* is boiled under reflux with aqueous sodium hydroxide, a clear solution, *B*, is obtained.
   (b) When solution *B* is distilled, the distillate, *C*, turns warm acidified sodium dichromate solution green.
   (c) When a portion of the residual liquid from (b) is neutralised with hydrochloric acid, and calcium chloride solution added, a white precipitate, *D*, is observed.
   (d) When the remaining liquid from (b) is acidified and then warmed with potassium permanganate solution carbon dioxide is evolved and a colourless solution results.
   (e) When *A* is shaken with 0·880 ammonia solution, a white solid, *E*, is formed.

   Suggest the most probable structures for *A* and explain the reactions described.
   How would the result of the reaction of iodine and alkali on distillate *C* affect the decision on the structure assigned to *A*? (S(S))

28 A neutral compound, *A*, has an empirical formula $C_8H_7O_2$. When 0·4824 g of *A* is dissolved in 40·20 g of benzene, the boiling point is raised by 0·119°C (the boiling point constant, *K*, for benzene = 2·67°C per 1000 g of benzene). Treatment of *A* with an excess of boiling aqueous potash gives a clear solution which when saturated with carbon dioxide affords a liquid turning to a low-melting solid, *B*, molecular formula C_6H_6O. *B* gives a purple colour with iron(III) chloride [ferric chloride] solution, and a white precipitate when treated with bromine water.

   Acidification (with mineral acid) of the aqueous solution after removal of *B* yields a solid acid, *C*, empirical formula $C_2H_3O_2$, and this gives a neutral compound, *D*, when it is boiled with methanol in the presence of a trace of acid. The empirical formula of *D* is $C_3H_5O_2$ and the vapour density, 73. Also when *C* is heated it loses water to give *E*, a neutral substance with molecular formula $C_4H_4O_3$.

   Deduce the structure of *A* and explain the formation of the substances *B*, *C*, *D* and *E*. (JMB(S))

29 Give an account of the general properties associated with the unsaturated groups present in each of the following aliphatic types:

   (a) R—CH=CH—R
   (b) $\begin{array}{c} R \\ R \end{array} \!\!\! C\!\!=\!\!O$
   (c) R—C≡C—H
   (d) R—C≡N

   Call attention to any points of comparison or contrast which you consider of special interest.

30 When a compound *P* was heated with concentrated aqueous sodium hydroxide solution and the reaction product acidified, equimolar amounts of *Q*, containing C, 77·8 per cent; H, 7·4 per cent; and *R*, containing C, 68·8 per cent; H, 4·9 per cent were obtained. Mild oxidation of *Q* gave *P*, but more vigorous oxidation gave *R*. When 25 cm³ of a solution of *R*, containing 11·0 g/dm³, were titrated with 0·10N sodium hydroxide, 22·6 cm³ were required for neutralisation (phenolphthalein). When *R* was heated with soda-lime it gave a hydrocarbon *S*, containing C, 92·4 per cent; H, 7·6 per cent.

Identify P, Q, R and S and give equations for the reactions involved in the above sequence.

Describe how you would make a pure crystalline derivative from P. (C(S))

31 Three isomeric compounds, A, B and C, have the molecular formula $C_8H_8O_2$.

(a) When heated for a long time with soda-lime A gave benzene, while both B and C gave methylbenzene.

(b) A was a neutral compound, whereas both B and C were monobasic acids.

(c) B and C were both readily chlorinated in sunlight; B gave a compound containing three chlorine atoms per molecule which was easily hydrolysed to a dibasic acid D which in turn gave an anhydride on heating. When D was heated with soda-lime it gave benzene. The chlorinated product from C contained two chlorine atoms per molecule and was a strong monobasic acid.

Identify A, B, C and D, write equations for the reactions, and indicate how you would prepare A from benzoic acid. (C(S))

32 Give examples of (a) addition reactions, and (b) condensation reactions, of the carbonyl group $>C=O$. Account for the difference in reactivity of the carbonyl group in (a) acids, (b) aldehydes, (c) ketones, (d) esters. (L(S))

33 A neutral white solid, P (C, 40.6 per cent; H, 5.1 per cent), was refluxed with an excess of aqueous sodium hydroxide and the reaction mixture was then distilled to give a distillate, Q, and a residue, R.

No reaction was observed when Q was treated with iodine and alkali, but Q was oxidised by acid dichromate solution to give a compound S (C, 40.0 per cent; H, 6.67 per cent) which reduced silver nitrate to silver and mercury(II) chloride to mercury(I) chloride and mercury.

Careful acidification of R, followed by suitable treatment, gave an anhydrous crystalline compound, T, 0.90 g of which required 20.0 cm³ of 1M (1N) sodium hydroxide for neutralisation or 40.0 cm³ of 0.1M (0.5N) acidified potassium permanganate for oxidation.

Compound P reacted with ethanolic ammonia to give U, empirical formula CH_2NO, and Q.

Identify compounds P to U, and write equations for all the reactions. (C(S))

34 Describe in outline the preparation of diethyl propanedioate from ethanoic acid. Starting from diethyl propanedioate how would you prepare:

(a) butanoic acid;

(b) 2-methylpropanoic acid.

35 A neutral solid A, $C_2H_3O_2$, slowly dissolved in water to give a solution which became progressively more acidic as the substance went into solution. After boiling A with two molecular proportions of sodium hydroxide, evaporation of the solution gave B, CO_2Na, which on heating with soda-lime evolved hydrogen. On treatment with ammonia, A gave a white infusible solid C, CH_2ON, which gave the biuret reaction. All the formulae given above are empirical.

Elucidate the above reactions.

36 Discuss from a practical point of view the synthesis and hydrolysis of esters. What explanation can you offer for the following observations:

(a) Esterification and hydrolysis are equilibrium processes,

$$RCOOH + R'OH \rightleftharpoons RCOOR' + H_2O,$$

yet in some reactions, nearly quantitative yields of ester may be obtained.

(b) Most esters show a molar freezing point depression of two when dissolved in concentrated sulphuric acid.

(c) The methyl ester of 2,4,6-trimethylbenzoic acid gives a fourfold freezing point depression in sulphuric acid? (C Schol.)

37 Three alternative structures were proposed for a compound A, $C_{16}H_{14}O_4$, namely

(i) $\text{Ph}-\text{O}-\text{CO}-\text{CH}_2-\text{CH}_2-\text{CO}-\text{O}-\text{Ph}$

(ii) $\text{Ph}-\text{O}-\text{CO}-\text{CH}_2-\text{CH}_2-\text{O}-\text{CO}-\text{Ph}$

(iii) $\text{Ph}-\text{CO}-\text{O}-\text{CH}_2-\text{CH}_2-\text{O}-\text{CO}-\text{Ph}$

A dissolved completely when it was boiled with aqueous sodium hydroxide and it was found possible to isolate the organic products liberated when the solution was acidified, in almost quantitative yields.

Given bromine water and either standard acid or alkali, describe the investigations you might make to assign the correct structure to A. (O Schol.)

38 A neutral oil A ($C_{15}H_{26}O_6$) can be extracted from butter. It is immiscible with water but it dissolves on boiling with aqueous sodium hydroxide. From the resulting solution a syrupy liquid B ($C_3H_8O_3$) and a salt C ($C_4H_7O_2Na$) can be isolated. Phosphorus pentachloride converts B into a liquid which contains 24·41 per cent carbon, 3·39 per cent hydrogen and 72·20 per cent chlorine.

Suggest formulae for the compounds A, B and C and if alternatives are possible state whether it is easy to choose between them in assigning a structure to A. Give your reasons. (O Schol.)

39 A neutral compound A, $C_{10}H_{18}O_4$, dissolved slowly in boiling sodium hydroxide solution to give ethanol and, after acidification, a solid B, $C_6H_{10}O_4$. The solid B decomposed smoothly at its melting point (118°) to give a liquid acid C, $C_5H_{10}O_2$. The acid C was optically inactive, but could be resolved with an optically active base. Compound B could not be resolved.

Suggest structures for A, B and C and explain the reactions involved.

(O Schol.)

Chapter 15 Isomerism

15.1 Introduction

Isomerism is said to occur when two or more compounds have the same molecular formula, but have different physical or chemical properties. The compounds that exhibit isomerism are said to be **isomers**. There are two principal subdivisions of isomerism—**structural isomerism** and **stereoisomerism**.

Structural isomerism occurs when two or more compounds have the same molecular formula but different structural formulae; that is, at least one atom is bonded to a different atom in one isomer as compared with another.

Stereoisomerism occurs when two or more compounds have the same molecular formula and the same structural formula, corresponding atoms being linked to the same atoms, but different spatial arrangements of their bonds.

15.2 Structural isomerism

An outline of structural isomerism was given earlier (1.7) and many examples of isomerism have been met in the previous chapters. We shall here describe briefly the convenient sub-divisions of the subject.

(a) **Chain isomerism**, concerned with the arrangement of the carbon atoms in the molecule. For example, butane and 2-methylpropane are chain isomers:

$$CH_3-CH_2-CH_2-CH_3 \qquad CH_3-\underset{\underset{CH_3}{|}}{CH}-CH_3$$
Butane   2-Methylpropane

(b) **Position isomerism**, exhibited by isomers in the same homologous series which have the same carbon skeleton but differ in the position of the functional group, for example:

$$CH_3-CH_2-\underset{\underset{OH}{|}}{CH_2} \qquad CH_3-\underset{\underset{OH}{|}}{CH}-CH_3$$
Propan-1-ol   Propan-2-ol

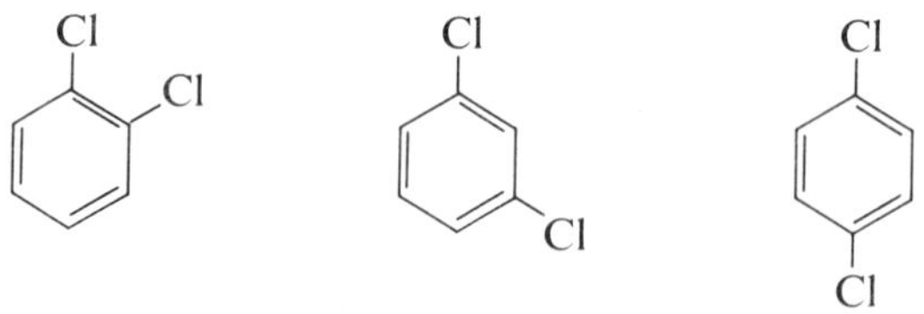

1,2-Dichlorobenzene   1,3-Dichlorobenzene   1,4-Dichlorobenzene

(c) **Functional group isomerism**, exhibited by isomers which have the same molecular formula but contain different functional groups, for example:

ISOMERISM

$$CH_3-CH_2-CH=O \qquad CH_3-\underset{\underset{O}{\|}}{C}-CH_3$$

Propanal   Propanone

(d) **Tautomerism**, exhibited by isomers which are in dynamic equilibrium. An example is ethyl 3-oxobutanoate; the two isomers rapidly interconvert, the equilibrium constant being about 0·11 (i.e. 90 per cent keto form and 10 per cent enol form):

$$CH_3-\underset{\underset{O}{\|}}{C}-CH_2-CO_2C_2H_5 \rightleftharpoons CH_3-\underset{\underset{OH}{|}}{C}=CH-CO_2C_2H_5$$

Keto form   Enol form

If a reagent is added which reacts with the carbonyl group (e.g. hydroxylamine), it removes the keto isomer and more of the enol form is then converted into the keto form to re-establish the equilibrium; eventually the entire mixture reacts with the reagent. The opposite occurs if a reagent is added which reacts with the enol form, such as bromine:

$$CH_3-\underset{\underset{OH}{|}}{C}=CH-CO_2C_2H_5 + Br_2 \rightarrow CH_3-\underset{\underset{O}{\|}}{C}-\underset{\underset{Br}{|}}{C}H-CO_2C_2H_5 + HBr$$

Eventually the entire mixture is converted into the bromo-compound.

15.3 Introduction to stereoisomerism

Stereoisomerism occurs when two or more compounds have both the same molecular formula and the same structural formula but differ in the spatial arrangement of the atoms. There are two subdivisions: **geometrical isomerism** and **optical isomerism**.

15.4 Geometrical isomerism

The essential requirement for the existence of geometrical isomers is the prevention of rotation around at least one bond in the compound. The commonest situation of this sort is when a carbon–carbon double bond is present; there can then be two geometrical isomers:

$$\underset{Y}{\overset{X}{\diagdown}}C=C\underset{Y}{\overset{X}{\diagup}} \qquad \underset{Y}{\overset{X}{\diagdown}}C=C\underset{X}{\overset{Y}{\diagup}}$$

cis isomer   *trans* isomer

Note that these isomers have the same structural formula, for in both isomers each carbon atom is attached to one group X and one group Y; however, the relative positions in space of these groups are different for the two isomers. There is a third compound with the same molecular formula ($C_2X_2Y_2$):

$$\underset{X}{\overset{X}{\diagdown}}C=C\underset{Y}{\overset{Y}{\diagup}}$$

but this is a structural isomer of the first two.

Typical examples of geometrical isomers are:

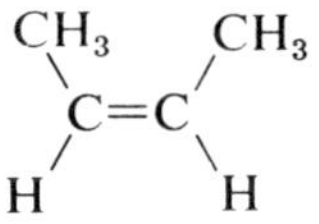

$$\underset{\text{cis-But-2-ene}}{\overset{CH_3 \quad CH_3}{\underset{H \quad\quad H}{C=C}}} \qquad \underset{\text{trans-But-2-ene}}{\overset{CH_3 \quad H}{\underset{H \quad\quad CH_3}{C=C}}}$$

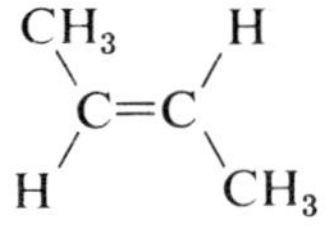

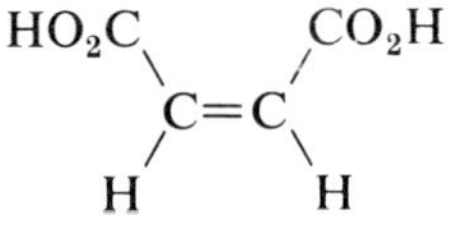

$$\underset{\substack{\text{cis-Butene-}\\\text{dioic acid}}}{\overset{HO_2C \quad CO_2H}{\underset{H \quad\quad H}{C=C}}} \qquad \underset{\substack{\text{trans-Butene-}\\\text{dioic acid}}}{\overset{HO_2C \quad H}{\underset{H \quad\quad CO_2H}{C=C}}}$$

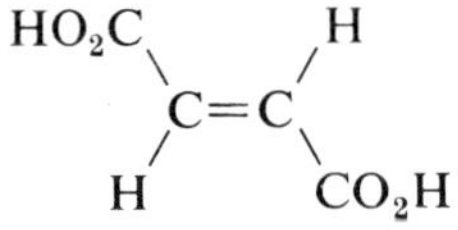

Oximes, which have a C=N bond, also exhibit geometrical isomerism, for example:

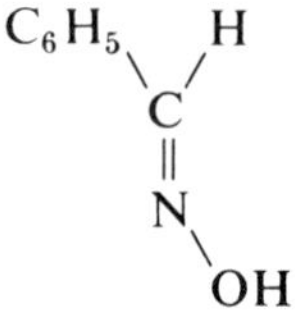

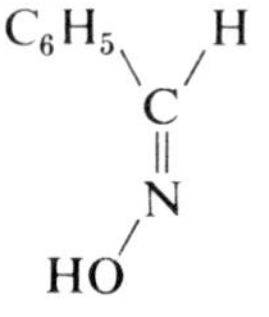

$$\overset{C_6H_5 \quad H}{\underset{\underset{OH}{N}}{C}} \qquad \overset{C_6H_5 \quad H}{\underset{\underset{HO}{N}}{C}}$$

Rotation is also prevented by the presence of a ring of atoms; for example, 1,2-dimethylcyclobutane exists in *cis* and *trans* forms:

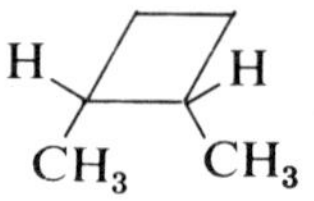

cis isomer 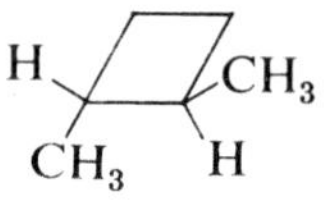   *trans* isomer

Geometrical isomers have different physical and chemical properties. For example, *cis*-butenedioic acid readily forms a cyclic anhydride on being heated:

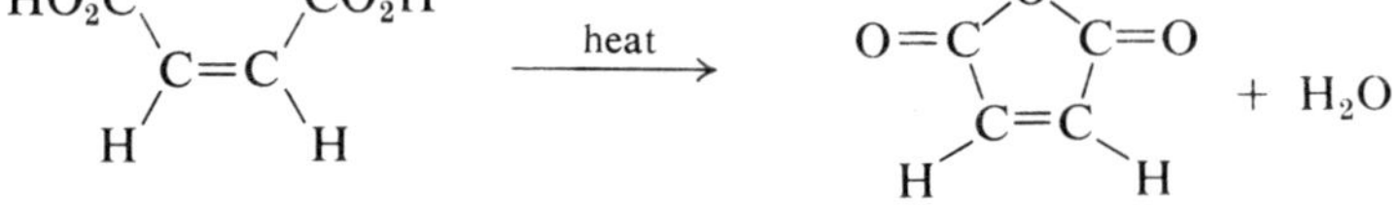

$$\overset{HO_2C \quad CO_2H}{\underset{H \quad\quad H}{C=C}} \quad \xrightarrow{\text{heat}} \quad \text{(cyclic anhydride)} + H_2O$$

but the *trans*-isomer is unable to do so.

15.5 Optical isomerism

When two compounds have the same molecular and structural formulae but one is not superimposable upon the other, they are described as **optical isomers**. The reason is that they differ in their optical properties.

For example, if a compound contains a carbon atom bonded to four different atoms or groups (e.g. 2-hydroxypropanoic acid, $CH_3CH(OH)CO_2H$; p. 191), the construction of models shows that it can

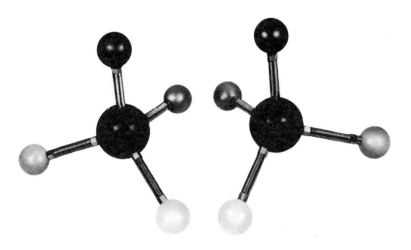

Plate 15.1. Molecular models of stereoisomers. The carbon atom is surrounded by four different groups of atoms. Each group is represented, for clarity, by a single coloured ball. The two stereoisomers are mirror images of one another which cannot be superimposed

exist in two forms which are not superimposable (Plate 15.1). The two forms are related as object and mirror image; if a mirror were placed between the models in the photograph, perpendicular to the page, then the reflection of the right-hand form in the mirror would be seen as identical with the left-hand form, and *vice-versa*. The two forms differ in their behaviour towards polarised light.

Introduction to polarised light

Light is a form of electromagnetic radiation, like radio waves and X-rays. It is a wave-motion, the wavelength of each wave being about 5×10^{-7} m, the exact value being characteristic of the colour of the light. In free space all these waves, irrespective of colour, travel with a velocity of 3×10^8 m per sec. When they impinge upon a transparent homogeneous material, however, they are slowed down in a constant ratio, so that:

$$\frac{\text{Velocity in free space}}{\text{Velocity in medium}} = \text{a constant,}$$

known as the 'refractive index', and usually denoted by the symbol μ; its value depends upon both the medium and the wavelength of the light.

Those who find difficulty in appreciating the properties of electromagnetic radiation may be helped by likening light waves to the material waves which can be made by regularly vibrating one end of a long string—which we shall assume to be stretched horizontally in what follows—although they must not push the analogy too far, nor expect an analogy to *prove* anything. Just as it is possible for the end of the string to oscillate vertically up and down, or horizontally to and fro, or even in a combination of these directions, so also the direction of oscillation of the light waves can be in any direction which lies at right angles to the direction of propagation of the light.

Many common media are not homogeneous, in either the optical or in any other sense; for example, the directional properties of wood are clearly recognisable in its 'grain'; the ease with which a paper-knife may be inserted between the pages of a closed book will depend upon whether its blade is parallel or perpendicular to the leaves of paper. In the optical analogues of such 'anisotropic' media, the velocity of light which is oscillating in one plane is not the same as that which is oscillating in the perpendicular plane, and therefore it will have two different refractive indices. Thus a ray of light which comprises oscillations in many planes in free space will travel on as two distinct rays when once it gets inside such a 'birefringent' medium. By subtle optical means it is possible to suppress one or other of these rays in such devices as Nicol Prisms (two crystals of calcite, $CaCO_3$, mounted together by Canada balsam) or Polaroids. The emergent ray then consists of vibrations in one plane only and it is said to be 'polarised' (Fig. 15.1).

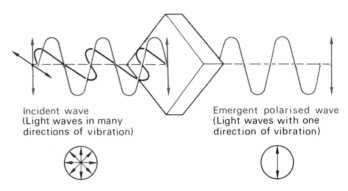

FIG. 15.1. *The passage of light through a Polariser to form plane polarised light*

To the human eye this polarised ray will be indistinguishable from an ordinary, non-polarised ray vibrating in many planes. But suppose a plane polarised ray, produced by a 'Polarising Prism', impinges subsequently upon another, similar 'Analysing Prism', several possibilities arise. If the axes of the two prisms are parallel to one another, the light will pass through them both without diminution in intensity, just as plane polarised waves on a string would pass unimpeded through parallel slits in two obstructing screens (Fig. 15.2).

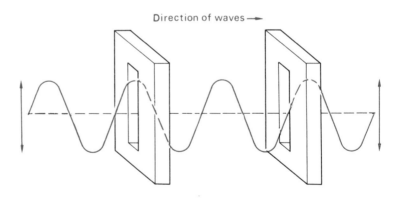

FIG. 15.2. *Transmission of plane polarised light when the axes of the Polariser and Analyser are parallel*

If, however, the second slit were turned through a right angle by rotating it about the direction of the plane of the waves in the string, the material waves would not be able to get through (Fig. 15.3). Light waves are similarly extinguished when the axes of the Polariser and the Analyser are perpendicular to one another, or 'crossed'. It can be shown that if the axes are at some intermediate angle, θ, then the intensity of the emergent ray is proportional to $\cos^2 \theta$.

FIG. 15.3. *The extinction of plane polarised light when the axes of the Polariser and Analyser are at right angles to one another*

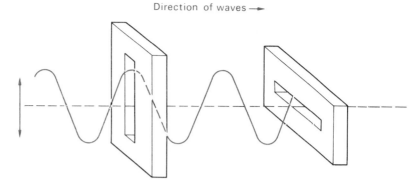

FIG. 15.4. *A Polarimeter*

A Polarimeter is an instrument for carrying out precise measurements of this effect. It consists of a fixed Polarising Prism adjacent to the source of monochromatic light, followed by a container for the substance under investigation, and finally an Analysing Prism carried on a circular scale which can be rotated about the axis of the whole apparatus (Fig. 15.4).

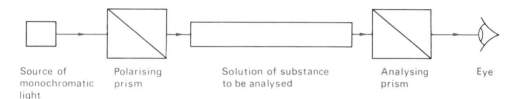

When an optically inactive solution is placed in the containing tube, light emerges at full intensity from the Analyser when its axis is parallel to that of the Polariser, and is completely extinguished when these two axes are perpendicular to one another; the latter situation can be more easily detected in practice, but even so it is subject to some uncertainty because the $\cos^2 \theta$ function changes only slowly as it reaches its minimum value.

If, however, a solution of an 'optically active' substance is used (this is explained in detail later), it will be found that the axis of the Analyser is no longer perpendicular to that of the Polariser when extinction occurs, but differs from that position by an angle through which the plane of polarised light has been rotated while the light has passed through the solution. Experiment shows that the angle of rotation is directly proportional to the length of the tube and to the concentration of the solution, for a given wavelength and at a fixed temperature.

The **specific rotation** of a compound, at a given temperature and for a given wavelength, is the rotation produced by a column of solution 100 cm long and of concentration 1 g per cm^3.

Introduction to optical activity

The ability to rotate the plane of polarised light is not confined to calcite crystals. Other inorganic crystals, such as sodium bromate and sodium iodate, have the same property. However, they lose the ability when they dissolve in water; it is the asymmetric arrangement of the atoms or ions in the crystal lattice which gives these crystals this property.

In contrast, any organic compound which contains a carbon atom attached to four different atoms or groups rotates the plane of polarised light both in the crystalline state and in solution. Such a carbon atom is described as an **asymmetric carbon atom** and is often shown as C*; an example is in 2-hydroxypropanoic acid. One form of the compound rotates the plane to the right and is said to be **dextrorotatory** (the (+) form), and the other (the mirror-image of the first) rotates it to the left and is said to be **laevorotatory** (the (−) form); the specific rotations of the two are equal but of opposite sign (by convention, the (+) form is defined as giving a positive rotation and the (−) form as giving a negative rotation). The two isomers are described as **enantiomers**. An equimolar mixture of the two does not rotate the plane of polarised light and is said to be optically inactive; it is also known as the **racemic form** ((±) form).

It is essential to realise that the phenomenon of optical activity can only be explained by assuming the tetrahedral arrangement for the molecule. Such an arrangement was proposed independently by van't Hoff and Le Bel in 1874, and is supported by two particular pieces of evidence: (a) there is only one form of such substances as dichloromethane CH_2Cl_2 (1.2), and (b) the phenomenon of optical isomerism.

Properties of enantiomers

1. The physical properties of enantiomers, such as melting point, boiling point and solubility, are identical except for the direction of rotation of polarised light. Note, however, that it is not possible to predict from the structural formulae of the enantiomers which will be the (+) and which the (−) form; this can only be determined by experiment once the actual structure has been found.

2. Enantiomers give crystals of the same type as each other, but the crystals of one are the mirror images of those of the other (p. 237), and the two are described as **enantiomorphs**.

3. The chemical properties of enantiomers are identical for their reactions with compounds which are not optically active; for example, (+) and (−)2-hydroxypropanoic acid undergo esterification with methanol at exactly the same rate as each other. However, enantiomers can react at different rates with an optically active reagent. This has especially important consequences when biological materials are involved, because enzymes, which are the catalysts for reactions in living systems, occur in optically active forms (18.2). Thus enantiomers often behave differently towards bacteria; for example, (+)2-hydroxypropanoic acid is consumed by penicillium glaucum but (−)2-hydroxypropanoic acid is relatively unaffected. Again, enantiomers can have different physiological properties; for example, (−)adrenalin is more active in contracting the blood capillaries than (+)adrenalin, and (−)nicotine is more poisonous than (+)nicotine.

Optical isomerism of compounds containing two asymmetric carbon atoms

The 2,3-dihydroxybutanedioic acid molecule contains two asymmetric carbon atoms:

$$HO_2C - \overset{H}{\underset{HO}{C^*}} - \overset{H}{\underset{OH}{C^*}} - CO_2H$$

2,3-Dihydroxybutanedioic acid exists in three stereoisomeric forms, which can be represented in two dimensions as follows:

(1)   (2)   (3)

Structures (1) and (2) are mirror images of each other but they are not superimposable on each other. Therefore, each is optically active and the two constitute a pair of enantiomers; one is dextrorotatory ((+)2,3-dihydroxybutanedioic acid) and the other is laevorotatory ((−)2,3-dihydroxybutanedioic acid).

Structure (3) is not superimposable on either (1) or (2), nor is it the mirror image of either. It is found that it is optically inactive, for the optical activity due to one of the asymmetric carbon atoms is counterbalanced by the optical activity due to the other; the top half of structure (3) is the mirror image of the bottom half, the dotted line representing a plane of symmetry. [In a sense, (3) has the structure of the top half of (1) and the bottom half of (2).]

Compounds which contain two or more asymmetric carbon atoms but are optically inactive are described as **internally compensated** and defined by the prefix *meso*; thus structure (3) is (*meso*)2,3-dihydroxybutanedioic acid. An equimolar mixture of (+) and (−)2,3-dihydroxybutanedioic acid is also optically inactive since the specific rotations of the two are equal in magnitude but opposite in sign; it is described as a **racemic mixture**, or as the (±)form, and is said to be **externally compensated**.

The (+) and (−)isomers have identical physical and chemical properties except in the direction of rotation of polarised light and in their reactions with other optically active substances However, the (*meso*)isomer has different properties, for example:

Physical property	(+)	(−)	(*meso*)
M.p. (°C)	170	170	140
Density (g per cm^3)	1·76	1·76	1·67
Solubility (g per 100 cm^3 H$_2$O)	139	139	125

When a compound contains two asymmetric carbon atoms of which one is not bonded to the same three groups as the other, four stereoisomers exist; for example, for the compound HO$_2$C—CH(OH)—CH(CH$_3$)—CO$_2$H:

ISOMERISM

The first two are non-superimposable mirror-images of each other and so constitute one pair of enantiomers, and likewise the other two constitute a second pair of enantiomers, with different properties from the first pair.

In general, a compound with n asymmetric carbon atoms exists in 2^n optically active forms, although when one set of substituents in the compound mirrors another, internal compensation occurs to reduce this number, as we have seen for 2,3-dihydroxybutanedioic acid.

Other optically active compounds

As we have seen, compounds containing an asymmetric carbon atom can exist in optically active forms. However, compounds which do not possess an asymmetric carbon atom can also exist in optically active forms provided that the *molecule* is asymmetric. Examples occur with certain derivatives of propadiene (CH_2=C=CH_2), such as 1,3-diphenylpropadiene: the central atom forms sp-hybridised σ-bonds with the other carbon atoms and provides two p electrons, one in each of two p orbitals which are mutually perpendicular, for forming π-bonds with the other carbon atoms. As a result, the substituents at one end of the molecule are in a plane which is perpendicular to that of the substituents at the other end, so that the compound exists in two forms which are non-superimposable mirror-images and are optically active (Fig. 15.5).

FIG. 15.5. *The stereoisomers of 1,3-diphenylpropadiene*

Substituted biphenyls exhibit optical isomerism when substituents in the 2- positions are large enough to prevent rotation about the bond joining the two benzene rings. For example, biphenyl-2,2′-disulphonic acid exists in two forms

which are non-superimposable mirror-images but do not interconvert at room temperature because the energy required to twist one ring through 180° relative to the other is too high (about 80 kJ per mol). This in turn is because, during the twisting process, the two sulphonic acid groups must come into very close proximity when the two benzene rings become coplanar and strong repulsive forces are introduced.

Resolution of enantiomers

The separation of a racemic mixture into the individual enantiomers is described as **resolution**. Three methods for resolving enantiomers are:

(a) **Crystal picking.** When sodium ammonium 2,3-dihydroxybutanedioate crystallises from solution below 28°C, the (+) and (−)isomers form crystals which are mirror images of each other (Fig. 15.6). The two types of crystals can be separated by hand.

This is a tedious method, and is in any case not always applicable (e.g. when the enantiomers are liquid). It is mentioned mainly for historical interest, for it was the first method to be employed, by Pasteur. In one sense

FIG. 15.6. *The crystal forms of (+) and (−)sodium ammonium 2,3-dihydroxybutanedioate*

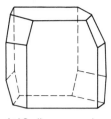
(+)Sodium ammonium 2,3-dihydroxybutanedioate

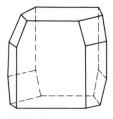
(−)Sodium ammonium 2,3-dihydroxybutanedioate

Pasteur was fortunate, because if his solution of sodium ammonium 2,3-dihydroxybutanedioate had crystallised *above* 28°C, he would have obtained only one type of crystal, so that no separation would have been possible. In this crystal, there is an ordered arrangement of the (+) and (−)isomers, and the state is called a **racemic compound**.

(b) **Chemical method.** This is the method of most general use. It is based on the principle that when each of the two enantiomers reacts with another compound which is optically active, the products are not mirror images and do not have identical properties. They are known as **diastereoisomers**. Usually one will be less soluble in a particular solvent than the other, and so the two can be separated by fractional crystallisation and then converted back into the individual enantiomers.

For example, a (±)acid ((±)A) can be resolved by making it into a mixture of two salts with an optically active base (say, (−)B),

$$(\pm)A + (-)B \longrightarrow \begin{array}{l} (+)A\text{-}(-)B \\ \text{Salt I} \\ \\ (-)A\text{-}(-)B \\ \text{Salt II} \end{array}$$

separating them, and then treating each with mineral acid to release (+)A and (−)A. A number of optically active bases suitable for the purpose occur naturally as alkaloids (e.g. quinine) in plants. Similarly, a (±) base can be separated by the use of an optically active acid.

Racemic mixtures which are not acids or bases can often be resolved by first making them into derivatives with acid groups and resolving the resulting mixture as above. For example, the enantiomers of butan-2-ol, CH_3—$\overset{*}{C}H(OH)$—C_2H_5, can be resolved by esterification with a dibasic acid (e.g. benzene-1,2-dicarboxylic acid) to give enantiomeric esters such as the (±) form of

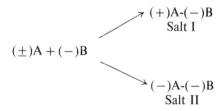

resolving with an optically active base, and then hydrolysing the individual (+) and (−)esters to give the (+) and (−)alcohols.

Diastereoisomers can sometimes be separated by chromatographic techniques. For example, it has been found (1969) that (±)amino-acids can be resolved by converting them into diastereoisomeric esters with an optically active secondary alcohol, separating these by gas-liquid chromatography, collecting the fractions and hydrolysing each separately to give the (+) and (−)amino-acids.

(c) **Biochemical method.** Bacteria will sometimes grow in solutions of racemates, and may feed on them by consuming one of the forms. Thus, (−)2-hydroxypropanoic acid can be prepared by allowing penicillium glaucum to feed on (±)2-hydroxypropanoic acid; it destroys the (+) acid.

Racemisation

Racemisation is the opposite of resolution; that is, it consists of the formation of equal amounts of a pair of enantiomers from either of the two. It can occur when a reaction takes place which breaks one of the bonds to the asymmetric carbon atom. For example, if a solution of one of the enantiomers of 2-iodobutane, say the (−)isomer, is treated with a solution of sodium iodide, an iodide ion in the solution displaces an iodide ion from the organic compound by attacking from the opposite side:

$$I^- + \begin{array}{c} C_2H_5 \\ H \cdots C - I \\ CH_3 \end{array} \longrightarrow \begin{array}{c} C_2H_5 \\ I - C \cdots H \\ CH_3 \end{array} + I^-$$

Consequently, the (+)isomer is formed. However, this too can react, in the reverse manner, to regenerate the (−)isomer, and eventually a point is reached when equal amounts of the two isomers are in dynamic equilibrium.

15.6 Summary

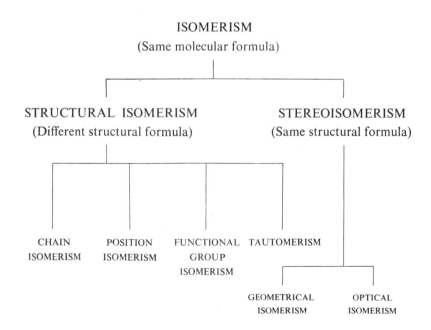

15.7 Practical work

Optical isomerism

Make ball and stick models of the following compounds, using different coloured balls to represent different functional groups:

(a) The stereoisomers of 2-hydroxypropanoic acid:

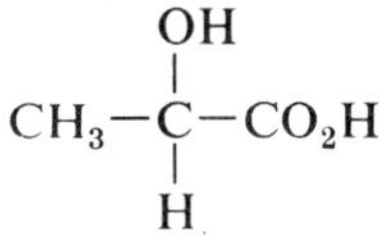

$$CH_3-\underset{\underset{H}{|}}{\overset{\overset{OH}{|}}{C}}-CO_2H$$

(b) The stereoisomers of 2,3-dihydroxybutanedioic acid:

$$\begin{array}{c} OH \\ | \\ H-C-CO_2H \\ | \\ H-C-CO_2H \\ | \\ OH \end{array}$$

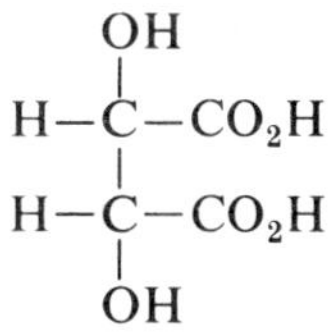

Make sure that you have constructed a model of (*meso*)2,3-dihydroxybutanedioic acid and that you can see why it is internally compensated.

(c) The stereoisomers of the acid:

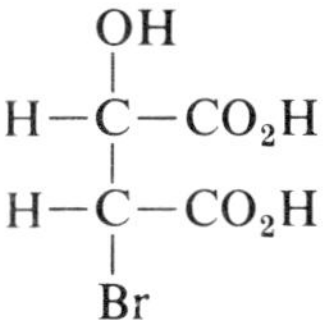

$$\begin{array}{c} OH \\ | \\ H-C-CO_2H \\ | \\ H-C-CO_2H \\ | \\ Br \end{array}$$

Show whether or not it is possible for this compound to have a (*meso*) form.

Geometrical isomerism

Dissolve 10 g of butenedioic anhydride in 12 cm^3 of boiling water in a test-tube to form a solution of *cis*-butenedioic acid:

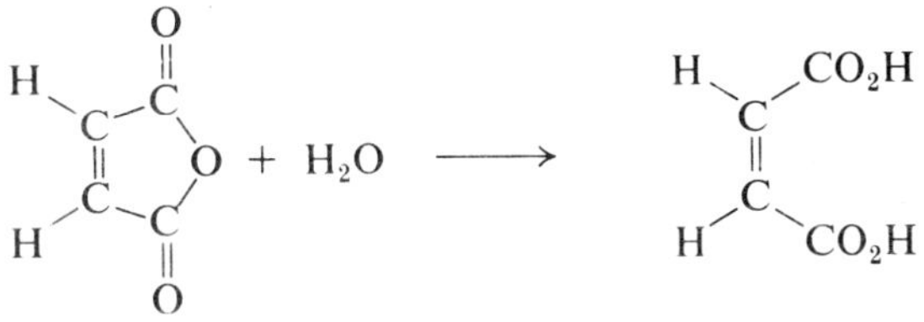

Cool the solution by placing the test-tube in a beaker of cold water. Filter the solid acid using a Buchner funnel (Fig. 2.8), but do not attempt to wash the solid. Dry it between pads of filter papers and determine its melting point.

Collect the filtrate, add 20 cm³ of *concentrated* hydrochloric acid and reflux the mixture for about 20 minutes. Crystals are formed which can be filtered, washed with water and dried between pads of filter paper. Determine the melting point. [M.p. of *cis*-butenedioic acid 130°C; m.p. of *trans*-butenedioic acid 287°C (sublimes).]

Hydrogen chloride adds on to the *cis*-butenedioic acid molecule to form an intermediate in which there is unrestricted rotation about the C—C bonds:

restricted rotation   unrestricted rotation   restricted rotation

15.8 Questions

1 Explain the terms (a) Empirical formula, (b) Molecular formula, (c) Structural formula, (d) Isomerism.

Three unsaturated dichloroalkenes of molecular formula $C_2H_2Cl_2$ exist and also two unsaturated dicarboxylic acids of molecular formula $C_4H_4O_4$. Account for the existence of these isomers discussing the stereochemical principles.

Indicate **one** method by which you could assign to each its structure if you were given pure specimens of the two unsaturated acids. (S)

2 (a) Distinguish between the terms empirical, molecular and structural formulae.

(b) Discuss the various types of isomerism which occur in organic chemistry, illustrating your answer by reference to the isomerism of the following compounds: (i) C_2H_6O, (ii) $C_2H_2Cl_2$, (iii) $C_3H_6O_3$ (only acids).

There are no isomers of compounds of the type CH_2X_2 where X = Cl, Br, etc.; what does this show? (O)

3 (a) What do you understand by (i) structural isomerism, (ii) optical isomerism and (iii) geometrical isomerism? Illustrate your answers with examples chosen from the chemistry of aldehydes and ketones and their derivatives.

(b) Three isomeric mononitrobenzoic acids are known to exist. Write down their structural formulae, and state, giving reasons, which of these would normally be formed on the nitration of benzoic acid. Suggest how you might attempt to prepare the other isomer(s). (L(X))

4 Discuss the various types of isomerism which occur in organic chemistry, illustrating your answer by reference to the isomerism that is shown by the following compounds: (a) $C_2H_2Br_2$; (b) CH_4N_2O; (c) $C_3H_6O_3$ (only acids); (d) C_3H_5N (no ring structures); (e) $C_6H_3Cl_3$ (only derivatives of benzene).

Describe **two** tests by which you would distinguish between two isomers in (d). (O(S))

5 Explain as fully as you can what is meant by the term 'isomerism', illustrating your answer by examples of your own choice.

How would you distinguish by not more than two chemical tests in each case between the substances in each of the following pairs?

(a) $(CH_3)_3C \cdot OH$ and $CH_3 \cdot CH_2 \cdot O \cdot CH_2 \cdot CH_3$,

(b) $CH_3 \cdot CH = CH \cdot CH_2 \cdot CH_3$ and $CH_3 \cdot CH = C \begin{smallmatrix} CH_3 \\ CH_3 \end{smallmatrix}$,

(c) $CH_3 \cdot CO \cdot O \cdot C_2H_5$ and $C_2H_5 \cdot CO \cdot O \cdot CH_3$. (L)

ISOMERISM

6 Describe, with diagrams, the type of isomerism shown by *cis*- and *trans*-butenedioic acids. Give examples of the differences in physical properties of these acids. How can these acids be distinguished by chemical methods?

The addition of hydrogen bromide to each of the two acids gives the same pair of isomers, while the addition of bromine to the two acids gives a total of three isomeric dibromo-acids. Draw the structures of the five brominated compounds and describe the type of isomerism involved. How might the two monobromodicarboxylic acids be separated?

7 An optically active compound A when oxidised with chromic acid yields a substance B which gives a condensation product with phenylhydrazine but no precipitate when warmed with a solution of potassium iodide in sodium hypochlorite. The empirical formulae of A and B are $C_6H_{14}O$ and $C_6H_{12}O$ respectively. When A is warmed with concentrated sulphuric acid it gives a hydrocarbon C, the percentage composition of which is C, 85·7; H, 14·3; and the vapour density, 42.

Treatment of a trichloromethane solution of C with trioxygen yields a compound which on decomposition with water yields two compounds, D and E, both having an empirical formula, C_3H_6O. Both D and E give precipitates with phenylhydrazine, but only E yields a yellow precipitate with a solution of iodine in sodium hydroxide. On the other hand the compound D is easily oxidised to an acid, $C_3H_6O_2$.

Deduce the structure of A, explaining briefly why it is optically active, and write equations for the formation of B, C, D and E.

8 What is meant by isomerism? Describe four different ways in which isomerism can arise in organic compounds. Where possible illustrate your answer by examples of each type.

9 Give an account of the consequences of the tetrahedral disposition of bonds about the carbon atom. Explain fully how the isomerism of *cis*- and *trans*-butene dioic acids differs from that of the 2-hydroxypropanoic acids.

10 What isomers of the following compounds exist:

(a) C_3H_8O
(b) $CH_3CH{=}CHCOOH$
(c) $CH_3CH(OH)COOH$?

Explain how and why the isomers differ in physical and chemical properties.

11 Show that you understand what is meant by the following terms: optical activity, geometrical isomerism, *meso* compound. Discuss the stereochemistry of the compound:

$$C_6H_5CH{=}C(C_6H_5){-}CH\genfrac{}{}{0pt}{}{\diagup CH_3}{\diagdown CH_2CH_3}$$

12 Define *isomerism*. What isomers exist for the following compounds:

$R_1R_2R_3SiCl$ (R_1, R_2, and R_3 represent different alkyl or aryl groups)
$HOOCCH{=}CHCOOH$
$Pt(NH_3)_2Cl_2$
$R_1R_2R_3R_4N^+I^-$ (R_1, R_2, R_3, and R_4 represent different alkyl or aryl groups)?
(C Schol.)

13 Show by means of examples what types of isomers are encountered in organic chemistry. Write the structures of the isomers of formula C_3H_5Cl. Indicate briefly how you might distinguish between them.

ISOMERISM

14 A compound is found to have the molecular formula $C_4H_8O_2$. Write down as many possible structures for this compound as you can *classifying* them according to the types of functional group present. Indicate briefly how you might distinguish between the various classes by chemical tests. (O Schol.)

15 Optically active 2-iodobutane

$$(CH_3.CH_2.CHI.CH_3)$$

racemises in acetone solution containing sodium iodide. When the experiment is carried out with labelled sodium iodide (prepared from ^{131}I, the radioactive isotope of iodine), the alkyl iodide is found to lose optical activity and to exchange its ordinary iodine for radioactive iodine. The rates of each of these reactions are proportional to

$$[\text{alkyl iodide}][I^-]$$

but racemisation is exactly *twice* as fast as isotopic exchange.

Discuss the stereochemical implications of these experiments.

What conclusions would you have drawn if the rates of racemisation and isotopic exchange had been independent of the concentration of iodide ion and exactly *equal*? (O Schol.)

16 Describe the different types of isomerism exhibited by organic compounds, illustrating your answer with specific examples.

Vigorous oxidation of a compound X $(C_2H_4)_n$ gives only ethanoic acid. With bromine X forms Y $(C_2H_4Br)_n$ and with hydriodic acid it forms Z, C_4H_9I.

Draw possible structures for compounds, X, Y and Z, and indicate the stereo-isomeric forms in which they could exist. (O Schol.)

Chapter 16

Amines

16.1 Introduction

Amines can be regarded as organic derivatives of ammonia. There are three classes of amines:

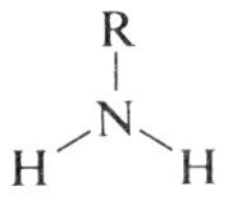

$$\underset{\text{Primary amine}}{\overset{R}{\underset{H}{\vphantom{R}}}\!N\!\overset{}{\underset{H}{\vphantom{R}}}} \qquad \underset{\text{Secondary amine}}{\overset{R}{\underset{R}{\vphantom{R}}}\!N\!\overset{}{\underset{H}{\vphantom{R}}}} \qquad \underset{\text{Tertiary amine}}{\overset{R}{\underset{R}{\vphantom{R}}}\!N\!\overset{}{\underset{R}{\vphantom{R}}}}$$

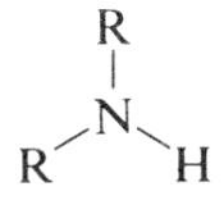

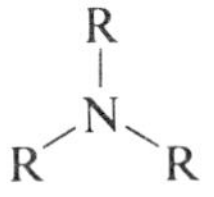

where R can be an alkyl group or an aromatic ring. The prefixes, primary, secondary and tertiary, define the number of carbon groups attached to the nitrogen atom and *not* (as with alkyl halides and alcohols) the number of such groups attached to the carbon atom which bears the functional group.

Aliphatic amines

Simple amines are usually named by adding the word amine to the names of the groups to which the nitrogen atom is attached (Table 16.1).

16.2 Nomenclature

Table 16.1. Some aliphatic amines

NAME	FORMULA	CLASS	B.P. (°C)
Methylamine	CH_3-NH_2	Primary	−7
Ethylamine	$CH_3CH_2-NH_2$	Primary	17
Dimethylamine	$CH_3-NH-CH_3$	Secondary	7
Methylethylamine	$CH_3CH_2-NH-CH_3$	Secondary	35
Diethylamine	$CH_3CH_2-NH-CH_2CH_3$	Secondary	56
Trimethylamine	$(CH_3)_2N-CH_3$	Tertiary	3
Triethylamine	$(CH_3CH_2)_2N-CH_2CH_3$	Tertiary	89

An alternative nomenclature employs the prefix **amino**; for example:

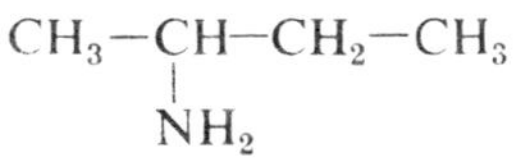

$$CH_3-\underset{\underset{NH_2}{|}}{CH}-CH_2-CH_3 \qquad CH_3-\underset{\underset{NH_2}{|}}{\overset{\overset{CH_3}{|}}{C}}-CH_2-CH_3$$

2-Aminobutane 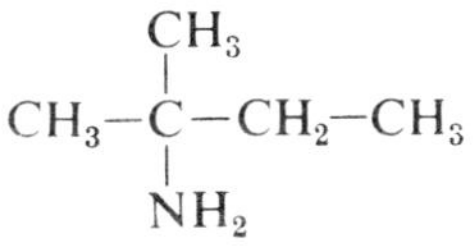   2-Amino-2-methylbutane

AMINES

Aromatic amines

The simplest, phenylamine, $C_6H_5-NH_2$, is sometimes called by its original name, aniline. Examples of substituted phenylamines are:

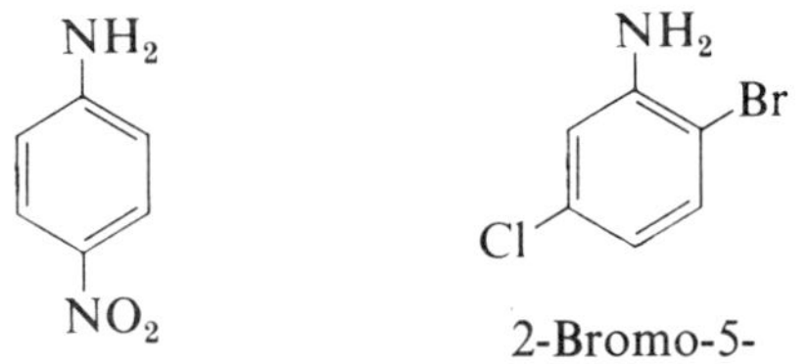

4-Nitrophenylamine

2-Bromo-5-chlorophenylamine

There are also amines which contain more than one aryl group, for example:

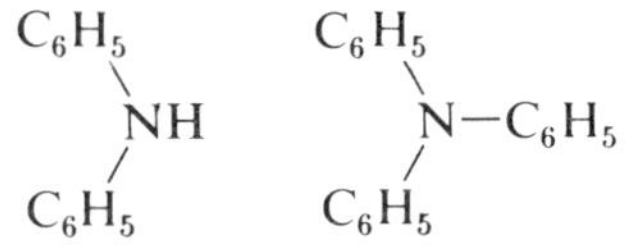

Diphenylamine   Triphenylamine

As will be seen, there are significant differences both in methods of preparation and in properties between aliphatic amines and those aromatic amines in which the nitrogen atom is attached to the benzene ring (aryl amines, e.g. phenylamine). (Phenylmethyl)amine, $C_6H_5CH_2NH_2$, although it contains the benzene ring, behaves like an aliphatic amine since the nitrogen atom is not attached directly to the ring.

16.3 Physical properties of amines

The lower aliphatic amines are gases or low-boiling liquids, their boiling points being lower than those of the corresponding alcohols. They have a smell rather like bad fish; indeed, decaying fish produce various amines. They are readily soluble in water and in organic solvents.

Aromatic amines are liquids or solids with high boiling points. They have a characteristic smell and are soluble in organic solvents but almost insoluble in water.

16.4 Methods of preparation of amines

Laboratory methods

1. Aliphatic amines can be prepared by the reaction between an alkyl halide and ammonia (9.3). However, the method is rarely used because a mixture of primary, secondary and tertiary amines and quaternary ammonium salts is obtained:

$$R-Br + NH_3 \rightarrow R-NH_2 + HBr$$
$$R-Br + R-NH_2 \rightarrow R_2NH + HBr$$
$$R-Br + R_2NH \rightarrow R_3N + HBr$$
$$R-Br + R_3N \rightarrow R_4N^+ \, Br^-$$

(The hydrogen bromide evolved reacts with ammonia and amines to form salts.)

Aryl amines cannot be made in this way because aryl halides are unreactive towards ammonia.

2. By the reduction of a nitro compound, giving a primary amine. The reducing agent can be hydrogen, catalysed by Raney nickel:

$$R-NO_2 + 3H_2 \xrightarrow{\text{Ni as cat.}} R-NH_2 + 2H_2O$$

or lithium aluminium hydride:

$$R-NO_2 \xrightarrow{\text{LiAlH}_4} R-NH_2$$

Tin and hydrochloric acid are used to reduce aromatic nitro compounds; the amine is produced as a complex salt from which it is liberated with alkali, for example:

$$2C_6H_5-NO_2 + Sn + 6HCl \rightarrow (C_6H_5-\overset{+}{N}H_3)_2SnCl_6^{2-} + 2H_2O$$

$$(C_6H_5\overset{+}{N}H_3)_2SnCl_6^{2-} + 8NaOH \rightarrow$$
$$2C_6H_5NH_2 + Na_2SnO_3 + 5H_2O + 6NaCl$$

3. Primary amines are also formed by the reduction of nitriles and amides with lithium aluminium hydride:

$$R-CN \xrightarrow{\text{LiAlH}_4} R-CH_2-NH_2$$

$$R-CO-NH_2 \xrightarrow{\text{LiAlH}_4} R-CH_2-NH_2$$

4. By the Hofmann degradation reaction, giving a primary amine (14.5). An amide is treated with bromine and alkali, and the resulting amine is distilled off:

$$R-CO-NH_2 + Br_2 + 4NaOH \rightarrow$$
$$R-NH_2 + Na_2CO_3 + 2NaBr + 2H_2O$$

It should be noted that an amide, $RCONH_2$, is reduced to the amine with the same number of carbon atoms, RCH_2NH_2, with lithium aluminium hydride but gives the lower homologue RNH_2 in the Hofmann reaction.

Manufacture

Primary aliphatic amines are manufactured by the reduction of nitroalkanes, the reaction between an alkyl halide and ammonia and the reaction between an alcohol and ammonia under pressure with a cobalt catalyst.

Phenylamine is manufactured by the reduction of nitrobenzene either with iron and hydrochloric acid or with hydrogen over a catalyst in the gas phase.

16.5 Chemical properties of amines

Reactions at the nitrogen atom

1. Amines, like ammonia, give alkaline solutions in water as a result of the equilibrium:

$$R-NH_2 + H_2O \rightleftharpoons R-\overset{+}{N}H_3 + OH^-$$

The strength of the base is described by the value of K_b where:

$$K_b = \frac{[R-\overset{+}{N}H_3][OH^-]}{[R-NH_2]}$$

This is explained further in Section 4.8.

Amines are weak bases; values of K_b are in Table 16.2.

Table 16.2. Dissociation constants of ammonia and some amines

COMPOUND	K_b AT 25°C
NH_3	1.8×10^{-5}
CH_3-NH_2	4.4×10^{-4}
$(CH_3)_2NH$	5.9×10^{-4}
$(CH_3)_3N$	6.3×10^{-5}
$C_6H_5-NH_2$	4.2×10^{-10}
$C_6H_5CH_2-NH_2$	2.2×10^{-5}

The aliphatic amines are approximately as basic as ammonia. However, the aryl amines such as phenylamine are much weaker bases.

This is because the unshared pair of electrons on the nitrogen atom in phenylamine is in an orbital which overlaps with the adjacent carbon p orbital, giving increased delocalisation in the aromatic, as compared with an aliphatic, amine. One of the delocalised π molecular orbitals is shown in Fig. 16.1.

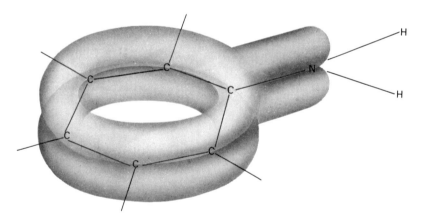

FIG. 16.1. A π molecular orbital in phenylamine

Therefore, relative to an aliphatic amine, phenylamine resists reaction with a proton since this process requires the use of the unshared pair of electrons on nitrogen in the formation of the new N—H bond and so results in the loss of the extra delocalisation (cf. the weaker basicity of an amide compared with an amine, p. 214).

On the other hand, an aromatic amine such as (phenylmethyl)amine does not contain a nitrogen atom adjacent to a carbon atom with a p orbital and is approximately as strong a base as an aliphatic amine.

The salts formed by amines with acids are analogous to ammonium salts. For example, amine hydrochlorides, like methylammonium chloride, $CH_3\overset{+}{N}H_3\ Cl^-$, are white crystalline solids with high melting points (above about 200 °C) which are soluble in water.

2. Primary and secondary amines react with acid chlorides and acid anhydrides to form substituted amides, for example:

$$CH_3-NH_2 + CH_3-CO-Cl \rightarrow CH_3-NH-CO-CH_3 + HCl$$
N-Methylethanamide

$$C_6H_5-NH_2 + CH_3-CO-O-CO-CH_3 \rightarrow$$
$$C_6H_5-NH-CO-CH_3 + CH_3CO_2H$$
N-Phenylethanamide

In these reactions the amine acts as a nucleophile:

$$R-\overset{H}{\underset{H}{N}}: \overset{R'}{\underset{O}{C}}\diagdown Cl \rightarrow R-\overset{H}{\underset{H}{\overset{+}{N}}}-\overset{R'}{\underset{O^-}{C}}-Cl \xrightarrow{-HCl} R-NH-CO-R'$$

3. Amines react with nitrous acid in a way which depends on whether they are primary, secondary or tertiary amines.

Primary aliphatic amines react to yield gaseous nitrogen. The reaction is complex; for example, propylamine gives nitrogen almost quantitatively, but a mixture of other compounds is also formed, among them being propene, propan-2-ol and a trace of propan-1-ol.

Primary aryl amines react to give aromatic diazonium salts, providing that the temperature is below about 10°C. For example, when phenylamine is treated with hydrochloric acid and sodium nitrite, it reacts with the nitrous acid formed to give benzenediazonium chloride:

$$C_6H_5-NH_2 + HNO_2 + HCl \rightarrow C_6H_5-\overset{+}{N}\equiv N\ Cl^- + 2H_2O$$
Benzenediazonium
chloride

Aromatic diazonium salts are of value in making other aromatic compounds and are discussed separately (16.8).

Secondary amines, both aliphatic and aromatic, react with nitrous acid to give nitroso compounds, which are yellow oils:

$$R_2NH + HNO_2 \rightarrow R_2N-N=O + H_2O$$
A nitrosamine

Tertiary amines react with nitrous acid to give solutions containing substituted ammonium nitrites. Since these are salts formed by a weak acid and a weak base, they are extensively hydrolysed; that is, the equilibrium

$$R_3N + HNO_2 \rightleftharpoons R_3\overset{+}{N}H\ NO_2^-$$

lies to the left-hand side.

4. Primary amines react with trichloromethane and a solution of potassium hydroxide in ethanol to form isocyano-compounds (carbylamine reaction; 9.4):

$$R-NH_2 + CHCl_3 + 3KOH \rightarrow R-\overset{+}{N}\equiv\overset{-}{C} + 3KCl + 3H_2O$$

For example, phenylamine forms isocyanobenzene, C_6H_5NC.

Substitution in the aromatic ring of aryl amines

The amino group in an aryl amine directs electrophilic reagents mainly to the 2- and 4-positions and renders the compound far more reactive than benzene towards these reagents. For example, whereas benzene requires a halogen carrier to react with bromine to form bromobenzene (8.3), phenylamine reacts rapidly with bromine water in the absence of a carrier at all three 2- and 4-positions:

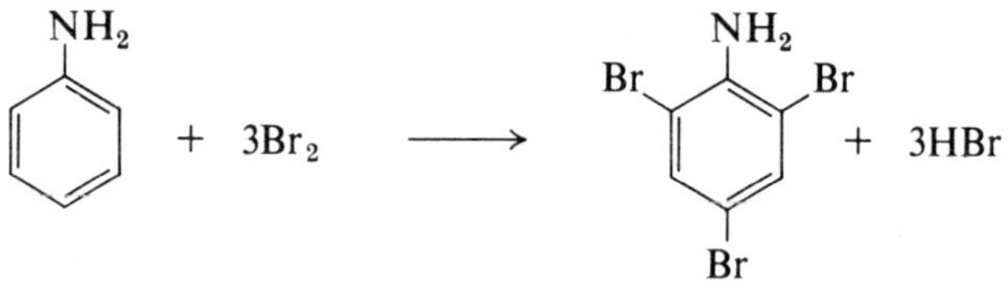

The reason is that the amino group stabilises the adducts formed by reaction at the 2- and 4-positions by sharing the positive charge (8.5). This can be represented by the structures:

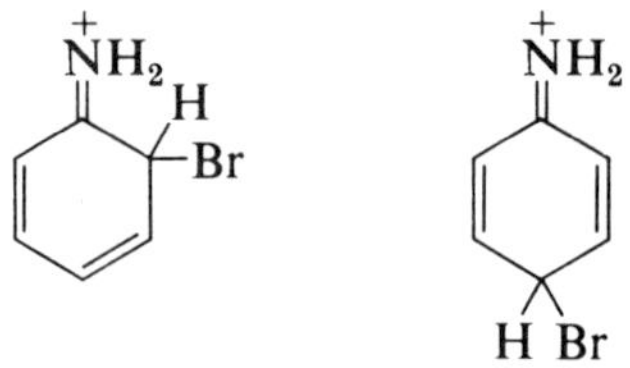

In order to obtain a monobromo derivative, it is necessary to reduce the reactivity of the aromatic ring. This can be achieved by ethanoylation of the amino group with ethanoic anhydride (p. 247); the substituent —NH—CO—CH$_3$ is 2-,4-directing, but much less strongly activating than —NH$_2$, and reaction with bromine gives N-2- and N-(4-bromophenyl) ethanamide. The ethanoyl group can then be removed by hydrolysis with a dilute solution of sodium hydroxide:

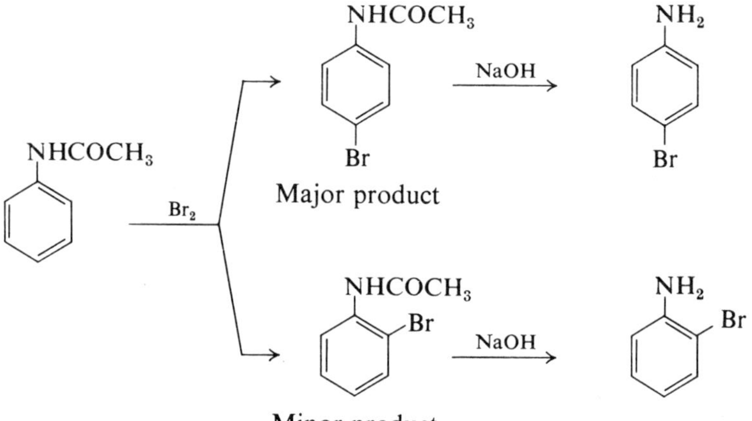

This use of the ethanoyl group is an example of **protection**: the aromatic ring is protected from the extensive reactions which occur with phenylamine itself. Protection is also required for the nitration of phenylamine, in this case not only because otherwise reaction would occur at all the 2- and 4-positions but also because the amino group is itself readily oxidised by nitric acid. Nitration of N-phenylethanamide gives the 2- and 4-nitro derivatives from which 2- and 4-nitrophenylamine can be obtained by hydrolysis:

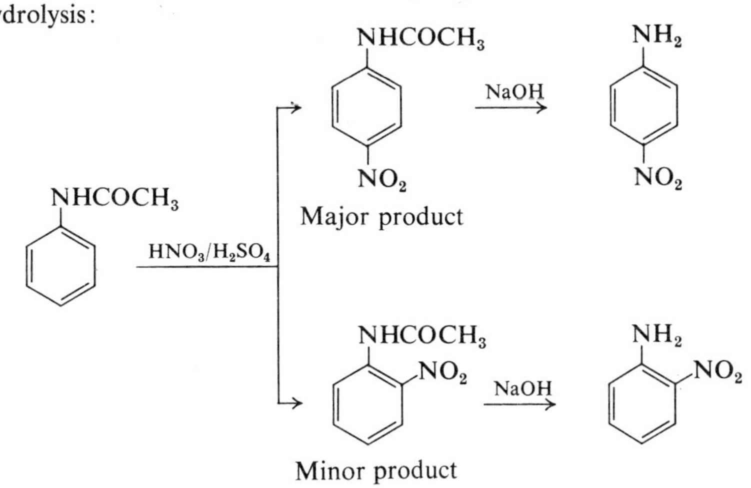

16.6 Uses of amines

1. *Plastics.* 1,6-Diaminohexane, $H_2N-(CH_2)_6-NH_2$, is used in the manufacture of nylon-6.6 (p. 320). Other amines are used in the production of isocyanates for polyurethane plastics (p. 317).

2. *Inhibitors.* Amines are effective at preventing the deterioration of rubber through oxidation by atmospheric oxygen.

3. *Dye-stuffs.* Primary aromatic amines are used to make azo-dyes (16.8).

4. *Medicines.* Amines are used in the manufacture of many pharmaceuticals (e.g. Mepacrine, an antimalarial drug).

16.7 Amino-acids

Amino-acids contain at least one amino and one carboxyl group. The most important are the 2-amino-acids, which are traditionally referred to as α-amino-acids and retain their original names, which will be used in this book; examples are aminoethanoic acid (glycine), 2-aminopropanoic acid (α-alanine) and 2,6-diaminohexanoic acid (lysine). Their importance stems from their being the constituents of the proteins (18.2). There are also 3-amino-acids (β-amino-acids), such as 3-aminopropanoic acid (β-alanine), 4-amino-acids (γ) and so on.

$$H_2N-CH_2-CO_2H \quad H_2N-\underset{\underset{\text{α-Alanine}}{|}}{\overset{\overset{CH_3}{|}}{CH}}-CO_2H \quad H_2N-CH_2-CH_2-CO_2H$$
$$\text{Glycine} \qquad\qquad\qquad\qquad\qquad \text{β-Alanine}$$

$$\underset{\text{Lysine}}{H_2N-\underset{\underset{\underset{\underset{NH_2}{|}}{(CH_2)_4}}{|}}{CH}-CO_2H} \qquad \underset{\text{Aspartic acid}}{H_2N-\underset{\underset{\underset{CO_2H}{|}}{CH_2}}{CH}-CO_2H}$$

Physical properties

Amino-acids are high-melting crystalline solids (e.g. glycine has m.p. 235°C) which are usually readily soluble in water but insoluble in organic solvents. In these respects, they resemble salts, and this is because they exist as internal salts, known as **zwitterions**, in which both cation and anion are held together in the same unit, for example:

$$\overset{+}{H_3N}-CH_2-CO_2^-$$

Amino-acids with an equal number of amino and carboxyl groups (e.g. glycine) are neutral in solution, but when the solution is acidified the carboxyl group is protonated,

$$\overset{+}{H_3N}-CH_2-CO_2^- + H^+ \rightarrow \overset{+}{H_3N}-CH_2-CO_2H$$

and when it is basified the amino group is freed:

$$\overset{+}{H_3N}-CH_2-CO_2^- + OH^- \rightarrow H_2N-CH_2-CO_2^- + H_2O$$

If the amino-acid contains more carboxyl groups than amino groups (e.g. aspartic acid), its solution in water is acidic. Conversely, if it contains an excess of amino groups (e.g. lysine), its solution in water is basic.

Most α-amino-acids have at least one asymmetric carbon atom and exist in optically active forms (e.g. α-alanine).

Preparation of α-amino-acids

1. By reaction of a 2-chloro-acid with concentrated ammonia solution, for example:

$$Cl-CH_2-CO_2H \xrightarrow{2NH_3} H_2N-CH_2-CO_2^- NH_4^+ \xrightarrow{HCl} H_2N-CH_2-CO_2H + NH_4Cl$$
Glycine

The disadvantage of this method is that the amino group which is introduced can react with a second molecule of the chloro-acid:

$$H_2N-CH_2-CO_2H + Cl-CH_2-CO_2H \longrightarrow HN\begin{matrix}CH_2-CO_2H\\ \\CH_2-CO_2H\end{matrix} + HCl$$

A mixture of products is therefore formed. This problem is overcome by using potassium benzene-1,2-dicarboximide as the source of the nitrogen atom:

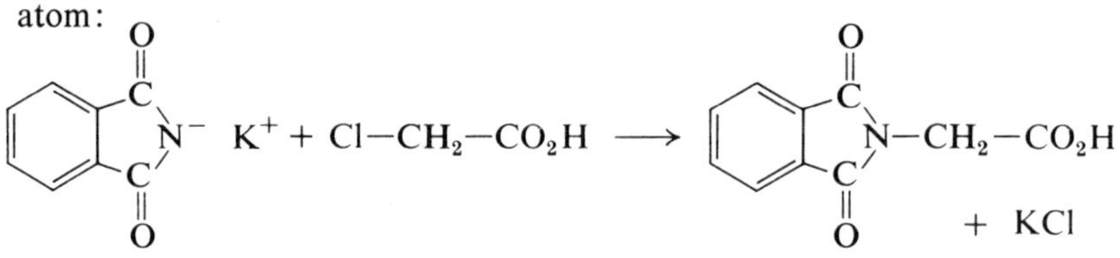

$$\text{(phthalimide)}N^-K^+ + Cl-CH_2-CO_2H \longrightarrow \text{(phthalimide)}N-CH_2-CO_2H + KCl$$

The resulting derivative is hydrolysed with acid:

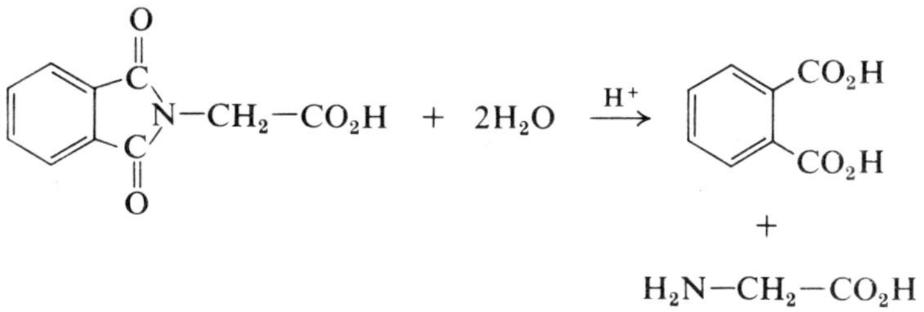

$$\text{(phthalimide)}N-CH_2-CO_2H + 2H_2O \xrightarrow{H^+} \text{(benzene)}(CO_2H)_2 + H_2N-CH_2-CO_2H$$

2. By reaction of an aldehyde with a mixture of potassium cyanide and ammonia:

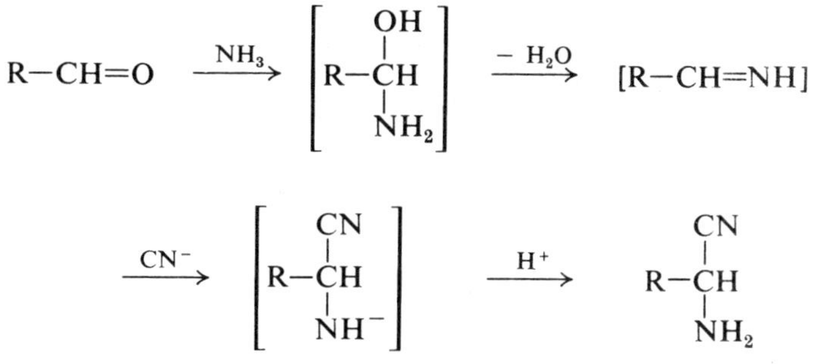

$$R-CH=O \xrightarrow{NH_3} \left[\begin{array}{c} OH \\ | \\ R-CH \\ | \\ NH_2 \end{array} \right] \xrightarrow{-H_2O} [R-CH=NH]$$

$$\xrightarrow{CN^-} \left[\begin{array}{c} CN \\ | \\ R-CH \\ | \\ NH^- \end{array} \right] \xrightarrow{H^+} \begin{array}{c} CN \\ | \\ R-CH \\ | \\ NH_2 \end{array}$$

followed by hydrolysis of the nitrile:

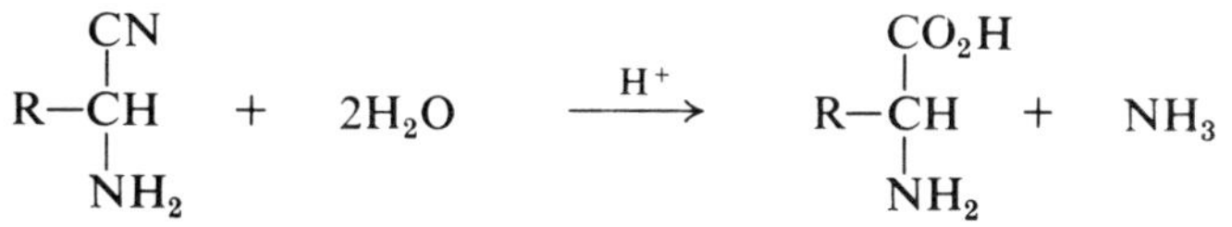

$$\begin{array}{c} CN \\ | \\ R-CH \\ | \\ NH_2 \end{array} + 2H_2O \xrightarrow{H^+} \begin{array}{c} CO_2H \\ | \\ R-CH \\ | \\ NH_2 \end{array} + NH_3$$

Chemical properties of α-amino-acids

α-Amino-acids show reactions of both acids and amines individually, as well as some reactions dependent on the presence of both groups. Typical reactions, illustrated for glycine, are:

Reactions of the carboxyl group

1. Formation of esters, for example:

$$H_2N-CH_2-CO_2H + CH_3-OH \xrightleftharpoons{H^+} H_2N-CH_2-CO-O-CH_3 + H_2O$$

2. Decarboxylation, when heated with soda-lime (cf. the decarboxylation of ethanoic acid, p. 193):

$$H_2N-CH_2-CO_2H + 2NaOH \rightarrow \underset{\text{Methylamine}}{H_2N-CH_3} + Na_2CO_3 + H_2O$$

Reactions of the amino group

1. Formation of acyl derivatives, for example:

$$H_2N-CH_2-CO_2H + CH_3-CO-Cl \rightarrow CH_3-CO-NH-CH_2-CO_2H + HCl$$

AMINES

2. Reaction with nitrous acid:

$$H_2N-CH_2-CO_2H + HNO_2 \rightarrow HO-CH_2-CO_2H + N_2 + H_2O$$
Hydroxyethanoic acid

Reactions dependent on the presence of both groups

1. When heated, a cyclic compound (diketopiperazine) is formed:

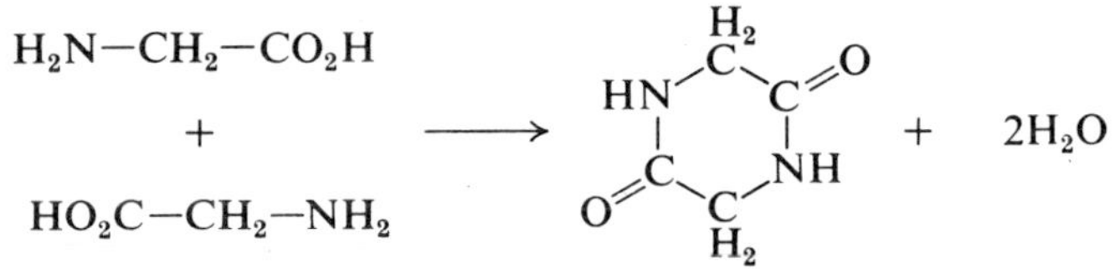

$$H_2N-CH_2-CO_2H + HO_2C-CH_2-NH_2 \rightarrow \text{diketopiperazine} + 2H_2O$$

2. When treated with copper(II) ion in water, a deep blue colour is produced due to the compound, copper(II)-glycine:

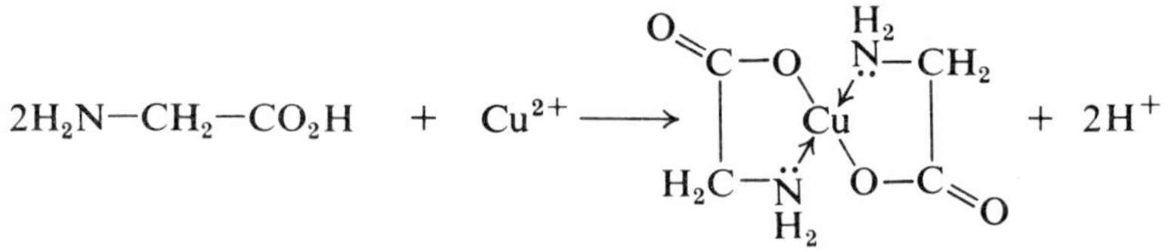

$$2H_2N-CH_2-CO_2H + Cu^{2+} \rightarrow \text{copper(II)-glycine} + 2H^+$$

Copper(II)-glycine

16.8 Aromatic diazonium salts

Primary aromatic amines react with nitrous acid in acid solution below about 10°C to form aromatic diazonium salts, for example:

$$C_6H_5-NH_2 + HNO_2 + HCl \rightarrow C_6H_5-\overset{+}{N}\equiv N\ Cl^- + 2H_2O$$
Benzenediazonium chloride

The process is known as **diazotisation** and was discovered by Griess in 1858.

Diazonium salts are stable in solution provided that the temperature is kept low. However, most of them are explosive in the solid state, and so they are usually not isolated from the aqueous solutions in which they are made. They are useful in synthesis because the substituent $-\overset{+}{N}\equiv N$ can be replaced by a variety of other groups by treating the aqueous solution of the salt with an appropriate reagent.

These reactions can be divided into two groups: (a) those in which the two nitrogen atoms are replaced, and (b) those in which the nitrogen atoms are retained.

(a) Reactions in which the nitrogen atoms are replaced

1. Replacement by the **hydroxyl** group. If the solution of the diazonium salt is warmed, a phenol is formed, for example:

$$C_6H_5-\overset{+}{N}\equiv N\ Cl^- + H_2O \rightarrow C_6H_5-OH + N_2 + HCl$$
Phenol

2. Replacement by a **halogen** atom.
(i) **Iodine.** When the solution of the diazonium salt is warmed with an aqueous solution of potassium iodide, an iodo compound is formed, for example:

$$C_6H_5-\overset{+}{N}\equiv N\ Cl^- + KI \rightarrow C_6H_5-I + N_2 + KCl$$
$$\text{Iodobenzene}$$

(ii) **Chlorine** and **bromine**. A solution of a copper(I) halide dissolved in the concentrated halogen acid is added to the solution of the diazonium salt, for example:

$$C_6H_5-\overset{+}{N}\equiv N\ Cl^- \xrightarrow{\text{CuCl/HCl}} C_6H_5-Cl + N_2$$
$$\text{Chlorobenzene}$$

$$C_6H_5-\overset{+}{N}\equiv N\ Cl^- + HBr \xrightarrow{\text{CuBr/HBr}} C_6H_5-Br + HCl + N_2$$
$$\text{Bromobenzene}$$

These are known as *Sandmeyer reactions*.

(iii) **Fluorine**. When a solution of potassium tetrafluoroborate, KBF_4, is added to the solution of the diazonium salt, a precipitate of a diazonium fluoroborate is formed, for example:

$$C_6H_5-\overset{+}{N}\equiv N\ Cl^- + KBF_4 \rightarrow C_6H_5-\overset{+}{N}\equiv N\ BF_4^- + KCl$$

This salt, which is more stable than diazonium chlorides, is filtered and dried. On being heated carefully to about 120°, the aryl fluoride is formed:

$$C_6H_5-\overset{+}{N}\equiv N\ BF_4^- \rightarrow C_6H_5-F + N_2 + BF_3$$
$$\text{Fluorobenzene}$$

3. Replacement by the **nitrile** group. Copper(I) cyanide dissolved in aqueous potassium cyanide is added to the solution of the diazonium salt, for example:

$$C_6H_5-\overset{+}{N}\equiv N\ Cl^- + CuCN \xrightarrow{\text{KCN}} C_6H_5-CN + N_2 + CuCl$$
$$\text{Benzonitrile}$$

4. Replacement by a **hydrogen** atom. Hypophosphorous acid is added to the solution of the diazonium salt, for example:

$$C_6H_5-\overset{+}{N}\equiv N\ Cl^- + H_3PO_2 + H_2O \rightarrow C_6H_6 + N_2 + H_3PO_3 + HCl$$

(b) Reactions in which the nitrogen atoms are retained

1. Diazonium salts react with phenols and tertiary aromatic amines to form bright-coloured **azo-compounds**.

The reaction with phenols is carried out in alkaline solution (that is, with the phenoxide ion), for example:

[Reaction scheme: $C_6H_5-\overset{+}{N}\equiv N$ + $C_6H_5-O^-$ → $C_6H_5-N=N-C_6H_4-OH$]

[Reaction scheme: $C_6H_5-\overset{+}{N}\equiv N$ + Anion from naphthalen-2-ol → azo-coupled product (1-phenylazo-naphthalen-2-ol)]

AMINES

The products formed with naphthalen-2-ol are insoluble, mostly red compounds. Thus, the formation of a red precipitate when an amine is treated with nitrous acid and the solution is poured into an alkaline solution of naphthalen-2-ol shows that the amine is a primary aromatic one; this provides a useful test.

The reaction of diazonium salts with tertiary aromatic amines is carried out in neutral solution. An example is:

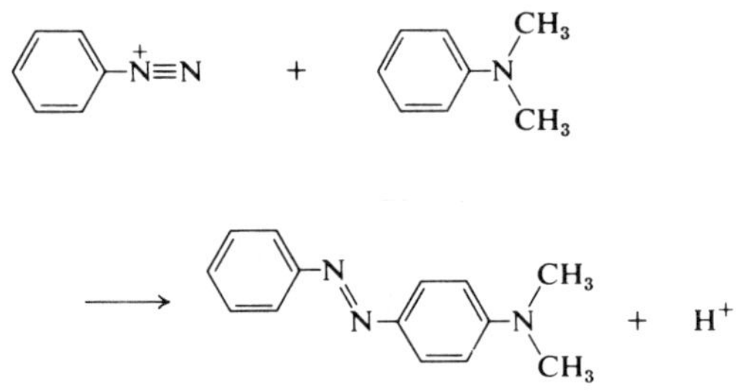

$$C_6H_5-\overset{+}{N}\equiv N + C_6H_5-N(CH_3)_2 \longrightarrow C_6H_5-N=N-C_6H_4-N(CH_3)_2 + H^+$$

Azo-compounds are coloured because the group —N=N— absorbs light; the group is known as a **chromophore**. The particular colour depends on what other substituents are present; these are called **auxochromes**. Many of the compounds are used as dyes; an example is Orange II, which is made by diazotising the sodium salt of 4-aminobenzenesulphonic acid and coupling the product to naphthalen-2-ol:

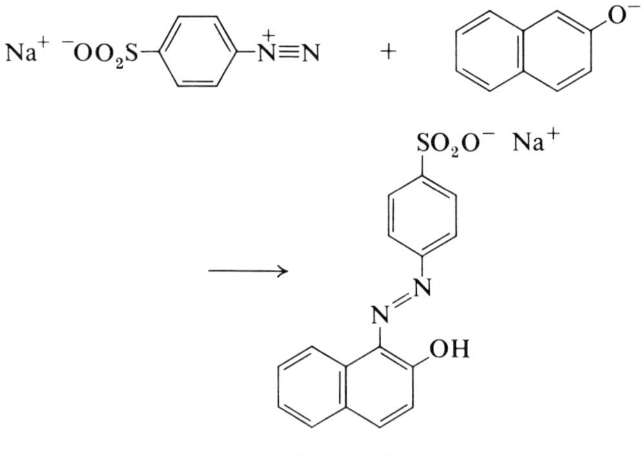

$$Na^+\ ^-OO_2S-C_6H_4-\overset{+}{N}\equiv N + \text{naphthalen-2-olate} \longrightarrow \text{Orange II}$$

Orange II

2. Diazonium salts are reduced by a solution of sodium sulphite to arylhydrazines, for example:

$$C_6H_5-\overset{+}{N}\equiv N\ Cl^- + 2Na_2SO_3 + 2H_2O \rightarrow$$
$$C_6H_5-NH-NH_2 + 2Na_2SO_4 + HCl$$
Phenylhydrazine

16.9 Practical work

Small-scale preparation of methylamine and methylammonium chloride: The Hofmann degradation

To 2 g of ethanamide and 2 cm³ of bromine in a boiling-tube, add 2 g of sodium hydroxide dissolved in 10 cm^3 of water, while shaking the tube under a stream of cold water. A solution of sodium N-bromoethanamide is formed, which is pale yellow.

In a separate boiling-tube, dissolve 4 g of sodium hydroxide in 10 cm^3 of water, and pour this solution into the flask (Fig. 16.2). Place the solution of sodium N-bromoethanamide in the dropping funnel.

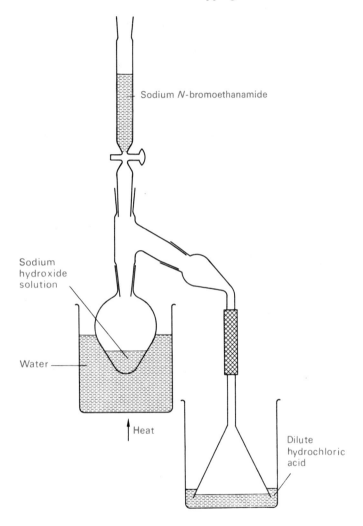

FIG. 16.2. *Preparation of methylammonium chloride by the Hofmann reaction*

Heat the flask to 70°C in a water-bath and add the sodium N-bromoethanamide solution dropwise. Keep the mixture at this temperature for a further 10 minutes. Remove the beaker of water and boil the solution, using a bunsen burner, to drive over methylamine. The base will be absorbed in the acid to yield methylammonium chloride, a salt.

Transfer the solution of the salt to an evaporating basin and evaporate the solution to dryness, using a water-bath.

Transfer a few crystals of methylammonium chloride to a test-tube and add dilute sodium hydroxide solution. Warm and test the gas evolved by (a) smell, (b) moist red and blue litmus papers.

If there is time, transfer the rest of the methylammonium chloride to a boiling-tube, add 5 cm³ of ethanol and warm the mixture in a beaker of boiling water. Filter (to remove any solid ammonium chloride which may have been formed from the alkaline hydrolysis of ethanamide), and allow the filtrate to cool. Filter off crystals of methylammonium chloride and dry them between pads of filter papers. Find the m.p.

Small-scale preparation of phenylamine

To 2 cm³ of nitrobenzene and 4 g of tin in a flask, add 10 cm³ of *concentrated* hydrochloric acid from the dropping funnel (Fig. 16.3). Shake the mixture, immersing the flask in cold water if the reaction becomes too vigorous.

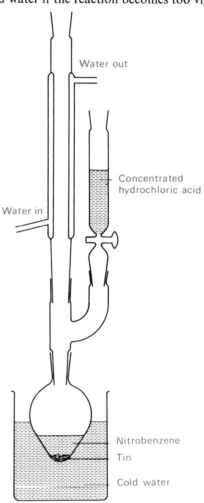

FIG. 16.3. *Preparation of phenylamine*

Remove the flask-head and heat the flask in boiling water for 30 minutes (to complete the reduction of nitrobenzene and to remove the excess of the nitro compound by steam distillation). If some tin remains, add *concentrated* hydrochloric acid dropwise until it has dissolved. Phenylammonium chloride and phenylammonium hexachlorostannate(IV) are soluble in water, but when excess of concentrated sodium hydroxide solution is added, phenylamine is liberated as an oil and the tin salts are precipitated and then dissolve to sodium stannate(IV).

Set up the flask for steam distillation (Fig. 2.6), keeping a low bunsen flame below the centre flask to prevent undue condensation of steam from the steam flask. Distil until no more oily drops of phenylamine can be seen distilling over. Transfer the distillate to a separating funnel and add about 2 g of sodium chloride and shake. (The salt reduces the solubility of phenylamine in water). Run off the organic layer into a test-tube and add some potassium carbonate to remove water. After standing for a few minutes, the turbidity should disappear.

Redistil phenylamine, using an air condenser (cf. Fig. 2.3), collecting the fraction boiling between 180 and 185°C.

Reactions of ammonia, methylamine, phenylamine and (phenylmethyl)amine

To compare their reactions as bases

1. (a) Place 2 cm^3 of an aqueous solution of ammonia, 2 cm^3 of an aqueous solution of methylamine, 3 drops of phenylamine and 3 drops of (phenylmethyl)amine in separate test-tubes, and test the solutions with Universal Indicator solution.

(b) To 3 drops of phenylamine and 3 drops of (phenylmethyl)amine in separate test-tubes, add 5 drops of water. Shake to see whether the amine dissolves. Then add dilute hydrochloric acid dropwise, with shaking.

2. To 1 g of methylammonium chloride in a test-tube, add 5 cm^3 of dilute sodium hydroxide solution. Boil gently. Test the vapour with moist red litmus paper. Note the smell of the gas.

Repeat the experiment with phenylammonium chloride.

Formation of amides

3. (a) To 5 drops of phenylamine in a test-tube, add 2 drops of ethanoyl chloride (**CARE**). A white precipitate of *N*-phenylethanamide can be seen.

Details of a small-scale preparation of *N*-phenylethanamide are given on p. 258.

(b) *Schotten-Baumann reaction*. To 3 drops of phenylamine in a test-tube, add 5 cm^3 of dilute sodium hydroxide solution. Shake to form a fine oily suspension and add 5 drops of benzoyl chloride. Fit a cork to the tube and shake for a minute. Cool the tube and contents under the tap, removing the cork periodically to release the pressure. Filter the residue of *N*-phenylbenzamide, using a small Buchner funnel and flask.

Transfer the residue to another test-tube and dissolve it in the smallest possible quantity of hot ethanol. Filter if necessary and allow the solution to cool. (If the crystals do not appear within 5 minutes, add a few drops of water, and 'scratch' the sides of the test-tube with a glass rod to 'seed'.) Filter the crystals, and dry them between pads of filter papers. M.p. 163°C.

Reactions with nitrous acid

4. Make up a solution of about 3 g of sodium nitrite in 10 cm^3 of water, and cool it to about 5°C.

In separate test-tubes, make up *solutions* of a few crystals of methylammonium chloride in water, a few drops of phenylamine in *concentrated* hydrochloric acid and of (phenylmethyl)amine in *concentrated* hydrochloric acid. Cool these solutions and add the cool solution of sodium nitrite to each. Observe what happens (a) in the cold, (b) when the phenylamine solution is warmed.

The reaction between phenylamine and nitrous acid is studied further on p. 258.

Reaction of the phenyl group

5. To 5 drops of phenylamine in a test-tube, add *concentrated* hydrochloric acid dropwise until the base dissolves. Add bromine water carefully until nothing more is seen to occur.

Compare the rate of, and the products formed by, the reaction of bromine and phenylamine with those of bromine with (a) benzene (p. 108), (b) phenol (p. 156).

Small-scale preparation of N-phenylethanamide

Cool a flask containing 4 cm^3 of glacial acetic acid and 4 cm^3 of ethanoic anhydride in a beaker of cold water, and add 4 cm^3 of phenylamine dropwise with gentle shaking.

Reflux the mixture for 30 minutes (cf. Fig. 2.1) and then pour the liquid into a beaker containing 100 cm^3 of water. Filter the crystals of N-phenylethanamide using a Buchner funnel and flask, and wash them with cold water.

Transfer the crystals to a boiling-tube, dissolve them in the minimum quantity of boiling water and allow the solution to cool. Filter the crystals again and dry between pads of filter paper. M.p. 114°C.

Small-scale preparation of benzenediazonium chloride solution

To 3 cm^3 of phenylamine and 10 cm^3 of water in a boiling tube, add 8 cm^3 of *concentrated* hydrochloric acid. Cork and shake the tube until the amine has dissolved.

Cool the solution in a beaker containing ice to about 5°C, and add a solution of sodium nitrite (3 g in 8 cm^3 of water), previously cooled to 5°C. Make sure that the temperature of the mixture does not rise above 10°C.

Reactions of benzenediazonium chloride solution

(a) *Replacement reactions*

1. Boil 2 cm^3 of the diazonium solution. Note the odour of phenol which separates as an oily liquid.

2. To 2 cm^3 of the diazonium solution at 5°C, add, drop by drop, 1 cm^3 of a 10 per cent solution of potassium iodide, previously cooled to 5°C. Allow to stand for 5 minutes and then gently boil. Observe oily drops of iodobenzene.

3. *Sandmeyer reaction.* Dissolve 1 g of copper(I) chloride in *concentrated* hydrochloric acid in a test-tube. Put the test-tube in a beaker of water at 60°C, and add 2 cm^3 of benzenediazonium chloride solution. Note whether the product is soluble in water and its smell.

(b) *Coupling reactions*

4. To 2 cm^3 of benzenediazonium chloride solution in a test-tube, add phenylamine (cooled to below 5°C) dropwise.

5. Dissolve 2 or 3 crystals of phenol in 2 cm^3 of dilute sodium hydroxide solution. Cool the solution in ice, and add the diazonium solution drop by drop. A yellow precipitate of a dye, the sodium salt of 4-hydroxyazobenzene, is obtained.

6. Repeat experiment 5 using napthalen-2-ol instead of phenol.

AMINES

Reactions of glycine

1. Test the solubility of glycine in (a) water, (b) ethanol, (c) ether.

(a) *Properties of the amino group*

2. Dissolve about 1 g of glycine in the minimum quantity of *concentrated* hydrochloric acid. Cool the mixture and observe whether a white crystalline solid is formed.

3. To about 1 cm^3 of an ice-cold solution of sodium nitrite in a test-tube, add 1 cm^3 of dilute hydrochloric acid. Some decomposition of the nitrous acid formed will occur. When the effervescence has subsided, introduce a few crystals (or a few drops of a concentrated aqueous solution) of glycine. Effervescence occurs again as nitrogen is evolved.

(b) *Properties of the carboxyl group*

4. To a solution of glycine in water, add some solid sodium hydrogen carbonate.

(c) *Properties due to both groups*

5. To a solution of 1 g of glycine in 10 cm^3 of water, add copper(II) carbonate until it is in excess. Filter the mixture using a Buchner funnel, and transfer the filtrate to a boiling-tube and allow it to stand. Observe whether crystals are formed, their colour and shape. Suggest what reaction has taken place.

16.10 Questions

1 Give **two** methods by which pure ethylamine may be prepared. How does ethylamine react with (a) iodoethane, (b) sodium nitrite and dilute hydrochloric acid, (c) ethanoic anhydride, (d) trichloromethane and ethanolic potash?

2 Outline **two** general methods which could be used to prepare a pure primary aliphatic amine, and state whether they could be applied to the preparation of phenylamine.
Give **three** types of reaction which both ethylamine and phenylamine undergo.
Describe **two** tests which may be used to distinguish between solutions of ethylamine and phenylamine in dilute hydrochloric acid.

3 An organic base A contains 61·01 per cent C, 15·25 per cent H, and 23·73 per cent N. When treated with nitrous acid A yields an alcohol B, and nitrogen is evolved. B contains 60·00 per cent C, and 13·33 per cent H, and on careful oxidation yields C, which has a vapour density of 29. C forms an oxime and an addition compound with sodium hydrogen sulphite, but does not react with Fehling's solution. Suggest structures for A, B and C, and indicate the course of the above reactions. (L)

4 A compound A gave on analysis C, 61·0 per cent; H, 15·2 per cent; N, 23·7 per cent. Treatment of A with acid and sodium nitrite yielded a compound B of molecular formula C_3H_8O. Oxidation of B with chromic acid gave C with a molecular formula C_3H_6O. The product C gave a positive iodoform reaction and formed a crystalline derivative with sodium hydrogen sulphite but did not react with ammoniacal silver nitrate. The compound B when treated with ethanoic anhydride gave a product D corresponding to $C_5H_{10}O_2$.
Deduce the identity of A, B, C and D. Explain the sequence of reactions by means of equations involving structural formulae for the organic molecules. (W)

5 A pungent-smelling liquid A was analysed and found to contain carbon, hydrogen, chlorine and possibly oxygen. Upon reaction with aqueous ammonia a neutral compound B was formed. When this was treated with bromine and potassium hydroxide a base C resulted. In acidic solution substance C reacted with sodium

AMINES

nitrite, giving nitrogen and an alcohol *D*. Upon mild oxidation, the alcohol *D* was converted into compound *E* of molecular formula C_2H_4O which gave a positive test with ammoniacal silver nitrate solution.

Describe the chemistry involved in these reactions and identify the compounds *A*, *B*, *C*, *D* and *E*. (L)

6 How would you prepare a specimen of phenylamine in the laboratory starting from nitrobenzene? Compare the reactions, if any, of phenylamine with those of methylamine towards the following reagents: (a) nitrous acid, (b) ethanoic anhydride, (c) water. (L)

7 Which of the following compounds are only sparingly miscible with water: ethyl ethanoate, nitrobenzene, diethyl ether, ethanol, propanone, 1,2-dibromoethane, tetrachloromethane?

Describe how you would purify a sample of phenylamine and determine the purity of the product. How would you demonstrate that phenylamine contains nitrogen? (W)

8 State, with equations, **three** methods by which primary amines can be prepared.

How, and under what conditions, do primary amines react with (a) nitrous acid, (b) ethanoic anhydride, and (c) dilute sulphuric acid?

Describe the chemical properties and reactions of glycine.

9 Phenylamine (b.p. 184°C) is prepared in the laboratory by the *reduction of nitrobenzene with tin and concentrated hydrochloric acid*. When the reduction is complete, this mixture is made alkaline with an *excess* of *sodium hydroxide* and *steam distilled*. *Salt is added* to the distillate, and then the resulting solution is extracted *twice* with *ether*.

Explain fully the reasons for the instructions in the above preparation which are printed in italics. (O)

10 What happens when an organic compound containing nitrogen is heated with copper(II) oxide?

Calculate the percentage of each of the elements in phenylamine, $C_6H_5NH_2$.

How does phenylamine react with (a) hydrochloric acid, (b) ethanoyl chloride, (c) bromine water, (d) nitrous acid? (AEB)

11 Define the term *base*. Arrange the following compounds in order of increasing base strength: ammonia, phenylamine, ethanamide, methylamine, ethylamine.

Compare the action of nitrous acid on phenylamine and ethylamine. Outline how ethanamide may be converted into (a) methylamine, (b) ethylamine (W)

12 Starting with ethanol, describe in outline how you would prepare propylamine (1-aminopropane)?

What is the action on propylamine of (a) iodomethane; (b) ethanoic anhydride?

Propylamine (*x* g) was dissolved in dilute hydrochloric acid and sodium nitrite solution added. The volume of gas evolved measured 896 cm³ (measured at 0°C and 760 mm mercury pressure). Calculate the value of *x*. (O and C)

13 How would you prepare, in the laboratory, a pure specimen of phenylamine from nitrobenzene?

Compare and contrast the behaviour of phenylamine and methylamine towards (i) nitrous acid, (ii) bromine, (iii) sulphuric acid, (iv) ethanoyl chloride.

14 *A* is a gaseous organic compound containing carbon and hydrogen only. A volume of *A* requires an equal volume of hydrogen for complete reduction to *B*, the volumes of gases being measured at the same temperature and pressure. *A* reacts with hydrogen chloride to produce a very volatile liquid *C*, with a composition C, 37·2 per cent; H, 7·75 per cent; Cl, 55·0 per cent. *C* reacts with potassium cyanide to produce *D*.

D reacts with sodium hydroxide to evolve an alkaline gas; and with dilute sulphuric acid to produce *E*, of which 1·00 g dissolved in water requires 27·0 cm³ of 0·5M (0·5N) sodium hydroxide for complete reaction.

AMINES

D reacts with zinc dust and an excess of concentrated hydrochloric acid to produce a solution, which, when made strongly alkaline with sodium hydroxide and warmed, evolves an alkaline gas *F*.

Identify the compounds *A* to *F* and write equations for the reactions in the above scheme. Outline the preparation of a sample of *A* in the laboratory.

(C(T))

15 Read the following instructions for the preparation of phenylamine. Explain, as fully as possible, the reasons for using the *apparatus, techniques, materials and conditions* printed in italics.

'Into a 250 cm³ *wide-necked* flask, fitted with an *air condenser*, place 8·4 cm³ of *nitrobenzene* and 18 g of granulated *tin*. Pour about 6 cm³ of *concentrated hydrochloric acid* down the condenser. *Shake thoroughly for 5 minutes.* Continue to add the acid in *5 cm³ portions at 5 minute intervals* with continued shaking until 40 cm³ has been added. During the addition of acid the flask can be immersed in a *cold water bath* but this should not be done *more than is judged to be necessary*. *Heat* the mixture on a boiling water bath for *30 minutes*. Cool. Add *gradually* a solution of 30 g of *sodium hydroxide* in 50 cm³ of water until the precipitate dissolves and the mixture, after shaking, is *alkaline*. Cool. *Steam distil* until the distillate is no longer *turbid*. Saturate the distillate with *salt*, transfer to a separating funnel and add about 20 cm³ of *ether*. Shake, releasing the tap momentarily. Allow to stand, separate the upper layer and place it in a corked conical flask with some *anhydrous magnesium sulphate*. Shake for several minutes. *Filter* and distil the *lower fraction* of the filtrate using a *warm water bath* and *water condenser*. With the lower fraction now removed, distil the higher fraction by *direct heating over a wire gauze* using an *air condenser*. The yield of phenylamine is 7·0 g.' (S)

16 Outline the preparation of a solution of a benzenediazonium salt. (Full practical details are not required.)

How does benzenediazonium chloride react with (i) copper(I) chloride, (ii) phenol, (iii) potassium iodide, and (iv) water? What is the industrial importance of diazonium salts?

17 A compound *Z* is boiled with excess alkali. Ammonia is expelled. The resulting mixture is evaporated to dryness and on prolonged heating, benzene is evolved. Acidification of the cold residue causes an effervescence of carbon dioxide.

Suggest a structure for *Z* to account for the reactions described and name the compounds involved.

Outline briefly reactions for converting *Z* into (a) phenylamine, (b) isocyanobenzene, (c) (phenylmethyl)amine, $C_6H_5.CH_2.NH_2$.

Explain why ammonia is a weaker base than trimethylamine. (S)

18 What is the diazo reaction? Illustrate the use of this reaction to prepare each of the following compounds starting from phenylamine: (a) benzoic acid, (b) 2,4,6-tribromobenzene, (c) methyl phenyl ether, (d) phenylhydrazine, (e) 4-hydroxyazobenzene. (W(S))

19 A compound *A* has a molecular formula C_8H_9NO. Ammonia is liberated when *A* is boiled with alkali, and the residue when acidified, gives a compound *B*, $C_8H_8O_2$, which effervesces with sodium carbonate. Treatment of *A* with bromine and alkali yields a compound *C*, C_7H_9N, which can be diazotised. Suggest possible structures for *A* and indicate what experiments you would perform to establish its identity. (W(S))

20 A compound *X* is believed to have the structure
$$H_2N.CH_2.CONH.CH_3.$$

How would you:

(a) show that *X* contains nitrogen;

(b) show that *X* contains an amine group?

How would *X* react with:

(i) soda-lime;

(ii) sulphuric acid;

(iii) ethanoyl chloride?

Suggest a series of reactions by which *X* could be prepared from ethanoic acid.

(C(S))

21 Illustrate, with examples, the difference in:

(a) the acidity of the —OH group in ethanoic acid, ethanol and phenol;

(b) the basicity of the —NH$_2$ group in ethanamide, phenylamine and methylamine.

How do you account for such differences?

How would you expect the basicity of (phenylmethyl)amine C$_6$H$_5$CH$_2$NH$_2$ to compare with that of phenylamine and why? (L(S))

22 A compound was found to have the structural formula illustrated below. From your knowledge of the typical groups it contains give an account of the principal chemical properties which you would expect it to possess.

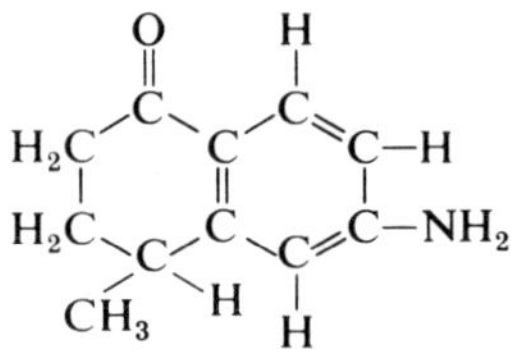

(L(X, S))

23 An optically active, acidic compound *A*, C$_5$H$_9$O$_3$N, on heating under reflux with 50 per cent sulphuric acid, gave two products, *B* and *C*. *C* was an acid of molecular weight 60 which, on treating with soda lime gave a combustible gas *D*. *B*, C$_3$H$_7$O$_2$N, was optically active, and, on heating with ethanol and sulphuric acid, gave an optically active base *E*, C$_5$H$_{11}$O$_2$N. Treatment of *E* with sodium nitrite and dilute hydrochloric acid gave an optically active neutral product *F*, C$_5$H$_{10}$O$_3$, which reacted with phosphorus pentachloride to give *G*, C$_5$H$_9$O$_2$Cl which was also optically active. Treatment of *G* with potassium cyanide and hydrolysis of the product gave an acid *H*, C$_4$H$_6$O$_4$, which was not optically active. *H* gave, on strong heating, another acid *J*, C$_3$H$_6$O$_2$, with loss of carbon dioxide. *J*, on heating with soda-lime, gave a combustible gas *K* which was different from *D*.

Assign structures to all the above compounds and explain the course of the reactions described. (L(S))

24 Describe degradative, analytical, and synthetic methods you would employ to establish the structure of 3-methylphenylamine. (W(S))

25 Describe one method by which glycine (aminoethanoic acid) may be prepared from ethanoic acid. How does glycine react with (a) hydrochloric acid, (b) sodium hydroxide, (c) sodium nitrite and dilute hydrochloric acid, (d) soda-lime?

26 An optically active compound *M*, C$_3$H$_7$O$_2$N, forms a hydrochloride, but dissolves in water to give a neutral solution. On heating with soda-lime, *M* yields *N*, C$_2$H$_7$N; both *M* and *N* react with nitrous acid, the former yielding a compound *P*, C$_3$H$_6$O$_3$, which on heating is converted to *Q*, C$_6$H$_8$O$_4$. Account for the above reactions and suggest how *M* may be synthesised.

27 A substance is stated to have the structure: HO.CH$_2$.CH$_2$.CH$_2$.NH$_2$. What experimental evidence would you require to verify the structure? How would this substance react with (a) nitrous acid, (b) ethanoic anhydride? (O. Schol.)

28 A compound *A*, C$_{14}$H$_{10}$N$_2$O, when heated with dilute sulphuric acid gave ammonium sulphate, a compound *B*, C$_8$H$_6$O$_4$, and a compound *C*, C$_6$H$_7$N (as its sulphate). Compounds *B* and *C* behaved as follows:

AMINES

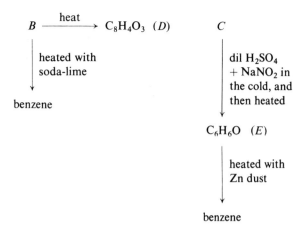

Suggest a formula for A and account for the above reactions. (C Schol.)

29 Explain the following changes, and deduce the nature of the compounds A to F:

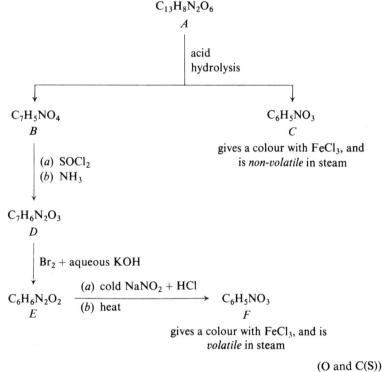

(O and C(S))

30 What are the principal reactions of the primary amino-group in an amine such as ethylamine? How do the properties of this group differ when it is present in an amide, e.g. ethanamide?

Is the structural formula $NH_2.CH_2.COOH$, satisfactory for glycine (aminoethanoic acid) which is soluble in water, very sparingly soluble in ether and benzene and which is still solid at 200°C? Give your reasons. (O Schol.)

31 An amino-acid NH_2RCOOH functions both as an acid and as a base in its reactions with water, according to the following equations.

(a) as an acid:

$$NH_2RCOOH + H_2O = NH_2RCOO^- + H_3O^+ \qquad K_a = 10^{-10}M.$$

(b) as a base:

$$NH_2RCOOH + H_2O = {}^+NH_3RCOOH + OH^- \qquad K_b = 2 \times 10^{-12} \text{ M}.$$

At the isoelectric point the concentrations of the cationic and anionic forms of the amino-acid are equal. Calculate the pH of an aqueous solution of the amino-acid at its isoelectric point from the acidic and basic dissociation constants given above and the fact that the ionic product of water is $10^{-14} M^2$.

The amino-acid is a crystalline solid which melts at 250°C. What do you deduce about its chemical structure in the crystal? (O Schol.)

Chapter 17

Nitro compounds

17.1 Introduction

General formula

$$R-NO_2$$

Nitro compounds have the structure

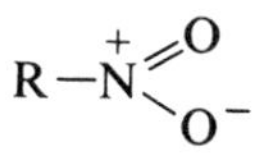

$$R-\overset{+}{N}\underset{O^-}{\overset{O}{\diagup\!\!\!\diagdown}}$$

where R is an alkyl group or an aromatic ring. They are isomeric with nitrites, R—O—N=O.

17.2 Nomenclature

Nitroalkanes are named by combining the prefix **nitro** with the name of the corresponding alkane, together with a number to indicate the position of the nitro group in the carbon chain where more than one position is possible.

Aromatic nitro compounds which contain the benzene ring are named as derivatives of nitrobenzene, for example:

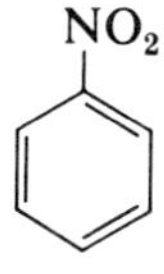

Nitrobenzene 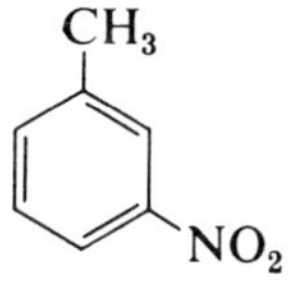   Methyl-3-nitrobenzene 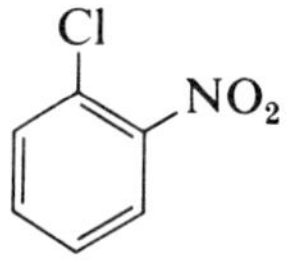   Chloro-2-nitrobenzene

Table 17.1. Some nitro compounds

NAME	FORMULA	B.P. (°C)
Nitromethane	CH_3-NO_2	101
Nitroethane	$CH_3CH_2-NO_2$	115
1-Nitropropane	$CH_3CH_2CH_2-NO_2$	132
2-Nitropropane	$(CH_3)_2CH-NO_2$	120
Nitrobenzene	$C_6H_5-NO_2$	210

17.3 Physical properties of nitro compounds

The lower nitroalkanes are colourless liquids which are sparingly soluble in water. Most aromatic nitro compounds are yellow crystalline solids (except for nitrobenzene and methyl-2-nitrobenzene which are yellow liquids) and are insoluble in water and may be purified by steam distillation.

17.4 Methods of preparation of nitro compounds

Laboratory methods

1. Nitroalkanes are prepared by the action of a solution of silver nitrite in ethanol on an alkyl halide, for example:

$$C_2H_5-Br + AgNO_2 \rightarrow C_2H_5-NO_2 + AgBr$$
$$\text{Nitroethane}$$

Ethyl nitrite (C_2H_5—ONO) is also formed (p. 118).

NITRO COMPOUNDS

2. Aromatic nitro compounds are prepared by the nitration of the aromatic compound with nitric acid. The choice of conditions depends on the reactivity of the aromatic compound. For benzene, a mixture of concentrated nitric acid and concentrated sulphuric acid is necessary (8.3):

$$C_6H_6 + HNO_3 \xrightarrow{H_2SO_4} C_6H_5-NO_2 + H_2O$$
$$\text{Nitrobenzene}$$

For compounds which are much less reactive than benzene (e.g. 1,3-dinitrobenzene), fuming nitric acid and fuming sulphuric acid are required, whereas for compounds which are much more reactive than benzene (e.g. phenol), dilute nitric acid is suitable.

Manufacture

Nitroalkanes are obtained by the reaction of alkanes with nitric acid in the vapour phase at about 350°C. Alkanes with more than two carbon atoms give mixtures of products which are separated by fractional distillation, for example:

$$CH_3-CH_2-CH_3 + HNO_3 \begin{cases} \rightarrow CH_3-CH_2-CH_2-NO_2 + H_2O \\ \quad\quad\text{1-Nitropropane} \\ \rightarrow CH_3-\underset{NO_2}{CH}-CH_3 + H_2O \\ \quad\quad\text{2-Nitropropane} \end{cases}$$

Aromatic nitro compounds are obtained industrially by the same methods as are used in the laboratory.

17.5 Chemical properties of nitro compounds

1. The most important general property of nitro compounds is their reduction to primary amines. This can be effected with hydrogen on Raney nickel:

$$R-NO_2 + 3H_2 \xrightarrow{\text{Ni as cat.}} R-NH_2 + 2H_2O$$

and with lithium aluminium hydride:

$$R-NO_2 \xrightarrow{LiAlH_4} R-NH_2$$

Aromatic nitro compounds are conveniently reduced with tin and hydrochloric acid (16.4), for example:

$$C_6H_5-NO_2 \xrightarrow{Sn/HCl} (C_6H_5-\overset{+}{N}H_3)_2SnCl_6^{2-} \xrightarrow{NaOH} C_6H_5-NH_2$$
$$\text{Phenylamine}$$

2. Nitroalkanes are hydrolysed by mineral acids to form a carboxylic acid and hydroxylamine:

$$R-CH_2-NO_2 + H_2O \xrightarrow{H_2SO_4} R-CO_2H + NH_2OH$$

3. Aromatic nitro compounds undergo electrophilic substitution in the

NITRO COMPOUNDS

aromatic ring. The nitro group reduces the reactivity of the ring and is 3-directing; for example, nitrobenzene undergoes nitration less readily than benzene and gives 1,3-dinitrobenzene (p. 105).

17.6 Uses of nitro compounds

Aliphatic nitro compounds

1. Solvents in industry, particularly for plastics and dyes. They are very useful as they have medium boiling points and do not have obnoxious smells.
2. Fuels for small engines.
3. As a starting material. For example, in the preparation of primary amines and carboxylic acids.

Aromatic nitro compounds

1. In the preparation of aromatic amines, used for the production of dyes (16.8).
2. A particularly important aromatic nitro compound is methyl-2,4,6-trinitrobenzene (trinitrotoluene, T.N.T.), a powerful explosive made by the nitration of methylbenzene with a mixture of fuming nitric and fuming sulphuric acids:

$$\text{C}_6\text{H}_5\text{CH}_3 + 3\text{HNO}_3 \xrightarrow{\text{H}_2\text{SO}_4} \text{C}_6\text{H}_2(\text{CH}_3)(\text{NO}_2)_3 + 3\text{H}_2\text{O}$$

T.N.T.

17.7 Practical work

Small-scale preparation of nitrobenzene

The experiment must be carried out in a fume-cupboard, and take great care not to breathe in benzene vapour. Benzene is highly toxic and must only be used under supervision.

To 7 cm³ of concentrated nitric acid in a flask, add slowly 9 cm³ of *concentrated* sulphuric acid, shaking and cooling the flask under a stream of running water.

Partly immerse the flask in a beaker of cold water and introduce, in small portions, 3 cm³ of benzene, using a dropping pipette. Shake the mixture gently after each addition and do not allow the temperature to rise above 50°C at this stage. When all the benzene has been added, maintain the flask at 60°C in a water-bath for 30 minutes. Gently swirl the contents of the flask occasionally, to ensure proper mixing of the reactants.

Allow the flask to cool, and transfer the contents to a separating funnel. Remove and discard the lower layer. Wash the nitrobenzene with successive 10 cm³ quantities of water, dilute sodium carbonate solution, and then water. In each case, separate and retain the lower layer.

Transfer the nitrobenzene to a stoppered test-tube, and add a few small lumps of anhydrous calcium chloride. When the liquid clears, decant and distil the nitrobenzene, using the condenser as an air condenser. Collect the fraction boiling between 207° and 211°C (cf. Fig. 2.3). Yield about 6 cm³.

Small-scale preparation of 1,3-dinitrobenzene

To 8 cm³ of concentrated nitric acid in a flask, add slowly 10 cm³ of con-

NITRO COMPOUNDS

centrated sulphuric acid, shaking and cooling the flask under a stream of running water.

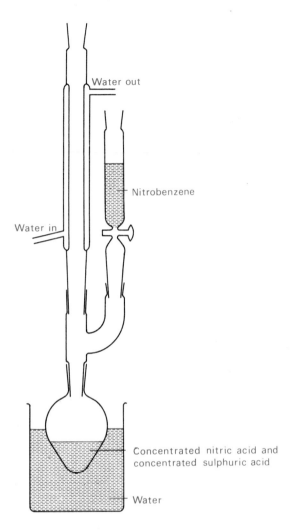

FIG. 17.1. *Preparation of 1,3-dinitrobenzene*

Arrange the apparatus (Fig. 17.1) and add 6 cm³ of nitrobenzene dropwise from the dropping funnel, gently shaking the flask.

Boil the water in the beaker, and heat the mixture for about 40 minutes.

Pour the mixture into 150 cm³ of cold water in a beaker and stir the contents from time to time for about 15 minutes. Filter off the solid and wash it with distilled water (Fig. 2.8).

Squeeze the crude product between filter papers to absorb any liquid impurities, and then place it in a boiling tube. Add about 20 cm³ of ethanol (or methylated spirit) and heat in a beaker of boiling water. (If there are any solid impurities, filter.) Allow the solution to cool and filter the product.

Dry the crystals between pads of filter paper (and if possible in an oven at 70°C). Find the m.p. of 1,3-dinitrobenzene.

Comparison of the rates of nitration of aromatic compounds

1. *Nitration of methylbenzene*

(a) Dissolve about 2 g of sodium nitrate in 10 cm³ of dilute sulphuric acid in a test-tube. Add 5 drops of methylbenzene, shake well and pour the mixture into a beaker containing about 10 cm³ of cold water. Observe whether nitration has occurred (a dense pale yellow liquid with a characteristic smell of almonds should be formed).

(b) Place 10 drops of *concentrated* nitric acid in a test-tube, and add 10 drops of *concentrated* sulphuric acid, shaking and cooling the test-tube under a stream of cold water. Add the mixed acids to 5 drops of methylbenzene in another test-tube. Shake the mixture under a stream of cold water and then pour it into a beaker containing about 10 cm³ of cold water. Note the colour and smell of the organic compound.

2. *Nitration of chlorobenzene*

(a) Repeat experiment 1 (a), using chlorobenzene instead of methylbenzene.

(b) Repeat experiment 1 (b), using chlorobenzene instead of methylbenzene. The mixture of chlorobenzene and the mixed acids should be gently warmed for 2–3 minutes, and then cooled before being poured into the beaker containing the cold water.

3. *Nitration of phenol*

Dissolve about 2 g of sodium nitrate in 10 cm³ of dilute sulphuric acid in a test-tube. Cool the solution by placing the test-tube in a beaker of ice.

In a second test-tube, dissolve 1 g of phenol in 2 cm³ of water (it may be necessary to warm the mixture). Add the solution of phenol dropwise to the solution of sodium nitrate, making sure that the temperature does not rise above 15°C. Allow the reaction mixture to stand for about 1 hour, decant the solution from the dark brown (black) solid formed, and wash it twice, in a separating funnel, with small amounts of water, discarding the aqueous layers.

Transfer the organic liquid to a small beaker and remove the last drops of water with a dropping pipette or the edge of a filter paper.

Identify the products by thin-layer chromatography (p. 27).

Reduction of aliphatic and aromatic nitro compounds

1. To 3 drops of nitrobenzene in a test-tube, add 1 cm³ of water, 1 cm³ of *concentrated* hydrochloric acid and 2 or 3 small pieces of tin. Warm. The phenylamine formed goes into solution in the form of its cation $C_6H_5-NH_3^+$.

Make alkaline with dilute sodium hydroxide solution, to liberate the phenylamine as an oil. Add enough alkali to dissolve the tin(IV) oxide formed as sodium stannate(IV).

Test for phenylamine by adding 1 drop of the oil to an aqueous suspension of bleaching powder. If phenylamine is present, a blue colour will appear.

2. To 1 cm³ of nitroethane in a test-tube, add 1 cm³ of dilute sodium hydroxide solution and a small quantity ($\frac{1}{2}$ cm on the end of a wooden splint) of powdered aluminium. Warm gently to start the reaction.

When the evolution of hydrogen ceases, warm the solution, note the characteristic smell of ethylamine and test the vapour with moist red litmus paper.

NITRO COMPOUNDS

17.8 Questions

1. Describe the preparation of a pure sample of 1,3-dinitrobenzene from nitrobenzene.
   How would you distinguish between the members of the following pairs by one chemical test in each case:
   (i) chlorobenzene and bromobenzene,
   (ii) nitrobenzene and phenylamine,
   (iii) bromoethane and 1,1-dibromomethane?

2. Describe in detail two experiments which you have seen or performed to illustrate the nitration of the benzene ring under varying conditions, and outline the purification of the products.

3. Describe briefly the preparation of a sample of pure nitrobenzene from benzene. Is it more difficult to nitrate nitrobenzene than to nitrate benzene?
   Draw the structural formulae of the compounds obtained when nitrobenzene is (a) heated with concentrated sulphuric acid; (b) reduced with a metal and acid (e.g. tin and hydrochloric acid).

4. Write structural formulae for the isomers corresponding to the molecular formula C_8H_{10} which contain a phenyl group. Give structural formulae for the different main mononitration products of each isomer. Assign possible structural formulae to $C_8H_8(NO_2)_2$ which can yield only two different nuclear monobromosubstitution products. (W)

5. A compound has the structural formula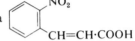
   Outline experiments to
   (a) confirm its molecular weight;
   (b) identify the functional groups present;
   (c) show how many acidic hydrogens are present;
   (d) convert the compound to

   (JMB Syllabus B)

6. Four bottles contain benzene, ethanonitrile, phenylamine and nitrobenzene, respectively, but the labels have fallen off and have become mixed up. What physical and chemical tests would you apply to these substances so that the bottles may be correctly re-labelled? (L)

7. Illustrate the directive influence of groups in aromatic nuclear substitution by giving the names and structural formulae of the main organic products formed by the reaction of a mixture of concentrated nitric and sulphuric acids on (a) methylbenzene, (b) benzoic acid, (c) methyl-2-nitrobenzene, (d) chlorobenzene.
   What is the active nitrating species in a mixture of concentrated nitric and sulphuric acids, and how may the reactivity of the acid mixture be increased?
   What procedure is adopted to protect the amino-group when phenylamine is nitrated with the concentrated mixed acids and what is the main product of nitration under such conditions? (W)

8. A compound A, $C_{11}H_{11}NO_4$, on treatment with aqueous sodium hydroxide yields ethanol and a compound B, $C_9H_7NO_4$, which is soluble in aqueous sodium carbonate with effervescence. Catalytic reduction of B gives C, $C_9H_{11}NO_2$, which can be diazotised. Vigorous oxidation of B gives a compound which can be prepared by the action of nitric and sulphuric acids on benzoic acid.
   Discuss these reactions.

9. A compound A contains C, 61·1 per cent, H, 5·1 per cent, N, 10·2 per cent, and has a vapour density of about 70. When A is treated with a mixture of concentrated sulphuric and nitric acids in the cold, a simple substance B is formed having the composition C, 46·1 per cent, H, 3·3 per cent, N, 15·4 per cent. Discuss the structures of A and B.

Chapter 18

Naturally occurring compounds

18.1 Introduction

Both animals and plants contain a large number of different types of organic compounds. Some constitute the structural materials of living organisms; for example, the carbohydrate **cellulose** is the principal constituent of the cell walls of plants, and **proteins** are the principal constituents of all living cells in animals and plants. Other organic compounds act as a source of energy, such as the carbohydrate **glucose**. Others act as catalysts for the essential chemical reactions which take place in living organisms; they are called **enzymes** and consist of proteins which in some cases are combined with other compounds. Finally, some organic compounds found in living systems are simply the end products of chemical processes and have no function; an example is **carbamide** (urea), $H_2N-CO-NH_2$, in human urine. The proteins and the carbohydrates are of special importance and are discussed in detail in this chapter.

18.2 Proteins

Introduction

A significant proportion of an animal is in the form of protein; for a human, the figure is about 15 per cent, of which muscle and blood account for 35 per cent, cartilage 20 per cent and hair 10 per cent. All animals require a constant supply of new protein, both for replacement of old protein and for growth.

Proteins are derived from α-amino-acids which are joined together *via* the amino group of one acid and the carboxyl group of the next, with the elimination of a molecule of water between the two. When two α-amino-acids are joined in this way, for example,

$$H_2N-CH_2-CO-NH-\underset{\underset{CH_3}{|}}{CH}-CO_2H$$

from glycine and α-alanine, the molecule is described as a **dipeptide**, the group —CO—NH— being known as the **peptide link**. A **tripeptide** is made up of three α-amino-acid molecules, and a **polypeptide** of up to about 40. A protein contains more than about 40 amino-acid residues although the distinction between a polypeptide and a protein is an arbitrary one (the term 'residue' is used for an α-amino-acid which has lost the elements of water in forming a peptide or protein).

Animals obtain α-amino-acids from proteins in the diet. These proteins are hydrolysed first to smaller units, e.g. tripeptides and dipeptides, and then to α-amino-acids by the enzyme (p. 275) pepsin in the stomach (under acid conditions) and by the enzymes trypsin and chymotrypsin in the intestines (pH 6–8). The amino-acids pass into the blood stream and then to the liver and other tissues where, under the influence of nucleic acids (p. 275), they are converted into proteins required by the body.

All plants need a supply of nitrogen in order to produce proteins. Some, for example those of the family *Leguminosae*, fix nitrogen directly from the air by means of bacteria in their root nodules; most other plants take in their nitrogen from the soil in the form of ions such as nitrate.

Naturally occurring proteins are made from a selection of about 20 different α-amino-acids. Of these, about 12 can be synthesised in the body from other amino-acids, for example:

$$\underset{\text{Phenylalanine}}{\text{C}_6\text{H}_5\text{—CH}_2\text{—CH(NH}_2\text{)—CO}_2\text{H}} \xrightarrow[\text{Enzyme}]{\text{'O'}} \underset{\text{Tyrosine}}{\text{HO—C}_6\text{H}_4\text{—CH}_2\text{—CH(NH}_2\text{)—CO}_2\text{H}}$$

However, eight are described as **essential** α-amino-acids; their residues must be present in the protein diet since they cannot be synthesised in the human body. Examples of α-amino-acids present in proteins, those which are underlined being essential, are:

$$\underset{\text{Glycine}}{\text{H}_2\text{N—CH}_2\text{—CO}_2\text{H}} \qquad \underset{\alpha\text{-Alanine}}{\text{H}_2\text{N—CH(CH}_3\text{)—CO}_2\text{H}} \qquad \underset{\underline{\text{Valine}}}{\text{H}_2\text{N—CH(CH(CH}_3\text{)}_2\text{)—CO}_2\text{H}}$$

$$\underset{\underline{\text{Phenylalanine}}}{\text{H}_2\text{N—CH(CH}_2\text{C}_6\text{H}_5\text{)—CO}_2\text{H}} \qquad \underset{\text{Cysteine}}{\text{H}_2\text{N—CH(CH}_2\text{SH)—CO}_2\text{H}} \qquad \underset{\underline{\text{Lysine}}}{\text{H}_2\text{N—CH((CH}_2\text{)}_4\text{NH}_2\text{)—CO}_2\text{H}}$$

Isolation and purification of proteins

Most proteins occur in mixtures with other proteins of closely similar properties and their separation is therefore difficult. Sometimes the careful adjustment of the pH of a solution of proteins results in one being precipitated. Other methods include column chromatography (2.6).

It is also difficult to determine whether a protein is pure or not, for they do not have sharp melting points but decompose on strong heating. One method of testing is by the use of the **ultracentrifuge**: the protein solution is spun at very high speeds, and the protein moves to the outer end of the spinning cell, owing to centrifugal force, at a rate which depends on its size. A special optical system enables the solution to be photographed during this process and reveals the moving protein. If impurities of different formula weight are present, they travel at different rates and consequently can be detected in the resulting photographs. The rate of sedimentation of the protein also enables its formula weight to be determined.

Structure of proteins

The physical and chemical properties of a protein are determined by its detailed structure, which can be subdivided into three aspects: the nature of the constituent amino-acid residues, the sequence of these residues and the shape of the molecule.

The shape is determined by the distances between certain functional groups, and two particular types of bond are important in determining it. One type is a covalent bond between two sulphur atoms, —S—S—; this can occur when the protein contains two or more cysteine residues, which can be linked by oxidation:

NATURALLY OCCURRING COMPOUNDS

$$\underset{\underset{NH}{|}}{\overset{\overset{CO}{|}}{CH}}-CH_2-SH + HS-CH_2-\underset{\underset{NH}{|}}{\overset{\overset{CO}{|}}{CH}} \rightarrow \underset{\underset{NH}{|}}{\overset{\overset{CO}{|}}{CH}}-CH_2-S-S-CH_2-\underset{\underset{NH}{|}}{\overset{\overset{CO}{|}}{CH}} \; (+2H)$$

This can happen even when the cysteine residues are separated by a large number of other residues, so that very large rings ('loops') are formed within the protein:

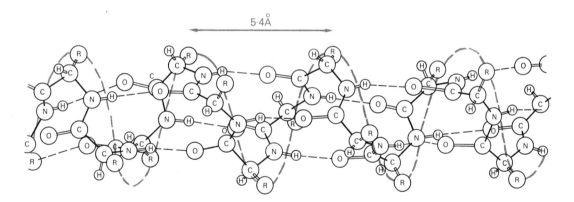

The second type of bond is the hydrogen-bond, which occurs between the N—H group of one residue and the C=O group of another:

$$\underset{/}{\overset{\backslash}{N}}-H---O=\underset{\backslash}{\overset{/}{C}}$$

This can lead to coiling of the protein chain, and the structure is described as an α-helix. In the diagram, the spacing between the turns is 5·4 Å.

FIG. 18.1. *α-Helix of a protein. The change in angle between one unit and the next occurs at the carbon atom to which the side group R is attached. The helix is held rigid by hydrogen-bonding*

The determination of the structure of a protein requires elucidating each of these three structural aspects. The first problem is to find out which α-amino-acid residues are present and the relative amounts of each, and this has become a relatively easy matter since the advent of chromatographic methods. The protein is hydrolysed with mineral acid and the resulting mixture of α-amino-acids is separated either by paper chromatography (2.6) or, better, by an ion-exchange resin.

The second problem, to find the sequence of the amino-acid residues, is very much more difficult and has so far been solved in only a small number of cases, and these for proteins of relatively low formula weight. One approach is based on chemical degradation: the protein is broken down into smaller units in various ways; these units are separately identified and, hopefully, it is shown that the various components could only have been arranged in the protein in one particular order. (To take a simple example, if a protein containing six amino-acids, A, B, C, D, E, F gives the units AB, CD, and EF by one method of degradation and the units BC, DE, A and F by another, its structure must be A—B—C—D—E—F.) By methods of this type, Sanger determined the structure of insulin (p. 275).

A second approach is to use X-ray diffraction analysis, and this is likely to prove the more important in the future.

The final problem is to determine the shape of the molecule. Here also X-ray diffraction is a powerful method, and the full details of the shapes of a small number of proteins have now been determined.

Fibrous proteins

The fibrous proteins (sometimes known as **structural proteins**) are essential constituents of skin, hair and muscle. Their peptide chains are generally coiled in the form of α-helices by hydrogen-bonding and are packed together to give a bundle of fibres. This structure accounts for their insolubility in water and their physical strength. In fibroin (silk), the chains are extended and the material cannot be stretched, but in keratin (in hair) the chains are folded and can be stretched to give a linear molecule; the molecule returns to its original form when the tension is released.

Globular proteins

Globular proteins are soluble in water (or can form colloidal solutions). Among examples are enzymes and protein hormones.

Enzymes are organic catalysts which enable complicated reactions to take place rapidly at body temperature. Some consist solely of a protein and others of a protein joined to another molecule (a **coenzyme**).

Enzymes are specific in their action, being able to catalyse only the breaking and making of one type of bond and, sometimes, one type of bond in a particular molecule. The shape of the protein enzyme is particularly important; it appears that the molecule in which reaction is to occur (the substrate) must fit exactly into the protein (often likened to a key fitting into a lock). Reaction then occurs within this complex of enzyme and substrate, after which the new molecule is released.

Protein hormones regulate growth and control metabolism. For example, insulin is required for the metabolism of carbohydrates; a deficiency of insulin leads to diabetes. Insulin is a protein of formula weight about 6000 which contains two polypeptide chains joined by —S—S— links. For his work on elucidating the structure of insulin, Sanger was awarded a Nobel Prize in 1958.

Biosynthesis of proteins

The synthesis of proteins within an animal—their **biosynthesis**—is controlled by **nucleic acids**.

Nucleic acids are polymers formed from **nucleotides**. A nucleotide is composed of a carbohydrate, a phosphate group and a nitrogen base, and in the nucleic acids the polymer chain consists of alternating carbohydrate and phosphate units; this is shown schematically in Fig. 18.2. There are only two general types of nucleic acid: **ribonucleic acid (RNA)**, in which the carbohydrate is ribose and the bases are cytosine and uracil (members of the pyrimidine group) and adenine and guanine (members of the purine group); and **deoxyribonucleic acid (DNA)**, in which the carbohydrate is deoxyribose and the bases are the same as in RNA except that thymine replaces uracil (Fig. 18.3). However, there is an enormous number of individual compounds in each of these two categories because of the variety of possible sequences for the four bases in each case.

In DNA, two nucleic acid chains, each in the form of a helix, are intertwined. The forces which hold the two helices together in this way are due to hydrogen-bonds between the bases in each, and these bonds are formed between specific pairs of bases: an adenine unit in one helix bonds to a thymine unit in the other, and likewise a guanine unit bonds to a cytosine unit (Fig. 18.4); Fig. 18.5 shows in detail the geometrical fit which results from these pairings. Thus, if the sequence of bases in one of the strands of DNA is adenine (A), guanine (G), thymine (T), cytosine (C), guanine (G), etc., then the sequence in the other strand is T—C—A—G—C—, etc. Watson, Crick and Wilkins were awarded a Nobel Prize in 1962 for elucidating the double-helix structure of DNA.

FIG. 18.2. *Alternating phosphate and carbohydrate residues in a nucleic acid. A base is attached to each carbohydrate. In (b), the bonding between the carbohydrate residues and the phosphate residues is shown. X represents H in deoxyribose and OH in ribose*

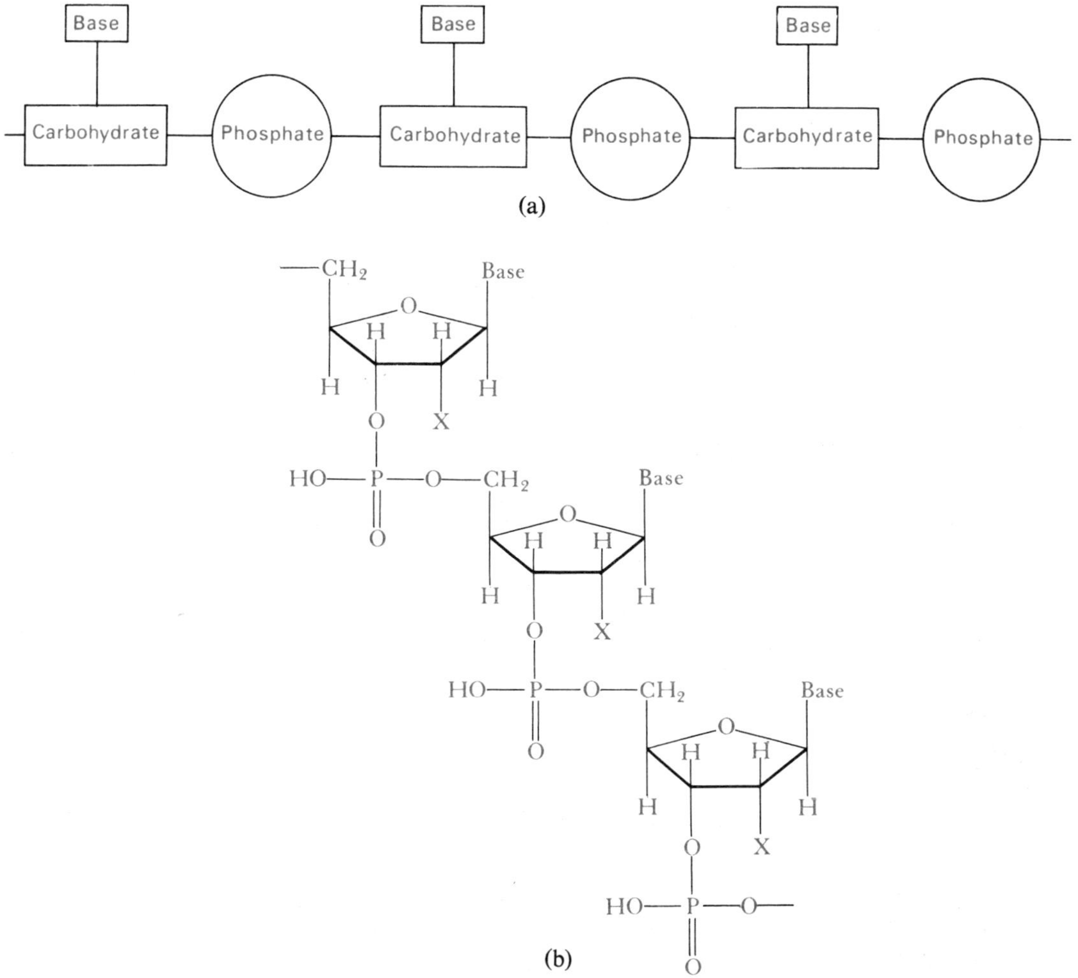

The DNA in the nucleus of a cell has two main functions. First, when the cell reproduces itself, by division, each molecule of a DNA produces two molecules of the same DNA. This is brought about by

FIG. 18.3. *The bases present in RNA and DNA. The bond that attaches the base to the carbohydrate is shown*

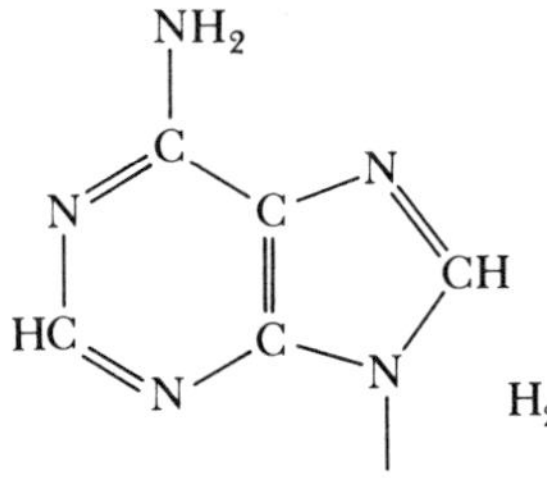

Adenine

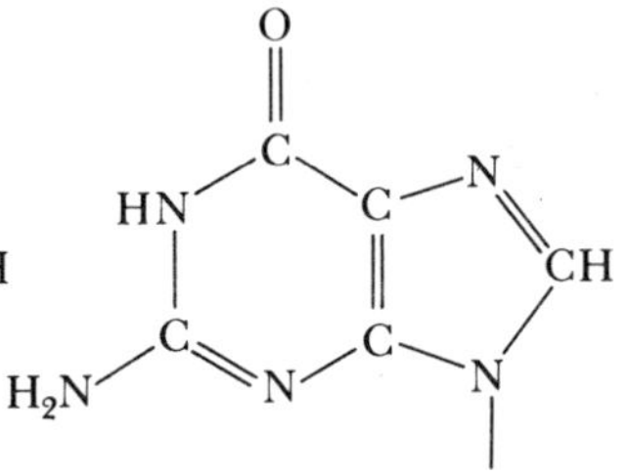

Guanine

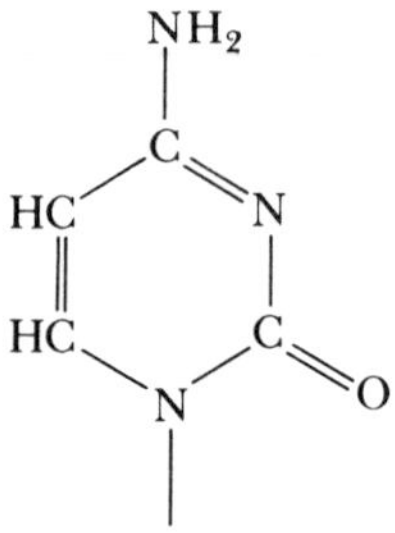

Cytosine

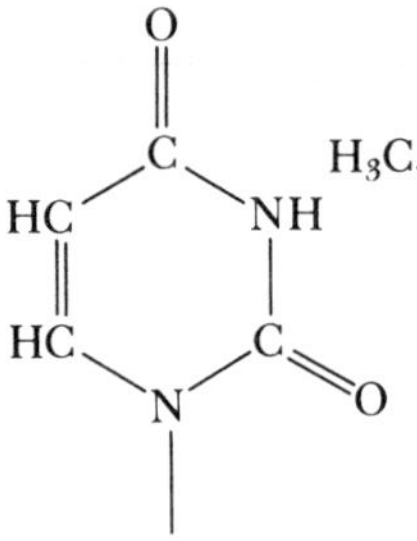

Uracil

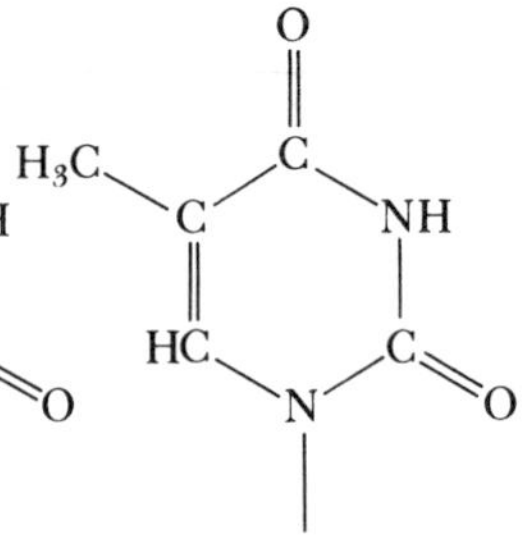

Thymine

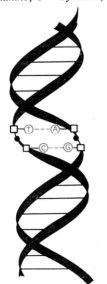

FIG. 18.4. *Part of the double helix of DNA. The helix consists of two strands of alternate carbohydrate (□) and phosphate (●) residues. To each carbohydrate residue, a base is attached. Hydrogen-bonding between two bases, one from each strand, holds the helix together. This is represented by the rungs of the ladder, and two of these rungs are shown in more detail (A = Adenine, T = Thymine, G = Guanine, C = Cytosine)*

the separation of the two helices of the DNA, each of which then acts as a template for the synthesis of a single strand of more DNA. This template action, like the method of intertwining of the helices in DNA, is dependent upon the specificity of the hydrogen-bonds formed by the base units; that is, nucleotides become associated with their 'partners' in the DNA chain and then link together to form a new strand of DNA. Thus, of two strands X and Y in the original DNA, X effects the synthesis of a new strand Y and Y effects the synthesis of a new strand X; the two new strands then yield a replica of the original molecule.

The second function of DNA is to control the synthesis of RNA. It effects this in essentially the same way, except that only one of the two strands of DNA acts as a template (since RNA has only one strand) and uracil plays the pairing role in the new RNA which thymine plays in DNA; hence, if the sequence of bases in the DNA strand is A—G—T—C—G—, etc., then the sequence in the new RNA is U—C—A—G—C—, etc. The length of an RNA chain is much shorter than that of a DNA chain, so that only a section of a DNA is required for determining the sequence of bases in an RNA, and one DNA, by the use of different sections, can be responsible for the generation of a number of different RNAs. A section of a DNA which controls the synthesis of a particular RNA corresponds to a **gene**.

Two types of RNA of special importance in protein synthesis are **transfer RNA** and **messenger RNA**; the formula weight of a transfer RNA is much less than that of a messenger RNA.

In protein synthesis, each of the α-amino-acids which is involved becomes attached, by a covalent bond, to a particular transfer RNA. In turn, the transfer RNA becomes associated with a particular part of a messenger RNA molecule. The specific hydrogen-bond pairs are again concerned in this association; a group of three adjacent bases in the transfer RNA fits on to the messenger RNA by the pairing of the appropriate base partners, a process referred to as **triplet coding**. For example, if the first six bases in the messenger RNA are A—G—U—C—G—A, the first transfer RNA which

FIG. 18.5. *The hydrogen-bonding between thymine and adenine and between cytosine and guanine*

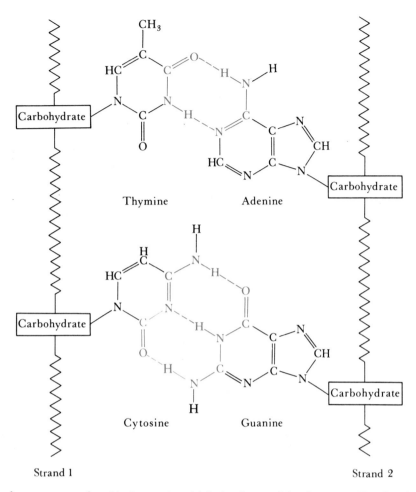

becomes associated is the one in which the three pairing bases are U—C—A and the second is the one in which the pairing bases are G—C—U. The two α-amino-acids attached to these two transfer RNA molecules then bond together by the formation of a peptide link. Further α-amino-acids are brought into position and then form peptide links with their neighbours in the same way, and eventually the complete protein is formed and released from the template.

In summary, the exact constitution of each protein which is synthesised in a living organism is governed by the precise structures of the messenger RNA molecules, and these in turn are governed by the natures of the DNA molecules within each cell. Hence, the DNA molecules, which are transmitted from one generation to the next, are responsible for determining the characteristics of the species.

Viruses

Viruses are composed of nucleic acid and protein, and some have been obtained in a crystalline form. Stanley was awarded a Nobel Prize in 1946 for his pioneering work which showed that tobacco mosaic virus is a crystalline compound of this type.

Viruses destroy living cells by injecting into them their nucleic acid. As a result, components of the cell are utilised in the synthesis of messenger

RNA characteristic of the virus DNA and thence of protein characteristic of the virus. One of the proteins so synthesised forms an enzyme which breaks down the cell and thereby destroys it.

It is debatable whether viruses should be described as living organisms. On one hand, they are not capable of independent existence in that they cannot, on their own, reproduce. On the other hand, they can make use of material from other organisms to generate exact replicas of themselves.

Denaturing of proteins

Globular proteins are very sensitive to heat and to acids and alkalis. For example, egg albumin changes its characteristics on mild heating, as can be seen when the white of an egg is boiled in water. The effect, known as **denaturation**, is due to the unfolding of the coiled chains which occurs when the hydrogen-bonds are broken. As a result, the proteins lose their physiological activity.

X-rays and other ionising radiations also denature proteins. Considerable research has been done in order to understand the changes that take place, so that methods of protecting living organisms from damage by irradiation can be found.

18.3 Carbohydrates

Carbohydrates are so called for the historical reason that the family of compounds was found to have the general empirical formula $C_x(H_2O)_y$—that is, to correspond to 'hydrates of carbon'. However, the name is misleading, and in any case several carbohydrates have since been found which do not have this general formula.

Carbohydrates are divided into two principal classes: the **sugars** and the **polysaccharides**. Sugars are crystalline solids which can be subdivided into two groups: the **monosaccharides**, most of which have the formula $C_5H_{10}O_5$ (the **pentoses**) or $C_6H_{12}O_6$ (the **hexoses**), and the **disaccharides**, which have the formula $C_{12}H_{22}O_{11}$. The polysaccharides are mostly insoluble powders, with the formula $(C_6H_{10}O_5)_n$ where n can range from 200 to several thousand.

Thus carbohydrates can be subdivided:

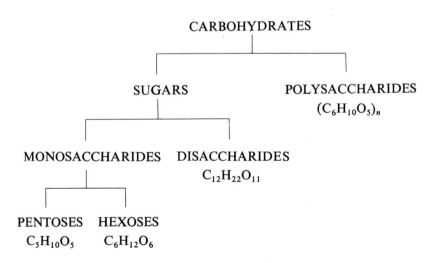

NATURALLY OCCURRING COMPOUNDS

Monosaccharides

The most important group of monosaccharides is the hexoses, with general formula $C_6H_{12}O_6$, and of these the most important are glucose and fructose:

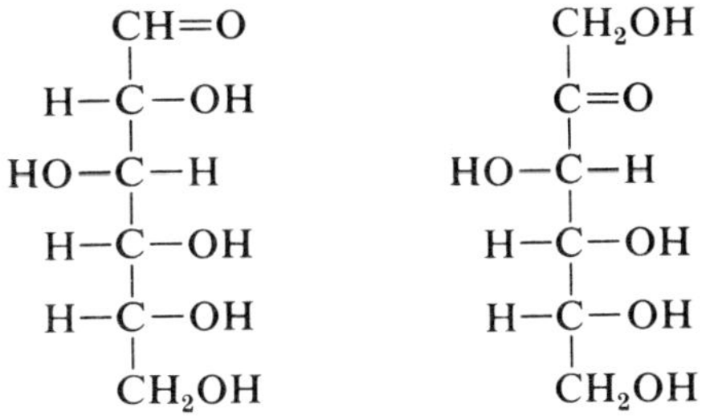

```
      CH=O                CH₂OH
       |                    |
    H—C—OH               C=O
       |                    |
   HO—C—H               HO—C—H
       |                    |
    H—C—OH               H—C—OH
       |                    |
    H—C—OH               H—C—OH
       |                    |
      CH₂OH               CH₂OH
     Glucose             Fructose
```

Both compounds, like all the other hexoses, possess a chain of six carbon atoms, but they differ in that glucose contains an aldehyde group and is described as an **aldose**, whereas fructose contains a ketone group and is described as a **ketose**. There are four asymmetric carbon atoms in glucose, so that (since there is no internal compensation) it is optically active and has $2^4 - 1 = 15$ stereoisomers (p. 236). Many of these, like glucose, occur naturally. Fructose, which has three asymmetric carbon atoms, is also optically active and has $2^3 - 1 = 7$ stereoisomers.

Glucose can be obtained in the solid state in two crystalline forms, which correspond to two different cyclic structures. When either form is dissolved in water an equilibrium is established between the two cyclic and one non-cyclic structures:

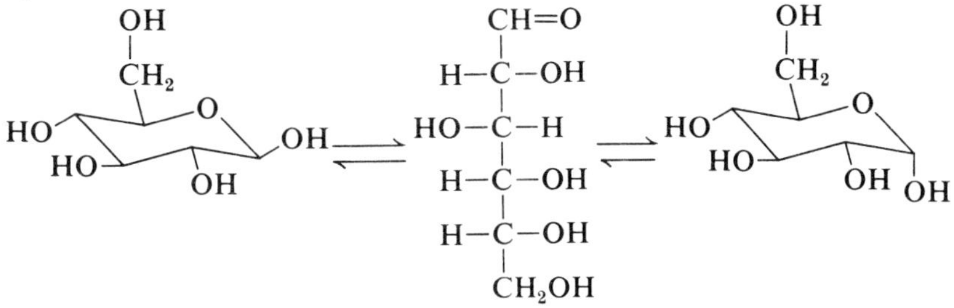

[cyclic form] ⇌ [open-chain form with CH=O, H—C—OH, HO—C—H, H—C—OH, H—C—OH, CH₂OH] ⇌ [cyclic form]

The process, known as **mutarotation**, can be followed with a polarimeter (p. 233) since the isomeric forms have different specific rotations.

The left-hand structure corresponds to the major component and the non-cyclic structure is present in the smallest proportion. However, equilibrium between the three is fairly rapidly established, so that if a reagent is introduced which reacts with aldehydes, for example Fehling's solution, it removes the non-cyclic isomer but more is then formed from the cyclic isomers to maintain the equilibrium; eventually the entire quantity of glucose reacts.

Preparation of glucose and fructose

Glucose is obtained by the hydrolysis of starch with dilute sulphuric acid at high temperatures under pressure:

$$(C_6H_{10}O_5)_n + nH_2O \xrightarrow{H^+} nC_6H_{12}O_6$$
$$\text{Starch} \qquad\qquad\qquad\qquad \text{Glucose}$$

Fructose is obtained by the hydrolysis of sucrose (p. 283) with dilute sulphuric acid:

$$C_{12}H_{22}O_{11} + H_2O \xrightarrow{H^+} C_6H_{12}O_6 + C_6H_{12}O_6$$
$$\text{Sucrose} \qquad\qquad \text{Glucose} \quad \text{Fructose}$$

Properties of glucose

Glucose is a white crystalline solid, soluble in water but insoluble in most organic solvents. It behaves like an aliphatic aldehyde in the following respects:

1. It reduces Fehling's solution to copper(I) oxide and ammoniacal silver nitrate to silver.

2. With weak oxidising agents, such as bromine water, it forms gluconic acid:

$$\begin{array}{c} \text{CHO} \\ | \\ (\text{CHOH})_4 \\ | \\ \text{CH}_2\text{OH} \end{array} \xrightarrow{\text{'O'}} \begin{array}{c} \text{CO}_2\text{H} \\ | \\ (\text{CHOH})_4 \\ | \\ \text{CH}_2\text{OH} \\ \text{Gluconic acid} \end{array}$$

With strong oxidising agents, such as nitric acid, the primary alcohol group is also oxidised:

$$\begin{array}{c} \text{CHO} \\ | \\ (\text{CHOH})_4 \\ | \\ \text{CH}_2\text{OH} \end{array} \xrightarrow{3\text{'O'}} \begin{array}{c} \text{CO}_2\text{H} \\ | \\ (\text{CHOH})_4 \\ | \\ \text{CO}_2\text{H} \\ \text{Saccharic acid} \end{array}$$

3. It forms a cyanohydrin:

$$\begin{array}{c} \text{CHO} \\ | \\ (\text{CHOH})_4 \\ | \\ \text{CH}_2\text{OH} \end{array} + \text{HCN} \xrightarrow{\text{OH}^-} \begin{array}{c} \text{HO}\diagdown\text{CH}\diagup\text{CN} \\ | \\ (\text{CHOH})_4 \\ | \\ \text{CH}_2\text{OH} \end{array}$$

4. It undergoes condensation reactions with hydroxylamine, phenylhydrazine, etc., for example:

$$\begin{array}{c} \text{CHO} \\ | \\ (\text{CHOH})_4 \\ | \\ \text{CH}_2\text{OH} \end{array} + \text{C}_6\text{H}_5-\text{NH}-\text{NH}_2 \rightarrow \begin{array}{c} \text{CH}=\text{N}-\text{NH}-\text{C}_6\text{H}_5 \\ | \\ (\text{CHOH})_4 \\ | \\ \text{CH}_2\text{OH} \end{array} + \text{H}_2\text{O}$$

With excess of phenylhydrazine, it forms an **osazone**:

$$\begin{array}{c}\text{CHO}\\|\\(\text{CHOH})_4\\|\\\text{CH}_2\text{OH}\end{array} + 3\text{C}_6\text{H}_5-\text{NH}-\text{NH}_2 \longrightarrow \begin{array}{c}\text{CH}=\text{N}-\text{NH}-\text{C}_6\text{H}_5\\|\\\text{C}=\text{N}-\text{NH}-\text{C}_6\text{H}_5\\|\\(\text{CHOH})_3\\|\\\text{CH}_2\text{OH}\end{array} + \text{C}_6\text{H}_5-\text{NH}_2 + 2\text{H}_2\text{O} + \text{NH}_3$$

Glucosazone

5. It is reduced, for example by sodium amalgam and water, to a polyhydric alcohol:

$$\begin{array}{c}\text{CHO}\\|\\(\text{CHOH})_4\\|\\\text{CH}_2\text{OH}\end{array} \xrightarrow{2\text{'H'}} \begin{array}{c}\text{CH}_2\text{OH}\\|\\(\text{CHOH})_4\\|\\\text{CH}_2\text{OH}\end{array}$$

Glucose as a source of energy

Glucose is required by the human body as a source of muscular energy and of heat. It is normally ingested in the form of starch and sucrose (cane sugar); these are hydrolysed to glucose by enzymes.

The mechanism by which glucose breaks down in the body to release energy is complex. It is first phosphorylated (i.e. converted into a phosphate ester) to give glucose-6-phosphate by reaction at its —CH$_2$OH group. Glucose-6-phosphate then isomerises to fructose-6-phosphate, and a second phosphorylation occurs to give fructose-1,6-diphosphate. This compound forms two 3-carbon molecules which, by oxidation, give 2-oxopropanoic acid, CH$_3$—CO—CO$_2$H, (which, at body pH, is ionised). Decarboxylation gives an ethanoyl group, CH$_3$CO—, which is attached to a coenzyme (Coenzyme A, or CoA). Further oxidation then gives carbon dioxide.

HO—CH$_2$—(CHOH)$_3$—CH(OH)—CHO →
Glucose

P—O—CH$_2$—(CHOH)$_3$—CH(OH)—CHO →
Glucose-6-phosphate

P—O—CH$_2$—(CHOH)$_3$—CO—CH$_2$OH →
Fructose-6-phosphate

P—O—CH$_2$—(CHOH)$_3$—CO—CH$_2$—O—P →
Fructose-1,6-diphosphate

P—O—CH$_2$—CH(OH)—CHO + HO—CH$_2$—CO—CH$_2$—O—P →

CH$_3$—CO—CO$_2^-$ → CH$_3$—CO—Coenzyme A → CO$_2$ + H$_2$O
2-Oxopropanoate ion

$$(P \text{ is } -\overset{\overset{\displaystyle O}{\|}}{\underset{\underset{\displaystyle O^-}{|}}{P}}-O^-)$$

The overall oxidation is accompanied by the release of 2870 kJ per mol of glucose and can be summarised as:

$$\text{C}_6\text{H}_{12}\text{O}_6 + 6\text{O}_2 \rightarrow 6\text{CO}_2 + 6\text{H}_2\text{O} + \text{Energy}$$

Any excess of glucose is converted into **glycogen**, a polysaccharide, which is stored in the liver and in muscles. The store in the liver is converted into glucose by the enzyme phosphorylase and enters the blood stream as required. The muscle supply is available for the provision of glucose for muscular activity, but if the activity is particularly intense, oxygen cannot be obtained rapidly enough for the energy-releasing oxidation described above. Instead, a reaction occurs which does not involve oxygen but nevertheless releases energy (although less energy, per mol, than the oxidative process):

$$C_6H_{12}O_6 \rightarrow 2CH_3\text{---}CH(OH)\text{---}CO_2H + \text{Energy}$$
$$\text{2-Hydroxypropanoic acid}$$

When the muscular activity has died down, the 2-hydroxypropanoic acid is reconverted into glycogen; this reaction requires energy, which is supplied by the oxidation of more glucose.

Properties of fructose

Fructose, an isomer of glucose, is a white crystalline solid which is very soluble in water but insoluble in most organic solvents. It behaves like a ketone in the following ways:

1. It forms a cyanohydrin with hydrogen cyanide, and gives condensation products with hydroxylamine, phenylhydrazine, etc.

It reacts with excess of phenylhydrazine to form the same osazone as glucose:

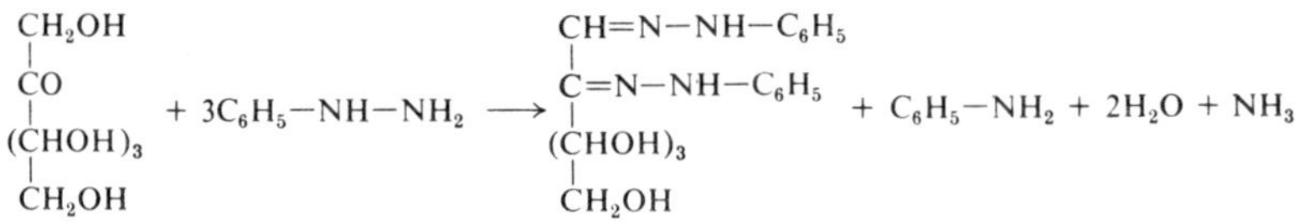

$$\begin{array}{l}CH_2OH \\ | \\ CO \\ | \\ (CHOH)_3 \\ | \\ CH_2OH\end{array} + 3C_6H_5\text{---}NH\text{---}NH_2 \longrightarrow \begin{array}{l}CH=N\text{---}NH\text{---}C_6H_5 \\ | \\ C=N\text{---}NH\text{---}C_6H_5 \\ | \\ (CHOH)_3 \\ | \\ CH_2OH\end{array} + C_6H_5\text{---}NH_2 + 2H_2O + NH_3$$

Fructose   Glucosazone

2. It is reduced to form a mixture of stereoisomeric alcohols:

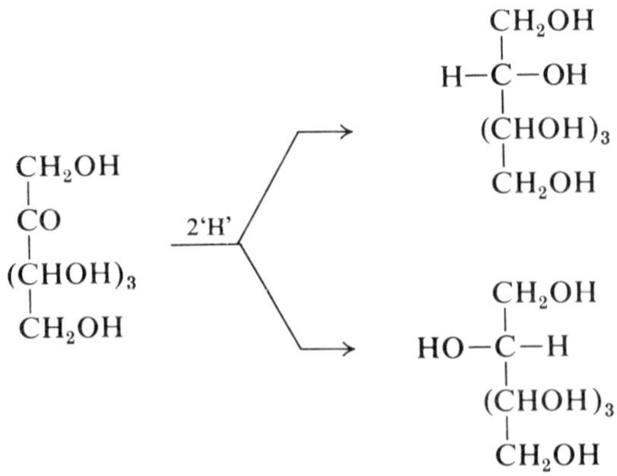

$$\begin{array}{l}CH_2OH \\ | \\ CO \\ | \\ (CHOH)_3 \\ | \\ CH_2OH\end{array} \xrightarrow{2\text{'H'}} \begin{array}{l}CH_2OH \\ | \\ H\text{---}C\text{---}OH \\ | \\ (CHOH)_3 \\ | \\ CH_2OH\end{array} \quad \text{and} \quad \begin{array}{l}CH_2OH \\ | \\ HO\text{---}C\text{---}H \\ | \\ (CHOH)_3 \\ | \\ CH_2OH\end{array}$$

Fructose differs from simple ketones in being a strong reducing agent; thus, it reduces Fehling's solution to copper(I) oxide and ammoniacal silver nitrate to silver.

NATURALLY OCCURRING COMPOUNDS

Disaccharides

Disaccharides have the molecular formula $C_{12}H_{22}O_{11}$, and consist of two monosaccharide molecules, $C_6H_{12}O_6$, joined together with the loss of a molecule of water. Three disaccharides occur naturally; these are maltose, lactose and sucrose. They all have similar physical properties, being white crystalline solids which are soluble in water.

Manufacture of sucrose

Sucrose is obtained from either sugar-cane or sugar-beet. The cane is cut into small pieces and crushed, and the juice is pressed out. The juice is made alkaline with calcium hydroxide, impurities being precipitated and filtered off. The proteins are precipitated by passing steam through the liquid, and the clear juice is concentrated by evaporation under reduced pressure. The syrup is cooled and some sugar crystallises out. The residual sugar remains in the thick liquid, known as **molasses**, and sucrose is recovered by dilution and recrystallisation.

The brown sugar obtained is dissolved in water, and is treated with calcium hydroxide and carbon dioxide, and more impurities are precipitated. The filtrate is decolorised by boiling with charcoal, and the solution is filtered and concentrated by vacuum distillation. The sugar is allowed to crystallise out and may be granulated or moulded into cubes. The molasses are used in the manufacture of ethanol by fermentation or for cattle foods.

Chemical properties of sucrose

1. Sucrose, like all disaccharides, is hydrolysed by dilute mineral acids to two monosaccharide molecules. Sucrose yields glucose and fructose:

$$C_{12}H_{22}O_{11} + H_2O \xrightarrow{H^+} \underset{\text{Glucose}}{C_6H_{12}O_6} + \underset{\text{Fructose}}{C_6H_{12}O_6}$$

Sucrose is dextrorotatory, and glucose and fructose are dextro- and laevo-rotatory, respectively. Although an equimolar mixture is formed, the resulting solution is laevorotatory as fructose has a higher specific rotation than glucose. This hydrolysis is known as the **inversion of sucrose**, and the resultant mixture is known as **invert sugar**. The reaction may also be effected by the enzyme invertase, present in yeast.

2. Concentrated sulphuric acid dehydrates sucrose, leaving almost pure carbon, known as sugar charcoal:

$$C_{12}H_{22}O_{11} \xrightarrow{\text{conc. } H_2SO_4} 12C + 11H_2O$$

3. Sucrose is a **non-reducing sugar**; it does not reduce Fehling's solution to copper(I) oxide or an ammoniacal solution of silver nitrate to silver.

Structure of sucrose

An examination of the chemical properties above suggests that sucrose is made up of a molecule of glucose and a molecule of fructose, with the elimination of a molecule of water. Further, the absence of reducing properties suggests that the glucose and fructose 'residues' are joined *via* the aldehyde group of glucose and the keto group of fructose. The structure of sucrose has been deduced, from this and other evidence, as:

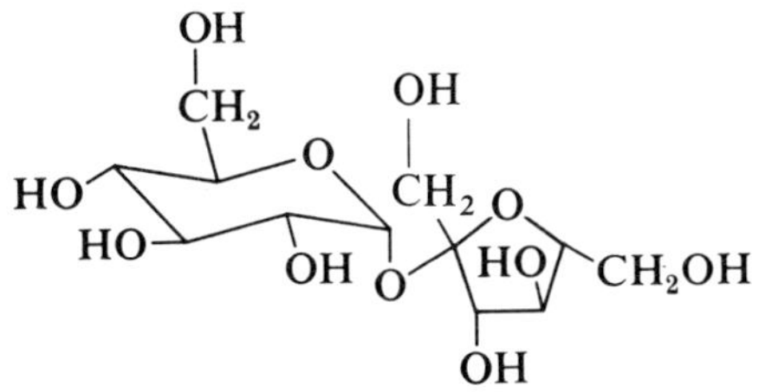

Chemical properties of maltose and lactose

1. They are hydrolysed by dilute mineral acids to two monosaccharide molecules:

$$C_{12}H_{22}O_{11} + H_2O \xrightarrow{H^+} 2C_6H_{12}O_6$$
Maltose → Glucose

$$C_{12}H_{22}O_{11} + H_2O \xrightarrow{H^+} C_6H_{12}O_6 + C_6H_{12}O_6$$
Lactose → Glucose + Galactose

2. Concentrated sulphuric acid dehydrates both disaccharides to charcoal.

3. Both disaccharides reduce Fehling's solution to copper(I) oxide and an ammoniacal solution of silver nitrate to silver. They are known as **reducing sugars**.

Structure of maltose and lactose

An examination of the chemical properties above suggests that both are made up of two monosaccharide 'residues'. However, their reducing properties suggest that the residues are *not* joined by aldehyde groups on both monosaccharides and that the cyclic structures are:

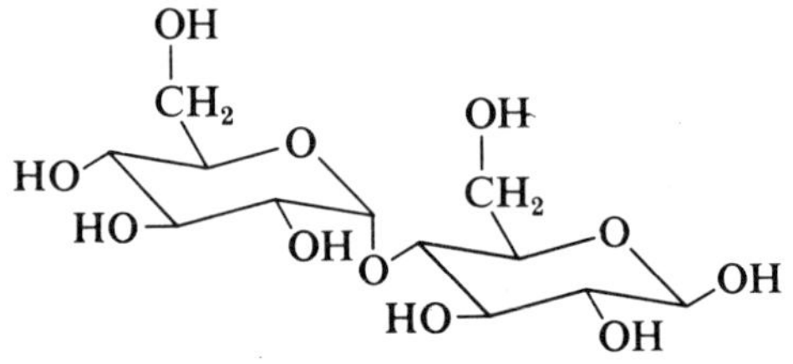

Maltose

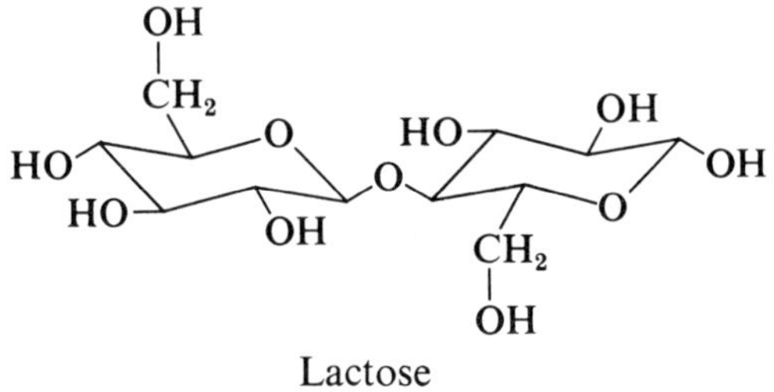

Lactose

The glucose unit in each will exist in solution as an equilibrium between two cyclic and one non-cyclic structure, the latter having a 'free' aldehyde group, which gives the molecule its reducing properties.

Polysaccharides

Polysaccharides are polymers, made up of monosaccharide units, which occur in both animals and plants. They are hydrolysed by mineral acid to monosaccharides:

$$(C_6H_{10}O_5)_n + nH_2O \xrightarrow{H^+} nC_6H_{12}O_6$$

The two most widely occurring polysaccharides are starch and cellulose.

Starch

Starch occurs in wheat, barley, rice, potatoes and all green plants. It is the main carbohydrate reserve of plants. It is also an important ingredient of animal foods since it provides a source of glucose; it is hydrolysed to glucose by enzymes in saliva.

Starch has two components: α-amylose and β-amylose (amylopectin).

α-Amylose is composed of long chains of glucose units:

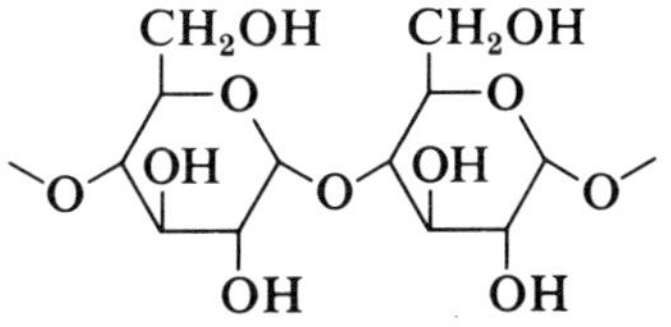

Its formula weight is between 10,000 and 60,000, depending on the degree of polymerisation. It is water-soluble and is used for making starch solutions.

β-Amylose also contains long chains of glucose units, but these are joined together at various points by other glucose chains and complex glucose derivatives, so that the structure is a complex three-dimensional network. Its formula weight is in the range 50,000–100,000 and it is insoluble in water.

Hydrolysis of starch

If a solution of starch is warmed to about 70°C in the presence of the enzyme amylase (present in human saliva), it is hydrolysed to the disaccharide, maltose:

$$2(C_6H_{10}O_5)_n + nH_2O \rightarrow nC_{12}H_{22}O_{11}$$
$$\text{Maltose}$$

Another enzyme which catalyses the hydrolysis is diastase (present in malt) (p. 286).

However, if starch is boiled with dilute sulphuric acid, it is hydrolysed to the monosaccharide, glucose:

$$(C_6H_{10}O_5)_n + nH_2O \rightarrow nC_6H_{12}O_6$$
$$\text{Glucose}$$

Presumably, starch is first hydrolysed to maltose, which, in turn, is hydrolysed to glucose.

Fermentation of starch

Wheat or barley are 'mashed' with hot water and then filtered to extract a solution of starch. The aqueous solution is heated to about 55°C with malt, which is germinated barley and contains a mixture of enzymes known as diastase. Starch is hydrolysed to maltose.

The liquid is cooled to 35°C, and yeast, which contains the enzyme maltase, is added to catalyse the hydrolysis of maltose to glucose:

$$C_{12}H_{22}O_{11} + H_2O \rightarrow 2C_6H_{12}O_6.$$

The yeast also contains a mixture of enzymes, zymase, which catalyses the decomposition of glucose to ethanol and carbon dioxide:

$$C_6H_{12}O_6 \rightarrow 2C_2H_5OH + 2CO_2$$

The fermented liquor contains up to 10 per cent ethanol, and is used in drinks (for example, beer from barley and malt) or is then purified to obtain pure ethanol. The purification by distillation is discussed on p. 141.

Cellulose

Cellulose is the principal constituent of the cell walls of plants. It consists of a three-dimensional network of chains of glucose units and some complex glucose derivatives.

Cotton is almost pure cellulose, but cellulose is usually manufactured from wood, which is a mixture of cellulose and a material known as lignin. In this process, wood shavings are heated with calcium hydrogen sulphite to dissolve the lignin and the cellulose is removed by filtration. It can be purified by dissolving it in a solution of ammonia containing a copper(II) salt and then adding mineral acid to precipitate the pure cellulose.

Rayon

Rayon is the name given to cover all fibres manufactured from cellulose. **Cellulose ethanoate**, celanese silk, is made by ethanoylating cellulose with ethanoic anhydride. Cellulose ethanoate does not burn readily, and is used to make films, lacquers and varnishes.

Viscose rayon is manufactured by treating cellulose with sodium hydroxide and carbon disulphide, and the reaction can be represented thus:

$$\underset{\text{Cellulose}}{X-OH} + NaOH + CS_2 \longrightarrow \underset{\substack{\text{Cellulose} \\ \text{xanthate}}}{S=C{\overset{O-X}{\underset{S^- Na^+}{\diagup\diagdown}}}} + H_2O$$

The viscose solution is forced through fine jets into a bath of dilute sulphuric acid, giving a fine thread of viscose rayon. The thread is spun and made into fabrics which are lustrous and supple, and can be easily dyed. Although

NATURALLY OCCURRING COMPOUNDS

rayon is cheaper than natural silk, it is not as strong or durable.

Cellophane is manufactured by forcing the viscose solution through slits into the acid.

If cellulose is treated with dilute nitric acid, cellulose mononitrate and dinitrate are formed; the mixture is known as **pyroxylin**. A solution of pyroxylin in ethanol, ether or propanone is known as **collodion**, and is used as an adhesive. If pyroxylin is heated with ethanol and camphor, **celluloid** is formed.

18.4 Practical work with proteins

Preparation of a protein solution

Beat an egg-white with about five times its volume of water, adding about 1 g of sodium chloride to help the albumen to dissolve. Filter carefully through a muslin cloth.

Reactions of the protein solution

1. *Denaturing*. Test 2 cm^3 portions for precipitation by (a) boiling, (b) adding M sulphuric acid, (c) adding 2M sodium hydroxide solution, (d) adding ethanol.

2. *Biuret test*. To a 2 cm^3 portion of protein solution, add 2 cm^3 of 2M sodium hydroxide solution, followed by 2 or 3 drops of copper(II) sulphate solution. Note the violet colour.

3. *Reactions catalysed by enzymes*.

(a) *The hydrolysis of carbamide (urea) catalysed by the enzyme urease.*

The hydrolysis of carbamide in neutral solution is catalysed by the enzyme, urease. Dissolve about 0·1 g of urea in 2 cm^3 of water in a test-tube, and add a few grains of urease powder.

Stand the test-tube in a beaker of water at 40°C, and test the gas evolved by (a) smell, (b) moist red litmus paper. It may take about 5 minutes to obtain enough gas to test with the litmus paper.

(b) *The hydrolysis of starch by enzymes present in saliva* (p. 288).

(c) *The decomposition of hydrogen peroxide catalysed by the enzyme, catalase.*

The decomposition of hydrogen peroxide solutions is catalysed by the enzyme, catalase. Catalase can be obtained as a pure powder and is also present in blood and in potatoes.

Set up an apparatus similar to that in Fig. 7.1, except with a boiling-tube as the receiver. The boiling-tube should first be calibrated in 10 cm^3 portions, by running in water from a burette and marking the heights for 10, 20, 30 and 40 cm^3 with a crayon or dab of paint.

Place 1 cm^3 of catalase solution (made up by dissolving about 0·01 g of catalase in 100 cm^3 of water) in the test-tube, and 2 cm^3 of 20-volume hydrogen peroxide in the dropping pipette. Squeeze the rubber teat so that all the hydrogen peroxide solution is added and collect the gas evolved in the boiling-tube.

Repeat the experiment using (a) a solution of 1 small drop of blood in 2 cm^3 of water, (b) a mixture made by chopping up a small piece of potato and water, instead of the solution of catalase in water.

If time, repeat the experiment using a small amount (ca. 0·1 g) of solid manganese(IV) oxide in place of the enzyme. By diluting the solution of catalase used above, see how active the enzyme is as a catalyst compared with manganese(IV) oxide.

NATURALLY OCCURRING COMPOUNDS

N.B. If blood is to be drawn, it is important that the correct procedure is used. Instructions are given in *Biology Teaching in Schools involving Experiment or Demonstration with Animals or with Pupils*, by J. J. Bryant (The Association for Science Education).

18.5 Practical work with carbohydrates

Reactions of glucose (a monosaccharide)

1. Heat about 1 g of glucose in a test-tube. It first melts, then gives off water of crystallisation and finally chars, smelling of burnt sugar.

2. Heat about 0·5 g of glucose with 2 cm^3 of concentrated sulphuric acid. Warm. Note charring with formation of carbon.

3. *Fehling's test* (p. 179). Use about 0·1 g glucose in place of the aldehyde.

4. *Silver mirror test* (p. 178). Use about 0·1 g glucose in place of the aldehyde.

5. *Formation of an osazone.* Dissolve 0·5 g of glucose in 5 cm^3 of water. Dissolve 1 g of phenylhydrazine in 1 g of ethanoic acid and dilute to 10 cm^3. Mix the solutions, and warm on a waterbath. Filter the crystals, wash with water and with ethanol. Recrystallise from ethanol. Study the crystalline structure of the osazone under a microscope. M.p. of the osazone is 204°C.

Reactions of fructose (a monosaccharide)

6. Repeat the experiments above with fructose in place of glucose.

Reactions of sucrose (a disaccharide)

7. Repeat experiments 1 and 2 above.

8. Dissolve 2 g of sucrose in 10 cm^3 of water and divide into two portions. With one, repeat experiment 3. Acidify the second portion with 5 drops of M sulphuric acid and boil for a few minutes. Neutralise by adding 2M sodium hydroxide solution dropwise and then test the solution with Fehling's solution.

Reactions of starch (a polysaccharide)

9. Make a paste with about 0·1 g of soluble starch and about 1 cm^3 of water. Pour 10 cm^3 of boiling water into the paste and shake the solution. Divide the solution into three parts.

To the first part, add 2 drops of iodine solution. Note the blue coloration.

Add 5 drops of M sulphuric acid to the second portion and boil for a few minutes. Neutralise the solution by adding 2M sodium hydroxide solution drop by drop. Divide the solution into two parts. Test one with iodine solution to show the absence of starch. Test the second with Fehling's solution to show the presence of a reducing agent (glucose).

Add a little saliva to the third portion and warm to about 70°C for a few minutes. Test the solution as in the previous experiment with iodine and with Fehling's solution.

NATURALLY OCCURRING COMPOUNDS

Small-scale preparation of cellulose ethanoate

Mix 20 cm^3 of ethanoic acid, 5 cm^3 of ethanoic anhydride and 2 drops of concentrated sulphuric acid in a beaker. Then add, with stirring, 0·5 g of shredded cotton wool.

Continue to stir so that the cotton wool is evenly distributed throughout the mixture and there are no air bubbles. Leave the mixture for about a day, stirring occasionally. The mixture gradually becomes less viscous, ending in a clear liquid.

Pour the liquid slowly into a large beaker containing 500 cm^3 of water. A precipitate of cellulose ethanoate separates out, which should be filtered using a small Buchner funnel (Fig. 2.8). Wash it with water and dry it between pads of filter paper.

Take about a quarter of the ester and dissolve it in about 15 cm^3 of trichloromethane in a boiling tube. If the solution is turbid, place some anhydrous calcium chloride in the tube and shake the solution. Decant the solution into a basin and allow it to evaporate in a fume cupboard. Remove the film of cellulose ethanoate.

18.6 Further reading

The Chemistry of Proteins. R. J. Taylor (1974). Unilever Educational Booklet, Advanced Series No. 3.
A Brief Introduction to Biochemistry. R. J. Light (1968). W. A. Benjamin, Inc.
The Double Helix. James D. Watson (1968). Weidenfeld and Nicolson. Paperback, Penguin 1970.

18.7 Films

The Structure of Protein. Unilever Ltd.
Biochemistry and Molecular Structure (CHEM—study series). Guild Sound and Vision Ltd.

18.8 Questions

1 What is meant by the term *sugar*? Give a brief account of the sources of sucrose and glucose, and their relationships with one another and with starch and ethanol.
How would you obtain a sample of ethanol from glucose and show it to be ethanol?

2 Give an account of the natural occurrence and properties of two types of polysaccharide. Write down structural formulae for glucose and fructose, and describe the principal reactions of these substances. Describe the preparation and chemical properties of sucrose.

3 How, if at all, do (a) glucose and (b) sucrose react with (i) dilute sulphuric acid, (ii) concentrated sulphuric acid, (iii) phenylhydrazine, (iv) hydrogen cyanide.

4 A hexapeptide gave, upon complete hydrolysis, two moles of glycine (Gly), and one mole each of alanine (Ala), leucine (Leu), serine (Ser), and tyrosine (Tyr). Partial hydrolysis under milder conditions gave, amongst other products, two dipeptides, glycylserine (Gly.Ser.) and tyrosylglycine (Tyr.Gly.), and a tripeptide, glycylalanylleucine (Gly.Ala.Leu.). Treatment of the hexapeptide with 2,4-dinitrofluorobenzene and alkali, followed by hydrolysis of the *N*-dinitrophenylpeptide, gave *N*-dinitrophenylglycine. Give structures for the hexapeptide consistent with these observations, and indicate how you could distinguish between them.
(*Note*: Structures for the amino-acid residues are *not* required. Use the abbreviations indicated in parentheses.) (O Schol.)

Chapter 19

Petroleum

19.1 Introduction

The term **petroleum** is used to describe the mixture of hydrocarbons in oil, including the gases above the liquid in oil wells and the gases and solids which are dissolved in the liquid.

Petroleum was formed in remote periods of geological time from the remains of living organisms. It is, therefore, a *fossil fuel*. Weathered rock material eroded from land masses and carried to the sea accumulated in layers over millions of years in subsiding basins, and the remains of large quantities of marine plant and animal organisms became incorporated in the sediment (Fig. 19.1). Owing to the great thickness of the sediments, high pressures built up which, probably in conjunction with biochemical activity, led to the formation of petroleum, although the detailed mechanism is obscure. Subsequent earth movements which caused uplift of the sedimentary basins also caused migration of the petroleum through pore spaces in the rocks, sometimes to areas distant from the formation zones. In the course of migration, some of the petroleum came to accumulate in *traps* where the porous rock was bounded by impermeable rock. The principal types of trap in oil fields such as those in the Middle East, North America and the North Sea are the *anticline* (an upfold in the strata), the *fault* trap and the *salt dome* (Fig. 19.2).

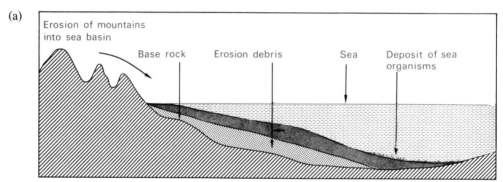
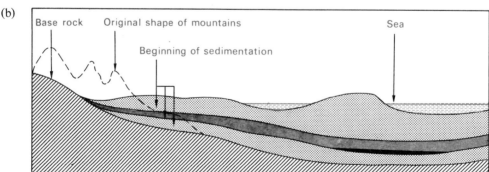

FIG. 19.1.(a) *Debris accumulates by erosion of the mountains, together with the bodies of marine organisms.* (b) *Pressure from the debris on the organisms contributes to the formation of oil*

The gas is principally methane, with small amounts of other alkanes. The liquid contains mainly straight-chain alkanes (with up to about 125 carbon atoms in the chain), with smaller amounts of cycloalkanes and

FIG. 19.2 *Structural traps.* (a) *Salt from an evaporated sea has been forced upwards. The resulting salt dome seals the oil-bearing porous rocks.* (b) *The crust of the earth may buckle. The up-folds are known as anticlines, and these may lead to an impermeable seal, thus keeping the oil in a confined space. On the right, the crust has been stressed causing a fault, thus trapping the oil*

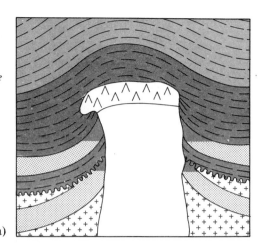

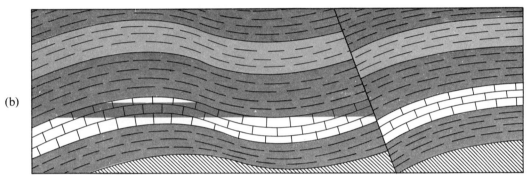

aromatic hydrocarbons. The relative amounts of the three classes of compound vary with the oil-field; for example, petroleum from the Middle East and Pennsylvania contains a high proportion of alkanes, whereas that from Venezuela is relatively richer in cycloalkanes and that from Mexico and California in aromatic hydrocarbons.

19.2 Natural gas

The discovery of natural gas fields under the North Sea in 1967 has focused attention on this very important material which serves both as a fuel and as a source of chemicals. Many fields all over the world are now exploited commercially, including those in Libya (from which gas is transported to the United Kingdom in the liquid state in specially constructed, refrigerated tankers), Italy, Holland and the U.S.A.

It is believed that natural gas is formed from the decomposition of petroleum or coal deposits. It contains principally methane (if it is more than 95 per cent methane it is known as 'dry' natural gas). It may contain larger amounts of ethane, propane and butane ('wet' natural gas). The large North Sea natural gas fields are 'dry', but those in the U.S.A. are 'wet', and the higher alkanes are recovered and used as fuels ('bottled' gases) and to make alkenes (20.4).

Natural gas also contains hydrogen sulphide, and although it is generally present only in small amounts, it is an important source of sulphur; the French field at Lacq contains about 15 per cent hydrogen sulphide. Certain American fields contain up to 5 per cent of helium, and these are now the most important sources of this gas.

Plate 19.1. An aerial photograph of the San Andreas fault in California. There is a sharp line between mountains and farmland (Esso Petroleum Co. Ltd.)

Plate 19.2. An aerial photograph of an anticline in Iran (Shell Petroleum Co. Ltd.)

19.3 Distillation of crude petroleum

Crude petroleum is first 'stabilised' by heating the oil to remove dissolved natural gas. This is usually done near the petroleum wells. The crude petroleum is then distilled into six principal fractions (Fig. 19.3).

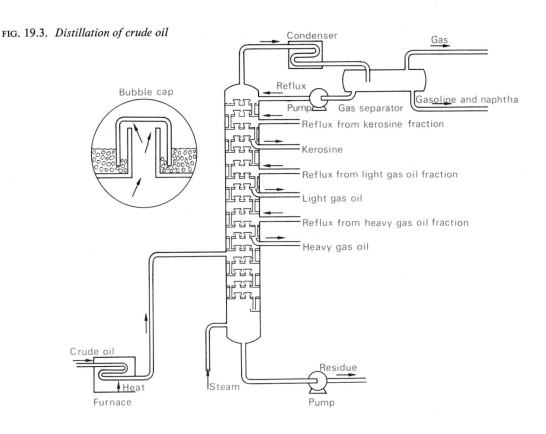

FIG. 19.3. *Distillation of crude oil*

	GAS	GASO-LINE	NAPHTHA	KEROSINE	LIGHT GAS-OIL	HEAVY GAS-OIL	RESIDUE (Reduced crude)
B.p. (°C)	<40	40–100	100–160	160–250	250–300	300–350	>350
% wt	3	7	7	13	9	9	52
Numbers of C atoms in the alkanes	<4	4–10		10–16	16–20	20–25	>25

All fractions derive from CRUDE PETROLEUM.

The values for percentage composition are approximate and vary widely, depending on where the crude petroleum is found (19.1). The residue, known as **reduced crude**, is distilled under vacuum to yield lubricating oils, paraffin waxes and a residue, bitumen.

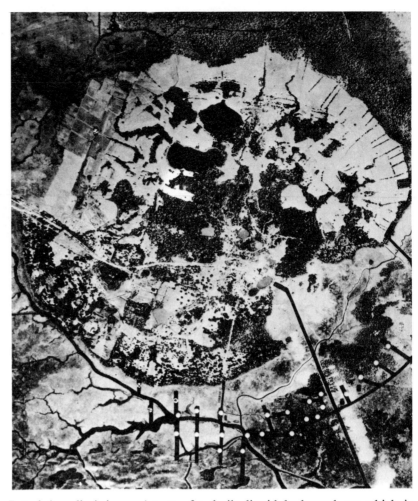

Plate 19.3. An aerial photograph of a salt dome in Louisiana. The salt has been forced upwards under high pressure and is now acting as a seal, preventing the escape of oil upwards from the porous rocks (Esso Petroleum Co. Ltd.)

19.4 The gasoline and naphtha fractions

Petrol (gasoline) is a mixture of volatile liquid hydrocarbons which is vaporised before entering the cylinder of the engine. The straight-chain alkanes, as obtained from crude petroleum, are not ideal as engine fuels because they do not burn uniformly in the cylinder, the process being known as 'knock'. This leads to both wear in the engine and wastage of petrol. However, the highly branched alkanes, as well as the cycloalkanes and the aromatic hydrocarbons, are less susceptible to knock. It is convenient to have a measure of the suitability of petrols as fuels, and for this purpose each compound is given a rating known as the **octane number**, which is determined by experiment. Two arbitrary reference points are used in the scale: heptane, octane number 0 (a poor fuel), and an isomer of octane 2,2,4-trimethylpentane, octane number 100 (a good fuel). The fraction obtained by the distillation of crude petroleum has an average octane number of less than 60, whereas modern car engines, which have high compression ratios, require petrol with an octane number of between 95 and 100 (for example, BP Super Plus, Super Shell, Esso Extra).

It is therefore necessary to enrich the petrol obtained from crude petroleum with branched-chain alkanes, cycloalkanes and aromatic hydrocarbons, and these compounds are themselves made from other fractions obtained by the distillation of crude petroleum. The methods for doing this include the **isomerisation** and **alkylation** of the smaller alkanes obtained from the gas fraction, the **reforming** of straight-chain alkanes obtained from the gasoline and naphtha fractions and the **cracking** of long-chain alkanes obtained from the kerosine and gas-oil fractions.

Isomerisation

Isomerisation is said to occur when a molecule undergoes a rearrangement to give an isomer. This process is used to convert straight-chain into branched alkanes over an acid catalyst. For example, butane is isomerised to 2-methylpropane:

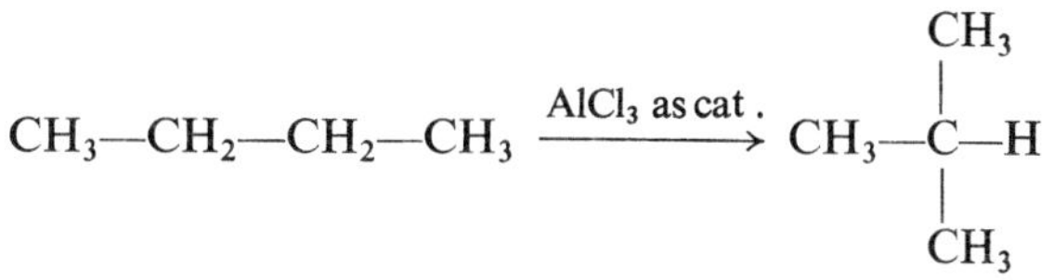

$$CH_3-CH_2-CH_2-CH_3 \xrightarrow{AlCl_3 \text{ as cat.}} CH_3-\underset{\underset{CH_3}{|}}{\overset{\overset{CH_3}{|}}{C}}-H$$

Reaction occurs *via* carbonium ions: initially, a secondary carbonium ion is formed and this, by rearrangement, gives the more stable tertiary carbonium ion (6.4):

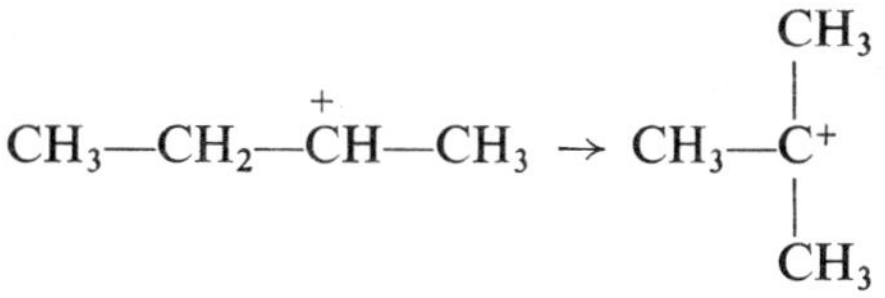

$$CH_3-CH_2-\overset{+}{CH}-CH_3 \rightarrow CH_3-\underset{\underset{CH_3}{|}}{\overset{\overset{CH_3}{|}}{C^+}}$$

Alkylation

Alkenes (obtained by cracking, p. 298) react with alkanes in the presence of acid to form larger alkanes. For example, 2-methylpropene reacts with 2-methylpropane (obtained by isomerisation as above) to give 2,2,4-trimethylpentane:

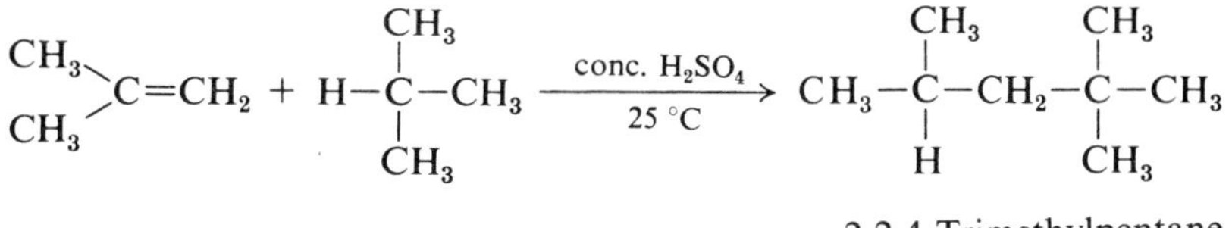

$$\underset{CH_3}{\overset{CH_3}{>}}C=CH_2 + H-\underset{\underset{CH_3}{|}}{\overset{\overset{CH_3}{|}}{C}}-CH_3 \xrightarrow[25\,°C]{\text{conc. } H_2SO_4} CH_3-\underset{\underset{H}{|}}{\overset{\overset{CH_3}{|}}{C}}-CH_2-\underset{\underset{CH_3}{|}}{\overset{\overset{CH_3}{|}}{C}}-CH_3$$

2,2,4-Trimethylpentane

Reforming

If the gasoline and naphtha fractions are passed over a catalyst above 500°C, the straight-chain alkanes undergo cyclisation and dehydrogenation, for example:

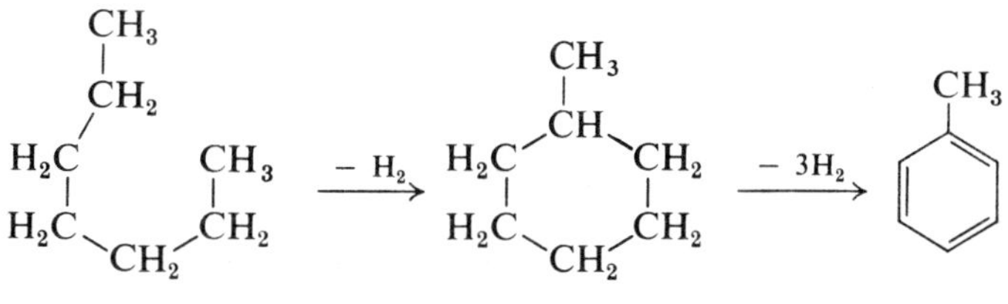

PETROLEUM

The process is known as **catalytic reforming**, and the catalyst is generally either finely divided platinum suspended on aluminium oxide (**platforming**) or finely divided molybdenum(VI) oxide suspended on aluminium oxide. They are known as **bifunctional catalysts**: the platinum or molybdenum(VI) oxide catalyses the dehydrogenation reaction while the aluminium oxide catalyses the rearrangement of the carbon skeleton. During the process, some of the alkane decomposes to carbon, but carbon formation can be suppressed by adding an excess of hydrogen (**hydroforming**). If carbon is formed, the catalyst is purified by removing it and burning off the carbon. In recent processes, the catalyst is continuously repurified and regenerated; this is known as a fluid-bed catalyst, and the method is similar to that used in cracking gas-oil.

As well as being employed for the enrichment of petrol, reforming methods are also now used for the production of aromatic hydrocarbons such as benzene, methylbenzene and the dimethylbenzenes (20.5).

Tetraethyllead

Although the combustion characteristics of modern high-grade petrols are considerably improved by their enrichment with a large proportion of branched-chain alkanes, cycloalkanes and aromatic hydrocarbons in the ways described above, still further improvement is necessary for use in engines with high compression ratios. Small amounts of a chemical, known as an 'anti-knock', are added of which the most effective are volatile compounds containing lead, such as tetraethyllead, $Pb(C_2H_5)_4$. This decomposes in the combustion chamber to give fine particles of lead or lead oxide which suppress certain combustion reactions which otherwise lead to knock.

However, legislation in the United States (and presumably later in the United Kingdom) may restrict the use of lead compounds in petrol in order to reduce pollution of the atmosphere. Manufacturers may therefore need to reduce the compression ratio of the cylinder, and petrol will need to be further enriched with reformed hydrocarbons; thus, processes such as alkylation and catalytic reforming may become even more important.

19.5 The kerosine fraction

The kerosine fraction boils between 160 and 250°C and contains straight-chain alkanes with 10–16 carbon atoms. Kerosine is used in space- and radiant-heaters; the 'blue flame stoves' burn the fuel without smell or smoke. The kerosine is marketed under such names as Aladdin Pink and Esso Blue.

Kerosine is the principal constituent of jet (gas-turbine) fuels. The fuel must have the correct viscosity (so that the fuel spray is atomised), volatility and freezing point, for the temperature of the air at 10,000 metres is below −90°C.

There is an excess of kerosine available from the distillation of petroleum and this is cracked (19.6).

19.6 The light gas-oil fraction

The light gas-oil fraction has a boiling point range of 250–300°C; the number of carbon atoms in the straight-chain alkane molecules varies from 16 to 20. Much of this fraction, and the kerosine not wanted for fuel, is heated in the presence of a catalyst, at high temperatures; the long-chain

FIG. 19.4. *A 'cat-cracker' using a fluid catalyst*

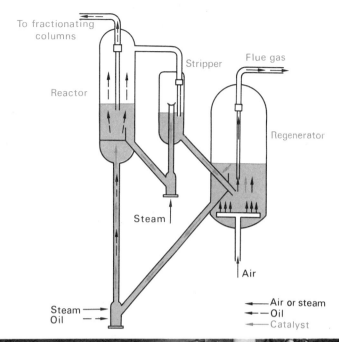

Plate 19.4. *A cat-cracker (a fluid catalytic cracking unit). The regenerator (A) and stripper (B) are in front of the reactor (C), which is partly hidden. On the left is a fractionating column (D) (Shell Petroleum Co. Ltd.)*

molecules are thereby broken into two or more smaller molecules. The process is known as **catalytic cracking** or **'cat-cracking'**. The gas-oil or kerosine vapour is passed through a fine powder made of silica and aluminium oxide at 400–500°C. The powder acts as a fluid and continuously flows out of the reactor into a second chamber through which air is passed. In this way, any carbon deposited on the catalyst is burnt off, thus reactivating the catalyst which then flows back to the reactor (Fig. 19.4).

The products from cat-cracking are: (a) a gas, known as **refinery gas**, of which the alkenes, ethene and propene, are major constituents; this is used to make many chemicals (20.4); (b) a liquid containing a high yield of branched-chain alkanes, cycloalkanes and aromatic hydrocarbons, which can be used as high-grade petrol; (c) a residue of high boiling point, used as a fuel-oil.

A variant of the process is called **hydrocracking**. The catalyst is a sodium aluminosilicate in which some of the sodium ions are replaced by platinum. In the presence of excess of hydrogen, no alkenes can be formed and all the products are saturated. For example, the gases produced contain a high proportion of 2-methylpropane used to make high-grade petrol by alkylation (p. 295).

19.7 The heavy gas-oil fraction

The heavy gas-oil fraction has a boiling-point range of 300–350°C, and the number of carbon atoms in the straight-chain alkanes varies from 20 to 25. It is used as a diesel fuel known as DERV (Diesel Engine Road Vehicle). In the diesel engine, the fuel is injected as a fine spray; the engine, which does not have a sparking plug, relies on the heat of compression to ignite the mixture of fuel and air.

19.8 The residue

The residue from the primary distillation of petroleum is known as **reduced crude** (p. 293). As this consists of over half of the crude petroleum and has often been transported over very large distances from the oil field to the refinery, it is important that as great a use is made of it as possible.

The reduced crude is distilled in vacuum to yield light and heavy lubricating oils, greases, waxes and a residue, bitumen.

The viscosity of lubricating oils changes with temperature; the oils are too thick when cold and too thin when hot. Recent developments in producing multigrade oils (for example, BP Viscostatic and Super Shell multigrade) have ensured that the viscosity changes relatively little with temperature. In particular, a polymer of 2-methylpropene is added whose viscosity is almost constant over a wide temperature range. Detergents are also added to keep any sludge formed as a fine dispersion.

The effect of engine wear has been studied with pistons which have been irradiated in an atomic pile and so contain a radioisotope of iron; the occurrence of wear leads to the enrichment of the oil with the radioactive iron and so can be followed by 'monitoring' the circulating oil. It has been found that most wear occurs while the engine is at rest; it results from attack on the metal by acids formed in the oxidation of oils. Chemicals, such as substituted phenols, are therefore added to inhibit the oxidation of oil.

Paraffin wax, deposited on distillation of the residue, is used for waterproofing paper cartons and in the manufacture of candles and polishes.

By cracking the wax in the presence of excess of steam at 500°C, straight-chain alkenes with terminal double bonds (i.e. $RCH=CH_2$) and 5–18 carbon atoms are formed; they are separated by fractional distillation. The lower members are used to make branched-chain alkanes for high-grade petrol (p. 295), and the higher members are used in the manufacture of detergents by a number of routes (13.10).

Bitumen (the residue from the vacuum distillation of reduced crude) is mainly used in making the surfaces for roads and in part for coating materials such as cables to give them water and electrical insulation.

19.9 Practical work

Distillation of crude oil

Pour in some crude oil to a depth of about 2·5 cm in a test-tube with a side arm and push down a plug of Rocksil to soak it up. Set up the apparatus

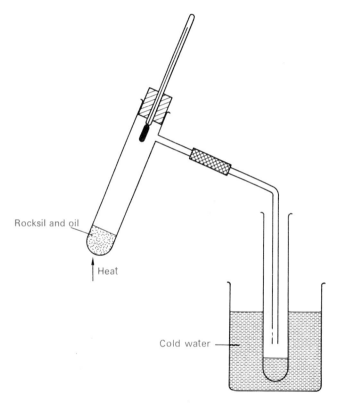

FIG. 19.5. *Distillation of crude oil*

as in Fig. 19.5 and distil the crude oil very slowly, collecting the fractions in different test-tubes:

Fraction 1—up to 70°C
Fraction 2—70–130°C
Fraction 3—130–180°C
Fraction 4—180–240°C
Fraction 5—the residue.

Pour each fraction on to a separate watch glass and light it with a burning match or splint.

Cracking of oil to form alkenes

Pour some paraffin oil in a test-tube to a depth of about 2·5 cm. Add

Rocksil until the oil has been soaked up, and then fill the tube to a depth of about 5 cm with porous pot chips (Fig. 19.6). Heat the chips and collect the gases in test-tubes, placing corks in them when they are full of gas.

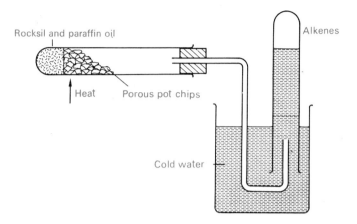

FIG. 19.6. *Cracking of oil to form alkenes*

Tests for alkenes

1. To one test-tube, apply a lighted splint. Note the colour of the flame and compare it with the colour of the flame of an alkane.

2. To another test-tube, add 2 drops of a solution of bromine in tetrachloromethane and shake.

3. To a third test-tube, add 2–3 drops of an alkaline solution of potassium permanganate (made by dissolving about 0·1 g of anhydrous sodium carbonate in 1 cm^3 of 1 per cent potassium permanganate solution). Shake the contents of the test-tube. A green solution may be obtained first [potassium manganate(VI)]. Subsequently, a brown precipitate of manganese(IV) oxide is formed.

19.10 Further reading

Our Industry. Petroleum. (1977). The British Petroleum Company Ltd.
The Petroleum Handbook (5th edition, 1966). Shell Education Service (Shell Mex House, Strand, London WC2R 0DX).
North Sea Heritage. D. Scott Wilson (1974). British Gas.
The Oil Venture. Shell Education Service.

19.11 Films

A wide range of films on prospecting for petroleum and natural gas can be obtained from Shell Film Library, British Petroleum Company Ltd and from the Gas Council. Films covering the distillation of petroleum can also be obtained from Shell Film Library and the British Petroleum Company Ltd.

Chapter 20

The petrochemical industry

20.1 Introduction

In 1920, less than 100 tonnes of the chemicals manufactured in the United States were derived from petroleum. By 1950, the output had grown to about 5 million tonnes and, by 1970, to about 30 million.

In the United Kingdom, the growth rate of the petrochemical industry was about 12–15 per cent per year during the Nineteensixties, compared with 6 per cent per year for the chemical industry as a whole and 3 per cent per year for the national economy. It is now smaller (about 6 per cent per year), though still ahead of the growth rate of the national economy.

Nowadays, over 80 per cent of all organic materials are derived from petroleum. In addition, one of the most interesting developments has been the production of inorganic compounds such as ammonia, sulphuric acid and hydrogen peroxide from petroleum sources; inorganic compounds now account for over 25 per cent of the chemicals derived from natural gas.

20.2 Primary sources

There are three important primary sources for the manufacture of petroleum chemicals: natural gas, refinery gas and naphtha. Natural gas and naphtha have been discussed in the preceding chapter. Refinery gas, which contains hydrogen, and alkanes and alkenes containing up to four carbon atoms, is obtained from the distillation of petroleum (19.3), catalytic cracking (19.6) and catalytic reforming (19.4).

20.3 Chemicals from alkanes

(a) Methane

Methane is obtained from natural gas. The sulphur compounds are first removed (20.8) and then the natural gas is fractionated to yield pure methane.

The principal use of methane is the formation of a mixture of **carbon monoxide** and **hydrogen** (known as **synthesis gas**) by reaction with steam (5.4). Methane is also used to make **ethyne** and, by direct oxidation, **methanol** (5.4).

Methane is the principal source of **hydrogen cyanide**:

$$2CH_4 + 2NH_3 + 3O_2 \xrightarrow[1000°C]{\text{Pt–Rh cat.}} 2HCN + 6H_2O$$

which is used in the manufacture of **propenonitrile** (p. 304), **hexanedinitrile** (p. 307) and **methyl 2-methylpropenoate** (p. 315), all essential for the plastics and fibres industry.

Chloromethane and **dichloromethane** are formed when an excess of methane is heated with chlorine at 400°C. Chloromethane is used in the manufacture of **silicones** (p. 323), and dichloromethane is an important solvent.

If methane is burnt with only a small amount of oxygen, and the flame impinges on a cold metal, finely divided carbon is formed, known as **carbon black**. This is compounded with rubber to make it resistant to wear, as required for tyres.

A more recent and increasingly important use of methane is in the manufacture of **ethyne** (5.4).

(b) Ethane

Ethane is obtained from natural gas and refinery gas, and is the principal source of **ethene** in the United States, high yields of the alkene being obtained when ethane is heated to about 700°C:

$$CH_3\text{---}CH_3 \rightarrow CH_2\text{=}CH_2 + H_2$$

Some ethane is chlorinated to form **chloroethane**, used to make **tetraethyllead** (19.4).

(c) Propane

Propane, like ethane, is obtained from 'wet' natural gas and from refinery gas. Propene is formed when propane is passed over a heated catalyst:

$$CH_3\text{---}CH_2\text{---}CH_3 \xrightarrow{\text{Cr}_2\text{O}_3 + \text{Al}_2\text{O}_3 \text{ as cat.}} CH_3\text{---}CH\text{=}CH_2 + H_2$$

If a catalyst is not used, the principal product is ethene.

Propane is also used to make **nitroalkanes**. When it is heated with nitric acid vapour at 350°C, a mixture of nitromethane, nitroethane, 1-nitropropane and 2-nitropropane is formed; the compounds are separated by distillation and used as solvents and as fuels for small engines.

(d) Butane

Butane is generally obtained from refinery gas (20.2). Some of it is now used to make **ethanoic acid**, the alkane being mixed with air under pressure in an inert solvent at 200°C in the presence of cobalt(II) ethanoate. Over half the acid produced in the United States is made by this method.

Another very important use is in the manufacture of **buta-1,3-diene**, an essential component in the synthetic rubber industry (21.4):

$$CH_3\text{---}CH_2\text{---}CH_2\text{---}CH_3 \xrightarrow[600°C]{\text{Cr}_2\text{O}_3 + \text{Al}_2\text{O}_3 \text{ as cat.}} CH_2\text{=}CH\text{---}CH\text{=}CH_2 + 2H_2$$

2-Methylpropane is used in the manufacture of 2,2,4-trimethylpentane for high-grade petrol (19.4).

(e) Higher alkanes

The naphtha fraction (C_4–C_{10} alkanes) is cat-cracked to form alkenes (20.4). It is also oxidised with air under high pressure at 200°C to form a mixture containing mainly **ethanoic acid**, with some methanoic and propanoic acids.

2-Methylbuta-1,3-diene, required for the synthetic rubber industry (21.4), is obtained in part from 2-methylbutane.

The **kerosine fraction** (C_{10}–C_{16} alkanes) contains both straight-chain and branched alkanes. These are separated by passing them through synthetic zeolites which 'sieve' the molecules, the straight-chain compounds (average diameter of 5 Å) being absorbed and the larger branched-chain compounds (diameter of about 5·5 Å) passing through. The straight-chain alkanes are then desorbed by washing the column with pentane, and these are converted into **detergents** (13.10).

20.4 Chemicals from alkenes

Production of ethene and propene

In the United States, ethene and propene are generally manufactured by cracking ethane and propane (p. 302). In the United Kingdom, they are manufactured from naphtha. Naphtha and steam are passed through pipes heated at 700–900°C to form a mixture of hydrocarbons which is distilled into the following fractions:

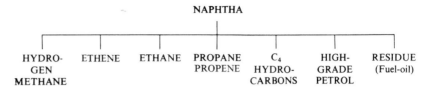

At 700°C, about 15 per cent of the feedstock is converted into ethene and a further 15 per cent into propene. When the temperature is raised to 900°C, the conversion into ethene increases to 30 per cent, but the amount of high-grade petrol (19.4) decreases from 30 to 25 per cent.

Ethene and propene are also obtained from refinery gas (p. 298).

Chemicals from ethene

The principal use of ethene is in the manufacture of the plastics, **poly(ethene)**, **poly(phenylethene)**, **poly(chloroethene)** and **poly(ethenyl ethanoate)**, which are discussed in Chapter 21. In addition, the Ziegler process for the manufacture of poly(ethene) (p. 312) has been adapted to make longer-chain alkenes and alcohols from ethene. The ethene is heated with triethylaluminium:

$$Al(C_2H_5)_3 + nC_2H_4 \rightarrow AlR^1R^2R^3$$

For the formation of alkenes, the resulting trialkylaluminium (in which R^1, R^2 and R^3 are alkyl groups larger than ethyl) is heated at 300°C under pressure, for example:

$$Al(CH_2CH_2CH_2CH_2R)_3 \rightarrow Al(C_2H_5)_3 + 3RCH=CH_2$$

For the formation of alcohols, the aluminium trialkyl is oxidised in air to an aluminium alkoxide which is subsequently decomposed with dilute sulphuric acid:

$$\underset{R^3}{\overset{R^1\ \ R^2}{\underset{|}{\overset{\diagdown\ \diagup}{Al}}}} \xrightarrow{\text{air}} \underset{OR^3}{\overset{R^1O\ \ OR^2}{\underset{|}{\overset{\diagdown\ \diagup}{Al}}}} \xrightarrow{H_2SO_4} R^1OH + R^2OH + R^3OH + Al_2(SO_4)_3$$

The alcohols are used in the manufacture of detergents (13.10).

Ethanol is made from ethene, either by direct hydration or *via* ethyl hydrogen sulphate (10.4), and is used to make ethanal. Ethanol is used as a solvent for lacquers and varnishes, cellulose nitrate (used to make cellophane) and in many cosmetic and toilet preparations.

Ethanal, which can be made from ethene *via* ethanol, is also manufactured directly from ethene by the Wacker process (12.3). Ethanal is then converted into **ethanoic acid** and **ethanoic anhydride** (14.4).

Ethene is oxidised directly to **epoxyethane** (11.6), which is used to make detergents (13.10) and **ethane-1,2-diol** (10.7). The diol is used as an anti-freeze and also to make **Terylene** (p. 321).

Chloroethane, used to make **tetraethyllead** (9.3), the antiknock agent for gasoline, is manufactured from ethene in the gas phase:

$$CH_2{=}CH_2 + HCl \xrightarrow[200°C]{AlCl_3 \text{ as cat.}} CH_3{-}CH_2Cl$$

and by the chlorination of ethane (9.3), while **1,2-dichloroethane** is made in the liquid phase by the addition of chlorine to ethene, with 1,2-dichloroethane itself as the solvent:

$$CH_2{=}CH_2 + Cl_2 \xrightarrow[50°C]{FeCl_3 \text{ as cat.}} Cl{-}CH_2{-}CH_2{-}Cl$$

1,2-Dichloroethane is subsequently heated under pressure to form **chloroethene** (p. 314):

$$Cl{-}CH_2{-}CH_2{-}Cl \xrightarrow[600°C]{\text{Pumice}} CH_2{=}CH{-}Cl + HCl$$

The hydrogen chloride can be used to manufacture further 1,2-dichloroethane by an 'oxidative chlorination' process:

$$CH_2{=}CH_2 + 2HCl + \tfrac{1}{2}O_2 \xrightarrow[250°C]{CuCl_2 \text{ as cat.}} Cl{-}CH_2{-}CH_2{-}Cl + H_2O$$

Chemicals from propene

By far the greatest use for propene is in the manufacture of raw materials needed to make plastics. For this purpose, most of it is converted into **propanone**. Two main processes are in current use, one *via* (1-methylethyl) benzene (10.8) and the other *via* **propan-2-ol** (12.3). The propan-2-ol is formed from propene either by direct hydration or through propyl hydrogen sulphate (10.4), and it is converted into propanone in part by catalytic dehydrogenation in the gas phase:

$$(CH_3)_2CH{-}OH \xrightarrow[500°C]{Cu \text{ as cat.}} (CH_3)_2C{=}O + H_2$$

and in part by oxidation in air in the liquid phase:

$$(CH_3)_2CH{-}OH + O_2 \xrightarrow{100°C} (CH_3)_2C{=}O + H_2O_2$$

The advantage of the latter method is that **hydrogen peroxide** is formed as a useful by-product.

Although propanone is used both as a solvent and in the manufacture of a variety of other products, one of the most important uses is in the manufacture of the thermosoftening plastic, **Perspex** (p. 315).

Some propene is used to make **propane-1,2,3-triol** (10.7) whose main use is in the manufacture of the group of plastics known as **glyptal resins** (p. 319). In addition, propene is now being used increasingly as the starting material for **propenonitrile**, used to make the plastic, poly(propenonitrile) (p. 315).

$$CH_3{-}CH{=}CH_2 + NH_3 + \tfrac{3}{2}O_2 \xrightarrow[450°C]{\text{Mo-based cat.}} CH_2{=}CH{-}CN + 3H_2O$$

Propene itself is polymerised to **poly(propene)** (p. 312).

Some propene is converted into **epoxypropane**, one of whose principal uses is in the manufacture of **polyurethane forms** (p. 317).

Butenes and pentenes

The butenes are obtained from refinery gas. The most important use of the straight-chain butenes (but-1-ene and *cis*- and *trans*-but-2-ene) is in the manufacture of **buta-1,3-diene**, which is required for the production of **synthetic rubbers** (p. 322).

2-Methylpropene is used in the production of a viscostatic lubricating oil, by polymerisation (p. 298), and to make **2,2,4-trimethylpentane** (p. 295).

Pentenes are used to make **2-methylbuta-1,3-diene**, the monomer for an **artificial rubber** (p. 323).

Higher alkenes

Higher alkenes are made by polymerisation of ethene (20.4) or the cracking of waxes (19.8). They are used principally for the manufacture of **detergents** (13.10) and long-chain alcohols [via the OXO process (20.7)] for making esters used as **plasticisers** (21.2).

20.5 Chemicals from aromatic compounds

Production of aromatic hydrocarbons

The traditional source of benzene, methylbenzene, the dimethylbenzenes and the polycyclic aromatic compounds such as naphthalene was coal-tar. However, a high proportion of all these hydrocarbons is now derived from petroleum; for example, over 80 per cent of benzene is obtained in this way.

The aromatic hydrocarbons are formed when naphtha is cracked to form alkenes (20.4) and during catalytic reforming (19.4). A number of physical processes are used to purify the mixtures of aromatic hydrocarbons. First, the aromatics are separated from other hydrocarbons by solvent extraction (bis-2-hydroxyethyl ether, $HOCH_2CH_2$—O—CH_2CH_2OH, is often used since aromatic hydrocarbons, unlike aliphatics, dissolve readily in it). Then, the aromatic hydrocarbons are carefully distilled. The first fraction contains **benzene**, and this is further purified by cooling, since it has a much higher freezing point than the other aromatic compounds which are present in the fraction. The second fraction contains **methylbenzene**, which is purified by further distillation. The third fraction, containing a mixture of the **dimethylbenzenes** and **ethylbenzene**, is purified by both fractional distillation and fractional recrystallization.

More methylbenzene and dimethylbenzenes are formed during the reforming processes than is required, and they are converted into benzene by a process known as **hydrodealkylation**. The vapours of the aromatics are mixed with hydrogen and passed over a catalyst under pressure at about 600°C. For example:

$$C_6H_5CH_3 + H_2 \xrightarrow[600\ °C]{Pt\ on\ Al_2O_3\ as\ cat.} C_6H_6 + CH_4$$

Chemicals from benzene

Most of the benzene which is manufactured is converted into materials needed for the plastics industry; examples are the formation of **phenylethene** and thence **poly(phenylethene)** (p. 313), and **cyclohexane** (p. 306), an intermediate in the manufacture of **nylon** (p. 320).

Benzene is also used to make **phenol** *via* (1-methylethyl)benzene (p. 150). The principal uses of phenol are in the manufacture of cyclohexanol (p. 153) used to make **nylon** (p. 320), thermosetting plastics such as **Bakelite** (p. 316), substituted phenols for **epoxy resins** (p. 319) and selective weed-killers such as **2,4-D** (2,4-dichlorophenoxyethanoic acid) (p. 154).

Chemicals from methylbenzene

Some methylbenzene is converted into **benzene** by hydrodealkylation, some is used in fuels for aeroplanes and cars and some is employed in the manufacture of the **polyurethane** plastic foams (p. 317).

Chemicals from dimethylbenzenes

1,2-Dimethylbenzene is used for making **benzene-1,2-dicarboxylic anhydride**, which is required for **plasticisers** (p. 311) and **glyptal resins** (p. 319):

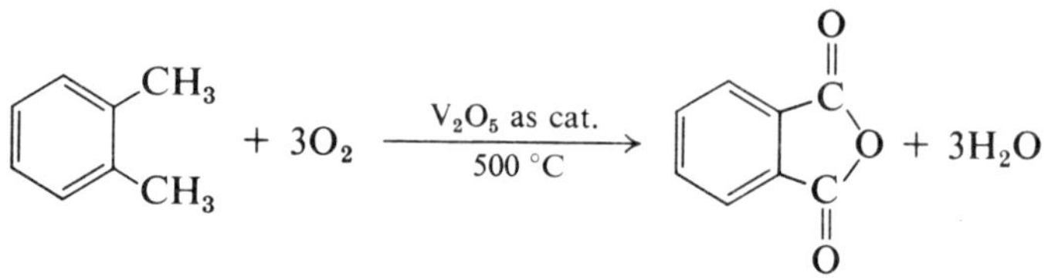

$$\text{o-C}_6\text{H}_4(\text{CH}_3)_2 + 3\text{O}_2 \xrightarrow[500\,°\text{C}]{\text{V}_2\text{O}_5 \text{ as cat.}} \text{benzene-1,2-dicarboxylic anhydride} + 3\text{H}_2\text{O}$$

Benzene-1,2-dicarboxylic anhydride

1,4-Dimethylbenzene is oxidised to **benzene-1,4-dicarboxylic acid**, required for **Terylene** manufacture (p. 321). One of the most recent methods for the oxidation is by passing air into the liquid hydrocarbon under pressure in the presence of a cobalt salt as catalyst:

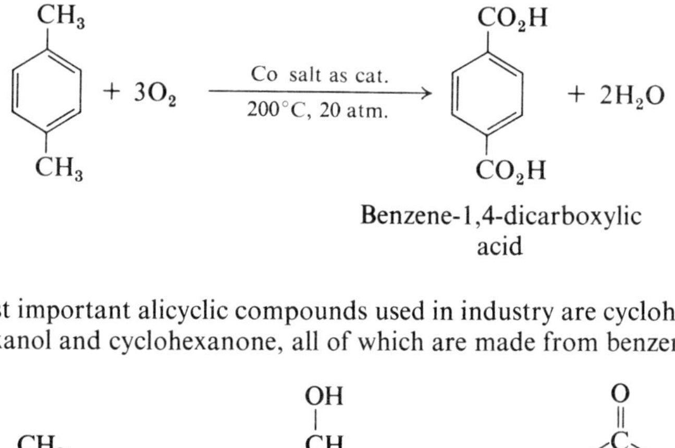

$$\text{p-C}_6\text{H}_4(\text{CH}_3)_2 + 3\text{O}_2 \xrightarrow[200\,°\text{C, 20 atm.}]{\text{Co salt as cat.}} \text{p-C}_6\text{H}_4(\text{CO}_2\text{H})_2 + 2\text{H}_2\text{O}$$

Benzene-1,4-dicarboxylic acid

20.6 Alicyclic compounds

The most important alicyclic compounds used in industry are cyclohexane, cyclohexanol and cyclohexanone, all of which are made from benzene.

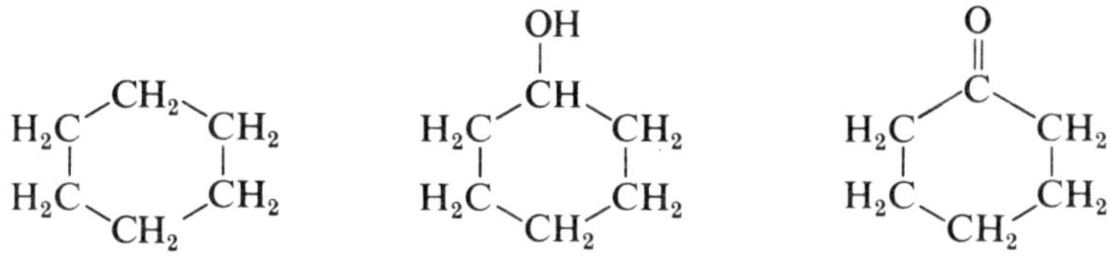

Cyclohexane   Cyclohexanol   Cyclohexanone

Cyclohexane is manufactured by passing hydrogen through liquid benzene in the presence of a nickel catalyst under pressure:

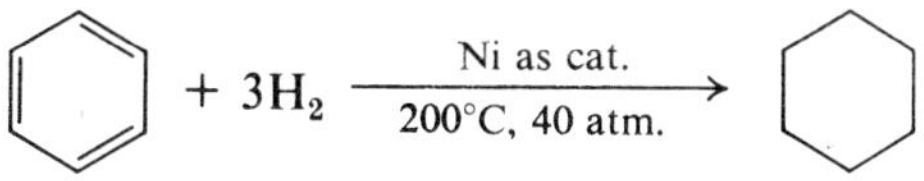

$$\text{C}_6\text{H}_6 + 3\text{H}_2 \xrightarrow[200°\text{C, 40 atm.}]{\text{Ni as cat.}} \text{C}_6\text{H}_{12}$$

Cyclohexane is oxidised by passing air through the liquid under pressure, in the presence of a catalyst (often a cobalt salt):

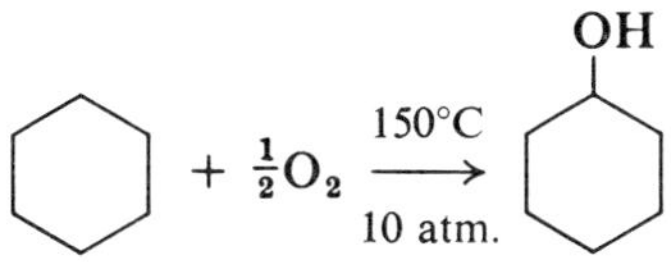

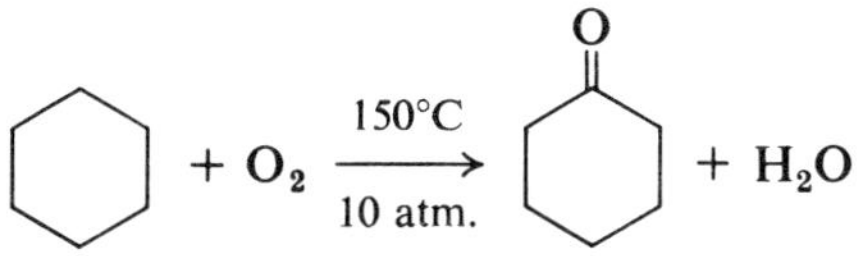

$$\text{C}_6\text{H}_{12} + \tfrac{1}{2}\text{O}_2 \xrightarrow[10\text{ atm.}]{150°\text{C}} \text{cyclohexanol}$$

$$\text{C}_6\text{H}_{12} + \text{O}_2 \xrightarrow[10\text{ atm.}]{150°\text{C}} \text{cyclohexanone} + \text{H}_2\text{O}$$

The mixture of cyclohexanol and cyclohexanone (known as 'mixed oil') can be oxidised to **hexanedioic acid** by heating with either moderately concentrated nitric acid or air:

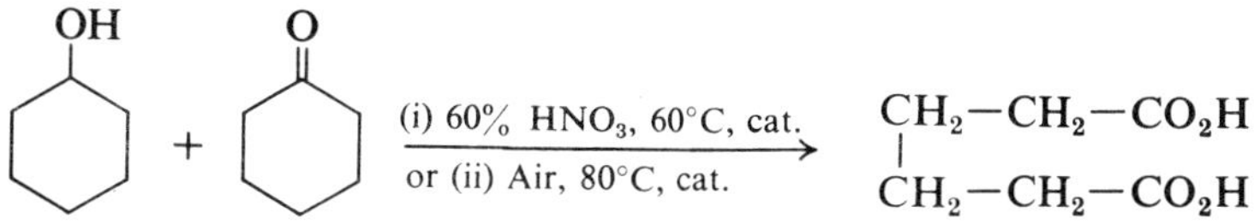

$$\text{cyclohexanol} + \text{cyclohexanone} \xrightarrow[\text{or (ii) Air, 80°C, cat.}]{\text{(i) 60\% HNO}_3\text{, 60°C, cat.}} \begin{array}{l}\text{CH}_2-\text{CH}_2-\text{CO}_2\text{H} \\ | \\ \text{CH}_2-\text{CH}_2-\text{CO}_2\text{H}\end{array}$$

Part of the hexanedioic acid is converted into **hexanedinitrile** which is reduced by hydrogen over nickel to **1,6-diaminohexane**. Hexanedioic acid and 1,6-diaminohexane are then used to make **nylon-6.6** (p. 320).

Another important polymer is **nylon-6**, for which pure cyclohexanone is required. When the mixed oil is heated under pressure with copper(II) and chromium(III) oxides, the cyclohexanol, which is a secondary alcohol, is dehydrogenated to the corresponding ketone:

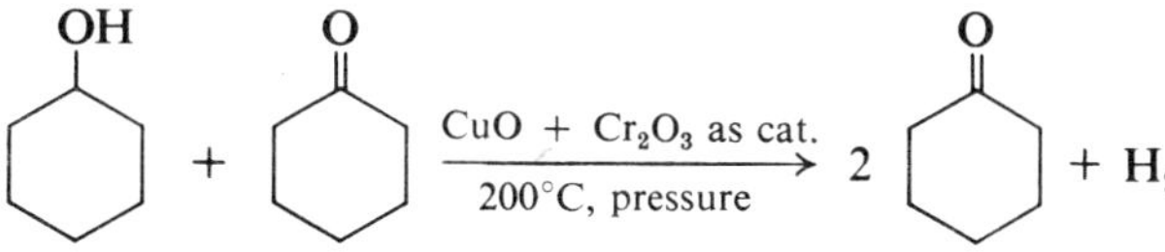

$$\text{cyclohexanol} + \text{cyclohexanone} \xrightarrow[200°\text{C, pressure}]{\text{CuO} + \text{Cr}_2\text{O}_3\text{ as cat.}} 2\,\text{cyclohexanone} + \text{H}_2$$

Cyclohexanone is then converted into **caprolactam**, from which nylon-6 is made (p. 320).

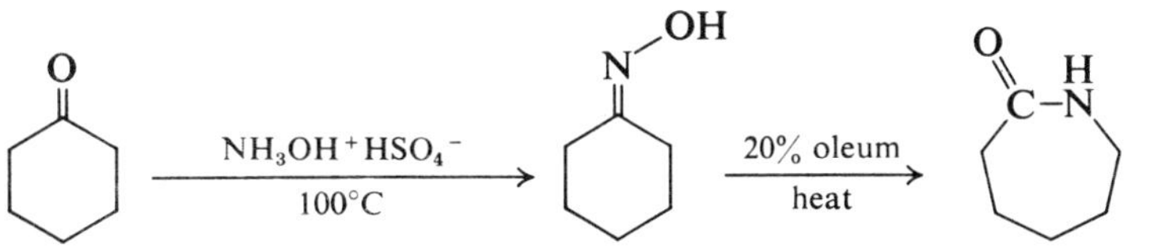

$$\text{cyclohexanone} \xrightarrow[100°\text{C}]{\text{NH}_3\text{OH}^+\text{HSO}_4^-} \text{Cyclohexanone oxime} \xrightarrow[\text{heat}]{20\%\text{ oleum}} \text{Caprolactam}$$

The isomerisation of the oxime to caprolactam by sulphuric acid is an example of the **Beckmann rearrangement** in which an oxime is transformed into an amide in the presence of acid:

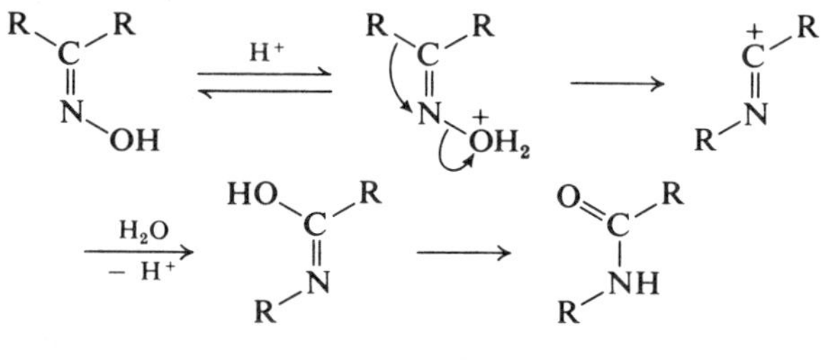

20.7 Synthesis gas

Manufacture of synthesis gas

'Synthesis gas' is a mixture of carbon monoxide and hydrogen. It is manufactured by heating methane or naphtha with steam on a nickel catalyst, for example:

$$CH_4 + H_2O \xrightarrow[900°C, 30 \text{ atm.}]{\text{Ni as cat.}} CO + 3H_2$$

Since methane contains a higher ratio of hydrogen to carbon than naphtha, it is the preferred feedstock when a high proportion of hydrogen is required in the product.

Uses of synthesis gas

1. When the mixture is passed over heated zinc and chromium(III) oxides under pressure, **methanol** is formed:

$$CO + 2H_2 \xrightarrow[300°C, 300 \text{ atm.}]{\text{ZnO} + Cr_2O_3 \text{ as cat.}} CH_3OH$$

Methanol is oxidised to **methanal**, used to make plastics (p. 316–317):

$$CH_3OH + \tfrac{1}{2}O_2 \xrightarrow[500°C]{\text{Ag as cat.}} CH_2=O + H_2O$$

2. **Higher alcohols** are manufactured by the OXO process. Synthesis gas and an alkene are heated under pressure in the presence of a cobalt(II) salt:

$$R-CH=CH_2 + CO + 2H_2 \xrightarrow{\text{Co(II) salt as cat.}} R-CH_2-CH_2-CH_2-OH$$

These alcohols are used in the manufacture of **detergents** (13.10).

3. The carbon monoxide and hydrogen in synthesis gas can be separated by passing the mixture through a solution of a copper(I) salt dissolved in aqueous ammonia. The carbon monoxide is absorbed, and, on regeneration, it can be converted into **carbonyl chloride**, $COCl_2$, which is used to make **polyurethane** foam plastics (p. 317).

4. Hydrogen is made by heating synthesis gas with an excess of steam:

$$CO + 3H_2 + H_2O \xrightarrow[400°C, 30 \text{ atm.}]{\text{Fe as cat.}} CO_2 + 4H_2$$

THE PETROCHEMICAL INDUSTRY

The gases are passed through an alkali (for example, a solution of potassium carbonate) to remove carbon dioxide. Traces of carbon monoxide can be removed by passing the mixture over nickel:

$$CO + 3H_2 \xrightarrow[300°C, 20 \text{ atm.}]{\text{Ni as cat.}} CH_4 + H_2O$$

Hydrogen is used to make **ammonia**:

$$N_2 + 3H_2 \xrightarrow[500°C, 150 \text{ atm.}]{\text{Fe as cat.}} 2NH_3$$

Nearly all ammonia is now made from petroleum sources and is used to make nitric acid and inorganic fertilisers such as **ammonium nitrate**. However, an organic nitrogen fertiliser, **carbamide**, is now the dominant fertiliser; it is made by heating ammonia and carbon dioxide under pressure:

$$2NH_3 + CO_2 \xrightarrow{200°C, 200 \text{ atm.}} H_2N\text{—}CO_2^- \, NH_4^+ \rightarrow$$
$$H_2N\text{—}CO\text{—}NH_2 + H_2O$$

Carbamide is also used to make plastics (p. 317).

20.8 Sulphur

Crude oil contains both hydrogen sulphide and organic sulphur compounds (which during the various processes in the refinery are converted into hydrogen sulphide). The refinery gas is passed through a solution of a weak base (an amine) which absorbs hydrogen sulphide, which is an acid. The solution is later boiled to release hydrogen sulphide in a concentrated form. Some natural gas also contains hydrogen sulphide and this is removed in the same way.

The removal of sulphur compounds is necessary to avoid pollution. However, it is also economically important since over a quarter of the sulphur now used in industry is recovered from petroleum sources. Hydrogen sulphide is burnt in air to form sulphur dioxide:

$$2H_2S + 3O_2 \rightarrow 2SO_2 + 2H_2O$$

Sulphur dioxide is mixed with excess of hydrogen sulphide and passed over aluminium oxide at 400°C:

$$2H_2S + SO_2 \rightarrow 3S + 2H_2O$$

Sulphur condenses out and is converted into **sulphuric acid** by the Contact Process. Sulphuric acid is used in many ways, for example, in the manufacture of fertilisers (superphosphate and ammonium sulphate), titanium(IV) oxide, rayon and cellophane (p. 286), explosives (p. 267), and detergents (p. 194).

20.9 Further reading

Heavy Organic Chemicals. A. J. Gait (1967). Pergamon Press.
Chemicals from Petroleum. A. L. Waddams (2nd ed. 1974). John Murray.
Petroleum Chemicals. BP Chemicals (U.K.) Ltd. (1970) (obtainable from Education Section, BP Chemicals (U.K.) Ltd., Devonshire House, Mayfair Place, Piccadilly, London W1.
Organic Chemicals from Petroleum. BP Chemicals (U.K.) Ltd. (1969).
Industrial Organic Chemistry. J. K. Stille (1968). Prentice-Hall, Inc.
Chemicals from Oil. Richard T. Nye (1970). Pergamon Press.
Modern Chemicals from Oil. Shell Education Service, Shell-Mex House, Strand, London, WC2R 0DX.

THE PETROCHEMICAL INDUSTRY

Advanced Reading

The Petroleum Chemicals Industry. R. F. Goldstein and A. L. Waddams (3rd ed. 1967). E. and F. N. Spon.

An Introduction to Industrial Organic Chemistry. P. Wiseman (1971). Applied Science Publishers.

20.10 Films

A large number of films on the petrochemical industry can be obtained from the Shell Film Library* and from the British Petroleum Company Ltd.
A film that describes catalysis is:
Catalysis. I.C.I. Film Library, Imperial Chemical Industries Ltd.*
* On free loan.

20.11 Questions

1 In the United States, the simple gaseous hydrocarbons are all cheaply available. Suggest routes by which certain of these gases could be used for the preparation of propanone, benzene, ethanol, ethanoic acid and poly(ethene).

2 Write an essay on petroleum as a raw material in the production of simple organic compounds of low formula weight.

3 Describe, with the aid of a diagram, the primary refining process used in the treatment of crude petroleum.
   Explain, quoting one example in each case, the meaning of: (a) catalytic cracking; (b) alkylation. Say why these processes were introduced into the petroleum industry.
   Describe how the following are made from refinery sources: (i) ethanol, (ii) ethane-1,2-diol, (iii) methylbenzene.

4 Write an account of the production of organic chemicals from crude petroleum.
(S(S))

5 Outline the means by which the following are industrially produced from petroleum: (a) petrol, (b) benzene, (c) propan-2-ol.
   By what means, in the petrochemical industry, are hydrocarbons containing about twelve carbon atoms changed into lower molecular weight homologues?
   Indicate all the steps involved in the production of benzene from a different natural source.
(S)

Chapter 21

Polymers

21.1 Introduction

When two or more molecules of a simple compound (a **monomer**) join together to form a new compound (a **polymer**), the process is described as **polymerisation**. In **addition polymerisation**, the polymer has the same empirical formula but a higher formula weight than the monomer; an example is the formation of poly(ethene) from the monomer ethene:

$$n\text{CH}_2=\text{CH}_2 \rightarrow \cdots-\text{CH}_2-\text{CH}_2-\text{CH}_2-\text{CH}_2-\text{CH}_2-\text{CH}_2-\cdots$$

In **condensation polymerisation**, polymerisation of the monomer is accompanied by the elimination of small molecules such as water or ammonia; an example is the formation of nylon-6.6:

$$n\text{H}_2\text{N}-(\text{CH}_2)_6-\text{NH}_2 + n\text{HO}_2\text{C}-(\text{CH}_2)_4-\text{CO}_2\text{H} \rightarrow$$
$$\text{H}[\text{NH}-(\text{CH}_2)_6-\text{NH}-\text{CO}-(\text{CH}_2)_4-\text{CO}]_n\text{OH} + (2n-1)\text{H}_2\text{O}$$

Natural polymers such as proteins (18.2) and carbohydrates (18.3) are discussed elsewhere. This chapter is concerned with synthetic polymers.

21.2 Plastics

Many synthetic polymers can be moulded into required shapes and are useful as **plastics** for the manufacture of a wide range of articles. The properties of plastics can be modified by the addition of compounds known as **plasticisers**. For example, esters of benzene-1,2-dicarboxylic acid, made from benzene-1,2-dicarboxylic anhydride (p. 306) with long-chain alcohols such as octanol, are added to PVC (poly(chloroethene)) (p. 314) to produce a softer and more easily worked material. **Dyes** and **pigments** are added to give the plastic colour, and it is sometimes possible to add a cheap material known as a **filler** to increase the bulk of the plastic and make it less expensive, without altering its desirable properties.

Plastics can be subdivided into two groups: the thermosoftening and thermosetting plastics.

Thermosoftening plastics

The characteristic of these plastics is that they become soft when heated and can then be moulded or remoulded. They are **linear** polymers; that is, they are of the general structure:

$$-\text{X}-\text{X}-\text{X}-\text{X}-\text{X}-\text{X}-\text{X}-$$

where X represents the monomer.

Many of the thermosoftening plastics are formed by the polymerisation of monomers which contain a C=C bond; examples are poly(ethene) from ethene, poly(propene) from propene, poly(chloroethene) from chloroethene and Perspex from methyl 2-methylpropenoate.

Poly(ethene) (polythene)

Two types of poly(ethene) are manufactured. One, with a low density and a formula weight ranging from 50,000 to 300,000, softens at a comparatively low temperature (about 120°C). It is made by compressing ethene under very high pressure at about 200°C, in the presence of a very small amount of oxygen:

$$n\text{CH}_2\!\!=\!\!\text{CH}_2 \xrightarrow[\substack{200°C,\ 1500\text{--}2000 \\ \text{atm.}}]{\text{O}_2 \text{ as cat.}} (\!\!-\!\text{CH}_2\!\!-\!\!\text{CH}_2\!\!-\!\!)_n$$

The second form is a high density material with a formula weight in the range 50,000 to 3,000,000 and a higher softening point (about 130°C). It is generally manufactured by a process developed by the Swiss chemist, Ziegler, in which ethene is passed under pressure into an inert solvent (an aromatic hydrocarbon) containing as catalyst triethylaluminium and titanium(IV) chloride. After polymerisation, the catalyst is decomposed by adding dilute acid, and the crystalline polymer is filtered off.

Poly(ethene) is used without a plasticiser or filler and can be readily coloured. The plastic is an insulator and is acid-resistant. It is easily moulded and blown, and is therefore used to make many household articles and, in particular, bottles. It is also readily rolled into thin film and used in sheets for packaging and as a coating.

Poly(propene)

While poly(ethene) has been manufactured since 1939, poly(propene) is a comparatively new material. Its manufacture did not begin in Great Britain until 1962, but the amount being produced is increasing by about 20 per cent per year. It can be used in place of poly(ethene), over which it has the advantages of being stronger and lighter and having a higher softening point; it is used particularly for making wrapping films and ropes.

It is manufactured by a method invented by the Italian chemist, Natta, in which propene is passed under pressure into an inert solvent (heptane) which contains a trialkylaluminium and a titanium compound:

$$2n\text{CH}_3\!\!-\!\!\text{CH}\!\!=\!\!\text{CH}_2 \xrightarrow[100°C,\ 10\ \text{atm.}]{\substack{\text{AlR}_3 + \text{TiCl}_3 \\ \text{as cat.}}} (\!\!-\!\!\overset{\overset{\displaystyle \text{CH}_3}{|}}{\text{CH}}\!\!-\!\!\text{CH}_2\!\!-\!\!\overset{\overset{\displaystyle \text{CH}_3}{|}}{\text{CH}}\!\!-\!\!\text{CH}_2\!\!-\!\!)_n$$

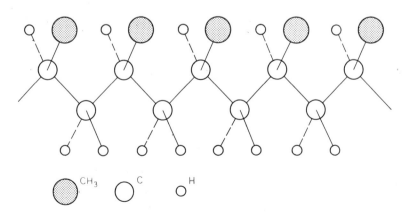

FIG 21.1. *A molecular model of poly(propene), an isotactic polymer*

This method produces a polymer in which the methyl groups on the alternate carbon atoms all have the same orientation. This can be represented as in Fig. 21.1. Such polymers are described as **isotactic**, and it is this regularity of structure which allows neighbouring molecules of poly(propene) to pack closely together to give a crystalline structure. In **atactic** polymers, the orientations of the side-chains are random and the compounds are non-crystalline.

Ziegler and Natta were awarded a Nobel Prize in 1963 for their work on polymerisation.

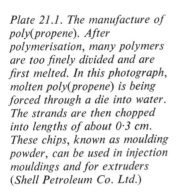

Plate 21.1. The manufacture of poly(propene). After polymerisation, many polymers are too finely divided and are first melted. In this photograph, molten poly(propene) is being forced through a die into water. The strands are then chopped into lengths of about 0·3 cm. These chips, known as moulding powder, can be used in injection mouldings and for extruders (Shell Petroleum Co. Ltd.)

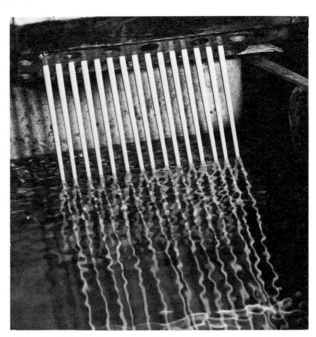

Poly(phenylethene)

Phenylethene, C_6H_5—CH=CH_2, is being produced on a rapidly expanding scale in order, principally, to make poly(phenylethene) and synthetic rubbers, especially SBR (p. 322). It is made largely by the reaction of benzene with ethene in the presence of aluminium trichloride:

$$C_6H_6 + CH_2\!\!=\!\!CH_2 \xrightarrow[300°C,\ 40\ atm.]{AlCl_3\ as\ cat.} C_6H_5\text{---}CH_2\text{---}CH_3$$
$$\text{Ethylbenzene}$$

followed by dehydrogenation of the resulting ethylbenzene at 600°C over a catalyst such as zinc, iron(III) or magnesium oxide supported on charcoal or alumina:

$$C_6H_5\text{---}CH_2\text{---}CH_3 \xrightarrow[600°C]{Fe_2O_3\ as\ cat.} C_6H_5\text{---}CH\!\!=\!\!CH_2 + H_2$$

The method by which phenylethene is polymerised is typical of that for a number of other monomers. It is treated with a compound, an **initiator**, which readily decomposes to form free radicals, for example di(benzoyl) peroxide on being heated:

$$C_6H_5\text{---}CO\text{---}O\text{---}O\text{---}CO\text{---}C_6H_5 \rightarrow 2C_6H_5\text{---}CO\text{---}O\cdot$$

One of the resulting radicals, R·, adds to the alkene, giving a new radical,

$$R\cdot + CH_2=CH-C_6H_5 \rightarrow R-CH_2-\overset{\cdot}{C}H-C_6H_5$$

which in turn adds to another molecule of phenylethene:

$$R-CH_2-\underset{\underset{C_6H_5}{|}}{CH}\cdot + CH_2=CH-C_6H_5 \rightarrow R-CH_2-\underset{\underset{C_6H_5}{|}}{CH}-CH_2-\underset{\underset{C_6H_5}{|}}{CH}\cdot$$

Further additions of this sort occur, giving long-chain radicals, but occasionally the chain is interrupted and polymerisation stopped by the chance collision of two radicals:

$$R-(CH_2-\underset{\underset{C_6H_5}{|}}{CH})_m-CH_2-\underset{\underset{C_6H_5}{|}}{CH}\cdot + \cdot\underset{\underset{C_6H_5}{|}}{CH}-CH_2-(\underset{\underset{C_6H_5}{|}}{CH}-CH_2)_n-R \rightarrow$$

$$R-(CH_2-\underset{\underset{C_6H_5}{|}}{CH})_m-CH_2-\underset{\underset{C_6H_5}{|}}{CH}-\underset{\underset{C_6H_5}{|}}{CH}-CH_2-(\underset{\underset{C_6H_5}{|}}{CH}-CH_2)_n-R$$

Collisions between radicals are very effective at yielding final products, but the concentration of radicals in the solution is very small, and it is much more likely that one radical will meet a molecule of phenylethene than another radical; in this way, long chains are built up. Note that the length that the growing polymer chain will have reached by the time that collision with another radical brings the growth to an end will vary from one chain to another, since it depends on chance collisions in the solution; therefore, the polymer does not have an exact formula weight, and formula-weight measurements give an average for all the chains. Note also that the phenyl groups occur at regular intervals in the chain except when two chains couple together in the example above. The regularity arises because the growing radical tends to add to the CH_2 group of phenylethene and not to the $CH(C_6H_5)$ group.

The groups at each end of a polymer depend on the nature of the initiator. For simplicity, the end groups are often omitted in representing the polymers.

The main uses of poly(phenylethene) are in making light-weight packaging materials and a wide variety of household goods such as egg boxes and the lining material for refrigerators.

Poly(chloroethene) (polyvinyl chloride, PVC)

Chloroethene is manufactured from ethene (p. 81) and from ethyne (p. 90), with the former route now becoming the more important. It is converted into poly(chloroethene) by heating in an inert solvent with di(benzoyl) peroxide to initiate polymerisation:

$$2nCH_2=CHCl \rightarrow \{CH_2-\underset{\underset{Cl}{|}}{CH}-CH_2-\underset{\underset{Cl}{|}}{CH}\}_n$$

PVC is a very tough polymer, and a plasticiser is added to soften it; esters of benzene-1,2-dicarboxylic acid are used for this purpose, sometimes accounting for up to 50 per cent of the total weight. PVC is easy to colour, it is resistant to weathering, fire and chemicals, and it is also a good electrical insulator. It is used as the insulator for cables and in the manufacture of artificial leather (e.g. car upholstery), household goods such as curtains and table cloths, gramophone records and floor coverings.

Poly(ethenyl ethanoate)

Ethenyl ethanoate is manufactured by passing a mixture of ethene, ethanoic acid vapour and oxygen over a noble metal at about 200°C:

$$CH_3-CO_2H + CH_2=CH_2 + \tfrac{1}{2}O_2 \rightarrow CH_3-CO-O-CH=CH_2 + H_2O$$

An older process, which is becoming relatively more expensive, is *via* ethyne. Ethyne and ethanoic acid vapour are passed over a catalyst of zinc ethanoate suspended on carbon at about 200°C:

$$CH_3-CO_2H + CH\equiv CH \xrightarrow[200°C]{(CH_3CO_2)_2Zn \text{ as cat.}} CH_3-CO-O-CH=CH_2$$

Ethenyl ethanoate is polymerised in the same way as phenylethene:

$$2n\,CH_3-CO-O-CH=CH_2 \rightarrow \begin{array}{c} \quad\;\; CH_3 \quad\quad\;\; CH_3 \\ \quad\;\; | \quad\quad\quad\;\; | \\ \quad\;\; CO \quad\quad\;\; CO \\ \quad\;\; | \quad\quad\quad\;\; | \\ \quad\;\; O \quad\quad\quad\; O \\ \quad\;\; | \quad\quad\quad\;\; | \\ -(CH_2-CH-CH_2-CH)_n- \end{array}$$

Poly(ethenyl ethanoate) is employed mainly as the essential constituent in plastic emulsion paints. Other uses are as "chewing gum base" and, in solution in propanone, to greaseproof paper.

Polyacrylic esters

Methyl 2-methylpropenoate is manufactured from propanone by addition of hydrogen cyanide in the presence of alkali to give the cyanohydrin (p. 167) followed by reaction with methanol in the presence of sulphuric acid:

$$(CH_3)_2C=O \xrightarrow[NaOH]{HCN} (CH_3)_2C\begin{array}{c} OH \\ CN \end{array}$$

$$\downarrow {98\%\ H_2SO_4}/{CH_3OH}$$

$$\left[\begin{array}{c} \quad OH \\ CH_3-C-CO_2CH_3 \\ \quad CH_3 \end{array} \right] \longrightarrow CH_2=C\begin{array}{c} CO_2CH_3 \\ CH_3 \end{array} + H_2O$$

Its polymer is formed in the same way as poly(phenylethene) and is sold under the name of "Perspex" (blocks and sheets) and "Diakon" (powders):

$$2n\,CH_2{=}C\!\begin{array}{c}CO_2CH_3\\[2pt]\\CH_3\end{array} \longrightarrow {+}CH_2{-}\underset{CH_3}{\overset{CO_2CH_3}{C}}{-}CH_2{-}\underset{CH_3}{\overset{CO_2CH_3}{C}}{\}_n}$$

The plastics are light, strong and transparent. Sheets are used where transparency is important (packaging, aeroplane windows, lenses, corrugated roof lights).

Polyoxymethylene

Polyoxymethylene is formed by the polymerisation of methanal; gaseous methanal is passed into an inert solvent (a hydrocarbon) which contains an amine as catalyst:

$$2n\,CH_2{=}O + H_2O \rightarrow HO{-}(CH_2{-}O{-}CH_2{-}O)_n{-}H$$

To prevent depolymerisation, the —OH groups at the end of each chain are esterified.

The plastic can be readily moulded, and is very hard, resistant to chemicals and remarkably resistant to abrasion. For example, a shaft running at 110 m.p.h. against a steel gear was found to have no wear after 10,000 miles. However, it tends to decompose above about 100°C.

Poly(tetrafluoroethene) (PTFE)

PTFE is made by heating tetrafluoroethene (p. 127) under pressure in the presence of ammonium peroxosulphate as catalyst:

$$2n\,CF_2{=}CF_2 \xrightarrow{(NH_4)_2S_2O_8 \text{ as cat.}} {+}CF_2{-}CF_2{-}CF_2{-}CF_2{\}_n}$$

The polymer is very hard, resistant to heat and chemicals and has a high softening point (above 320°C). These properties make it suitable for making seals and gaskets which are subject to heavy wear at high temperatures. Its 'anti-stick' properties make it useful as a surface coating for cooking equipment.

The anti-stick properties of PTFE are probably due to its structure, which is a helix, with the fluorine atoms on the surface of an inner chain of carbon atoms. There is a very smooth surface, and molecules can readily flow over it.

Thermosetting plastics

These plastics contain three-dimensional networks of bonds and are moulded during the polymerisation stage of their manufacture. Unlike thermosoftening plastics, they cannot be remoulded.

Phenol-methanal plastics (Bakelite)

In 1910, Baekeland patented a process for producing resins from phenol and methanal which are now known as Bakelite. The resins are moulded, together with a filler (such as wood shavings) and a pigment, to form a wide range of articles. The electrical resistance of Bakelite makes it especially useful for electric plugs, switches and tools.

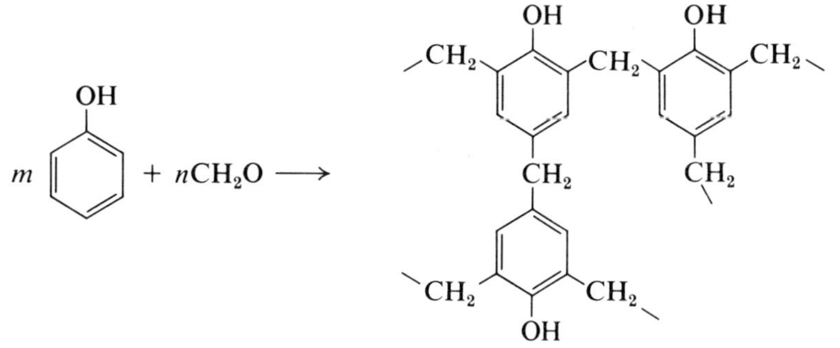

$$m\text{ C}_6\text{H}_5\text{OH} + n\text{CH}_2\text{O} \longrightarrow \text{[cross-linked phenol-methanal polymer]}$$

Carbamide-methanal and melamine-methanal plastics

Carbamide and methanal, when heated in a dilute solution of acid, give a resin:

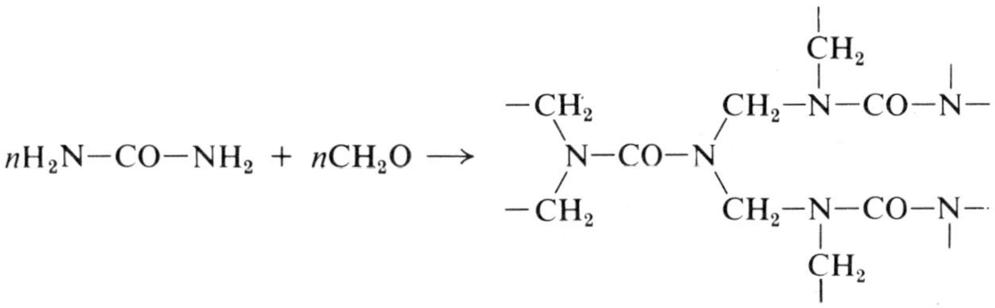

$$n\text{H}_2\text{N}-\text{CO}-\text{NH}_2 + n\text{CH}_2\text{O} \longrightarrow \text{[cross-linked carbamide-methanal polymer]}$$

Melamine can be used instead of carbamide. Melamine is manufactured from calcium cyanamide and has the structure:

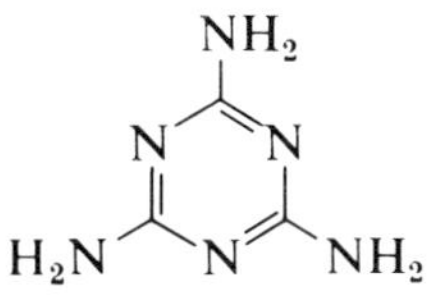

[melamine structure: 1,3,5-triazine ring with three NH$_2$ groups]

Both carbamide-methanal and melamine-methanal polymers absorb dyes easily, and the powders can be moulded under pressure to form tableware (e.g. Melaware), trays and many household utensils. They are used, in solution, to strengthen paper and to improve the shrink resistance of cotton, wool and rayon. Both are used as ion-exchange resins to demineralise water, as the free NH_2 groups combine with the acidic groups in the water.

Polyurethanes

Polyurethane resins are made by polymerisation of a mixture of a di-isocyanate and a molecule containing three hydroxyl groups (a triol):

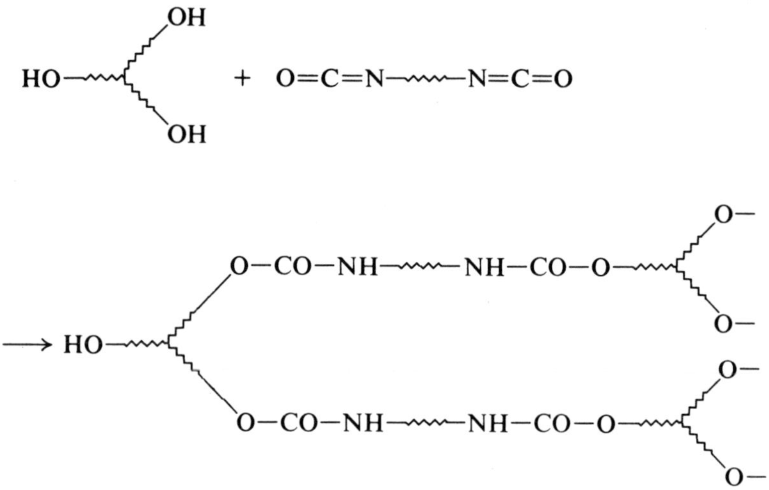

POLYMERS

$$HO-\!\!\!\!\sim\!\!\!\!\!<^{OH}_{OH} \quad + \quad O\!=\!C\!=\!N\!-\!\!\!\sim\!\!\!-N\!=\!C\!=\!O$$

$$\longrightarrow HO-\!\!\!\!\sim\!\!\!\!\!<^{O-CO-NH\!-\!\!\sim\!\!-NH-CO-O-\!\!\sim\!\!\!<^{O-}_{O-}}_{O-CO-NH\!-\!\!\sim\!\!-NH-CO-O-\!\!\sim\!\!\!<^{O-}_{O-}}$$

A triol often used is made from propane-1,2,3-triol and epoxypropane:

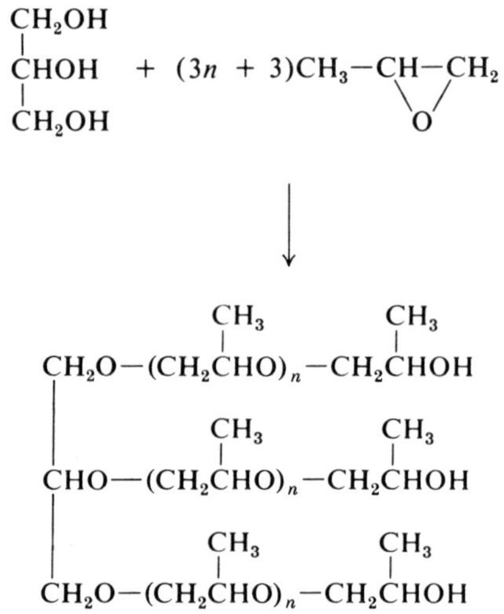

$$\begin{array}{c} CH_2OH \\ | \\ CHOH \\ | \\ CH_2OH \end{array} \quad + \quad (3n+3)CH_3\!-\!\underset{\underset{O}{\diagdown\!\diagup}}{CH\!-\!CH_2}$$

↓

$$\begin{array}{l} CH_2O\!-\!(CH_2\overset{CH_3}{\underset{|}{C}}HO)_n\!-\!CH_2\overset{CH_3}{\underset{|}{C}}HOH \\ | \\ CHO\!-\!(CH_2\overset{CH_3}{\underset{|}{C}}HO)_n\!-\!CH_2\overset{CH_3}{\underset{|}{C}}HOH \\ | \\ CH_2O\!-\!(CH_2\overset{CH_3}{\underset{|}{C}}HO)_n\!-\!CH_2\overset{CH_3}{\underset{|}{C}}HOH \end{array}$$

The di-isocyanate is made from methylbenzene:

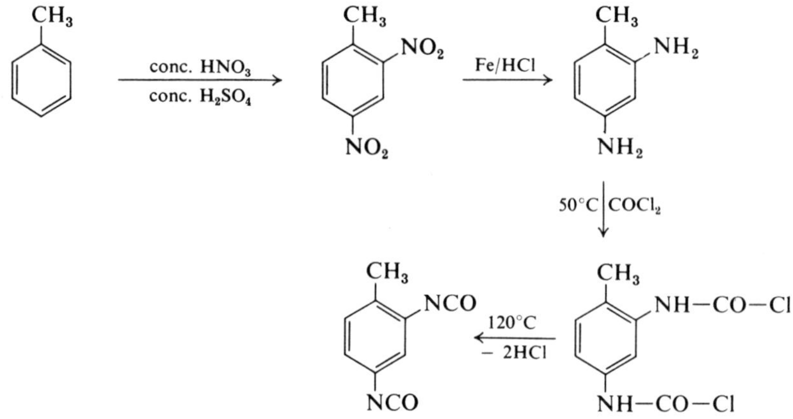

methylbenzene $\xrightarrow{\text{conc. HNO}_3 / \text{conc. H}_2\text{SO}_4}$ 2,4-dinitrotoluene $\xrightarrow{\text{Fe/HCl}}$ 2,4-diaminotoluene

$\xrightarrow{50°C \; COCl_2}$ bis(carbamoyl chloride) $\underset{-2HCl}{\overset{120°C}{\rightleftharpoons}}$ 2,4-diisocyanatotoluene

318

The polymers can be made into the well-known foam-polyurethane plastics by treatment with water. This converts some of the terminal isocyanate groups into amino groups,

$$-N=C=O + H_2O \rightarrow -NH_2 + CO_2$$

these react with other isocyanate groups to extend the polymer chain,

$$-NH_2 + O=C=N- \rightarrow -NH-CO-NH-$$

and the carbon dioxide liberated in the first step is embedded in the polymer and causes the characteristic 'foam', as used in making cushion pillows and padding. Polyurethane fibres (such as Lycra) are used in stretch fabrics.

Epoxy resins

The epoxy resins are made from the epoxide derived from 3-chloropropene and substituted phenols, for example:

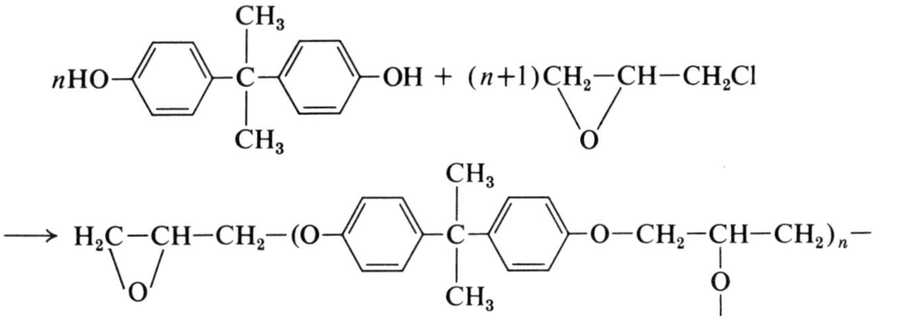

$$n\text{HO}-\text{C}_6\text{H}_4-\text{C}(\text{CH}_3)_2-\text{C}_6\text{H}_4-\text{OH} + (n+1)\text{CH}_2-\text{CH}-\text{CH}_2\text{Cl (epoxide)}$$

$$\rightarrow \text{H}_2\text{C}-\text{CH}-\text{CH}_2-(\text{O}-\text{C}_6\text{H}_4-\text{C}(\text{CH}_3)_2-\text{C}_6\text{H}_4-\text{O}-\text{CH}_2-\text{CH}-\text{CH}_2)_n-$$

The value of n can be controlled so as to give a range of resins varying from viscous liquids to solids with high melting points. The resins are used as adhesives (e.g. Araldite), surface coatings and electrical insulators.

Glyptal resins

The glyptals are polyesters formed from propane-1,2,3-triol (**glycerol**) and benzene-1,2-dicarboxylic (**phthal**ic) anhydride. Each of the three hydroxyl groups in glycerol forms an ester linkage with the anhydride, giving a three-dimensional, thermosetting polymer:

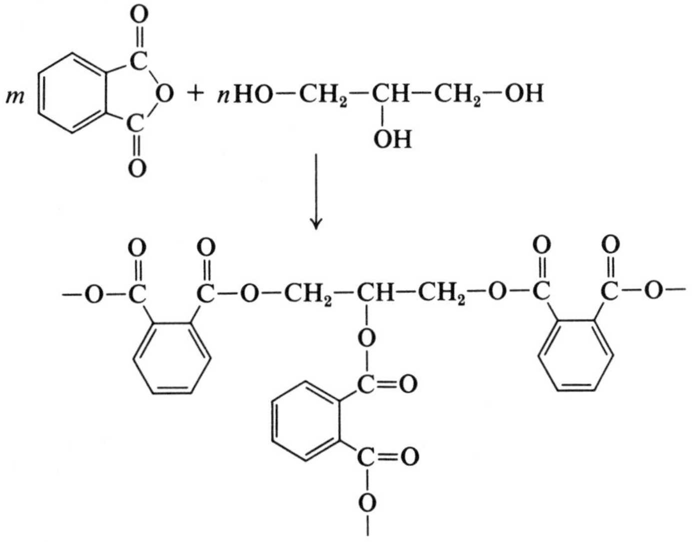

$$m\,(\text{phthalic anhydride}) + n\text{HO}-\text{CH}_2-\text{CH(OH)}-\text{CH}_2-\text{OH}$$

$$\downarrow$$

$$-\text{O}-\text{CO}-\text{C}_6\text{H}_4-\text{CO}-\text{O}-\text{CH}_2-\text{CH}-\text{CH}_2-\text{O}-\text{CO}-\text{C}_6\text{H}_4-\text{CO}-\text{O}-$$

21.3 Synthetic fibres

Natural fibres include carbohydrates such as cotton (which contains more than 90 per cent of cellulose) and proteins such as wool; the former are obtained from plants and the latter from animals. Synthetic linear polymers can also be made into fibres, and the following are important examples.

Nylon

Nylon was the first synthetic fibre to be prepared by polymerisation. It was discovered by Carothers in the U.S.A. in 1935 and first manufactured in 1940.

The nylon made by Carothers is now called **nylon-6.6** because it is made from two components each of which contains six carbon atoms. The components are hexanedioic acid and 1,6-diaminohexane (p. 307), which are heated together to give the polymer:

$$n\text{HO}_2\text{C}-(\text{CH}_2)_4-\text{CO}_2\text{H} + n\text{H}_2\text{N}-(\text{CH}_2)_6-\text{NH} \rightarrow$$
$$\text{HO}\!-\![\text{CO}-(\text{CH}_2)_4-\text{CO}-\text{NH}-(\text{CH}_2)_6-\text{NH}]_n\!-\!\text{H} + (2n-1)\text{H}_2\text{O}$$

There are other forms of nylon, each of which contain the peptide link —CO—NH— as in the naturally occurring protein fibres such as wool.

Nylon-6.10, prepared from decanedioic acid [$\text{HO}_2\text{C}-(\text{CH}_2)_8-\text{CO}_2\text{H}$] and 1,6-diaminohexane, has similar properties to nylon-6.6. Its preparation from decanedioyl chloride is described on p. 325.

Nylon-6 is softer and has a lower melting point than either nylon-6.6 or nylon-6.10. It is prepared from caprolactam, whose manufacture has been discussed (p. 307):

$$n \begin{pmatrix} \text{caprolactam ring} \end{pmatrix} \xrightarrow{260\,°\text{C}} \{\text{NH}-(\text{CH}_2)_5-\text{CO}\}_n$$

Plate 21.2. Molten terylene being forced through a metal disc. The filaments are then stretched to give them greater strength (Imperial Chemical Industries Ltd.)

The method for making the crude nylon polymer into fibres is to melt it and then force it through fine jets. It can be bleached with a dilute solution of peroxoethanoic acid. The finer threads are woven for clothing and the thicker ones are used for articles such as brushes, tarpaulins and tyre cord.

Polyester fibres

Polyester fibres are formed by condensation of polyhydric alcohols and polybasic acids. A particularly important one is **Terylene** (called **Dacron** in the U.S.A.) which is manufactured by the polymerisation of the dimethyl ester of benzene-1,4-dicarboxylic acid (p. 306) and ethane-1,2-diol (p. 146):

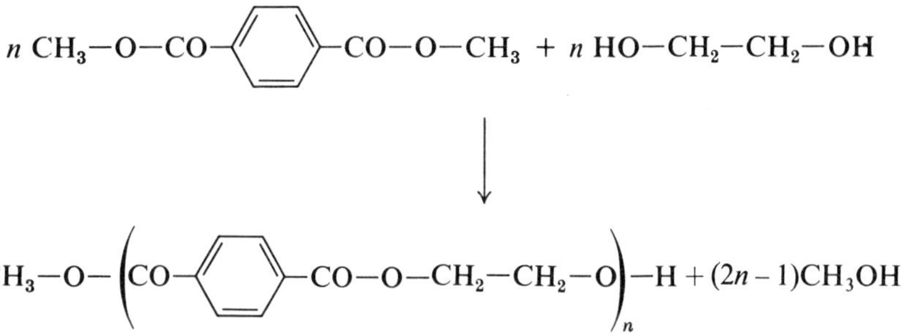

$$n\ CH_3-O-CO-\langle\bigcirc\rangle-CO-O-CH_3 + n\ HO-CH_2-CH_2-OH$$

$$\downarrow$$

$$CH_3-O-\left(CO-\langle\bigcirc\rangle-CO-O-CH_2-CH_2-O\right)_n-H + (2n-1)CH_3OH$$

The molten polymer is extruded to form fibres which are used for making material for clothes. It has the useful property of being able to form permanent creases for trousers and skirts.

Propenonitrile fibres

Propenonitrile is mainly manufactured from propene (p. 304). It is polymerised to form a fibre which is sold under the name **Orlon** for making clothes:

$$2nCH_2=CH-CN \rightarrow -(CH_2-\underset{\underset{CN}{|}}{CH}-CH_2-\underset{\underset{CN}{|}}{CH})_n-$$

Propenonitrile is also made into a **copolymer** by polymerisation together with ethenyl ethanoate. The copolymer contains a more or less regular alternation of the individual monomers:

$$2n\ CH_2=CH-CN + 2n\ CH_2=CH-O-CO-CH_3$$

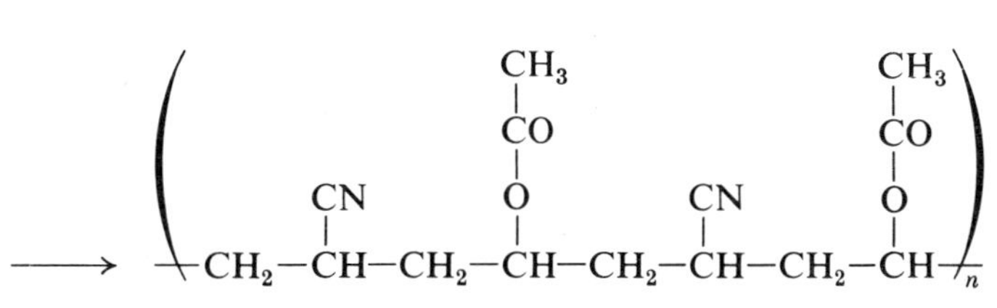

$$\longrightarrow \left(-CH_2-\underset{\underset{CN}{|}}{CH}-CH_2-\underset{\underset{\underset{CH_3}{|}}{\underset{CO}{|}}{\underset{O}{|}}}{CH}-CH_2-\underset{\underset{CN}{|}}{CH}-CH_2-\underset{\underset{\underset{CH_3}{|}}{\underset{CO}{|}}{\underset{O}{|}}}{CH}-\right)_n$$

The copolymer is used as a fibre for making materials under the name **Acrilan**.

21.4 Synthetic rubbers

Natural rubber is obtained from **latex**, which is an emulsion of rubber particles in water found in the bark of many tropical and sub-tropical trees. The latex slowly extrudes from the bark when the tree is cut, and it is coagulated by the addition of ethanoic acid.

Rubber has the structure

$$\underset{CH_3}{\overset{CH_2}{}}C=C\underset{H}{\overset{CH_2}{}}\underset{CH_2}{\overset{CH_3}{}}C=C\underset{CH_2}{\overset{H}{}}\underset{CH_3}{\overset{CH_2}{}}C=C\underset{H}{\overset{CH_2}{}}$$

←---------- 8·1 Å ----------→

and can be regarded as a polymer of 2-methylbuta-1,3-diene, $CH_2=C(CH_3)-CH=CH_2$. In its crude form, it is not a strong enough or elastic enough material for use, but these properties are improved by heating it with a few per cent of sulphur (**vulcanisation**). The exact mechanism of vulcanisation is not known, but it probably consists of the introduction of links between neighbouring chains of the polymer, corresponding to introducing the rungs into a ladder.

Synthetic rubbers mostly resemble natural rubber in having a series of double bonds in their polymer chain. For this reason, buta-1,3-diene is usually used in their manufacture; when it is incorporated into a polymer, one of the two double bonds is retained.

Buta-1,3-diene is made by the dehydrogenation of butane and butenes. They are passed over a catalyst consisting of aluminium and chromium(III) oxides at 600°C, for example:

$$CH_3-CH=CH-CH_3 \xrightarrow[600°C]{Al_2O_3 + Cr_2O_3 \text{ as cat.}} CH_2=CH-CH=CH_2 + H_2$$

The most widely used synthetic rubbers are made by the co-polymerisation of buta-1,3-diene with another monomer, phenylethene (p. 313).

The polymer is used in the manufacture of car tyres, hoses, shoe soles and waterproof boots; it is vulcanised with sulphur and carbon black is added to strengthen it.

Copolymerisation of buta-1,3-diene with propenonitrile gives **nitrile rubber**:

$$2nCH_2=CH-CH=CH_2 + nCH_2=CH-CN \rightarrow$$

$$+(CH_2-CH=CH-CH_2)_2-CH_2-\underset{\underset{CN}{|}}{CH}+_n$$

Nitrile rubber is very resistant to chemicals and so is used in oil seals and gaskets and for making flexible fuel and storage tanks.

Another rubber, known as ABS, is made by copolymerising propenonitrile (30 per cent), buta-1,3-diene (20 per cent) and phenylethene (50 per cent), and is used in car bodies.

Derivatives of buta-1,3-diene are often used to make rubbers of special

POLYMERS

qualities. One, known as **Neoprene rubber**, is resistant to organic solvents and is strong. It is used to make hoses and gaskets where oil resistance is needed. Neoprene is made from 2-chlorobuta-1,3-diene:

$$n\ CH_2=CH-\underset{\underset{Cl}{|}}{C}=CH_2 \xrightarrow{S_2O_8^{2-}\ as\ cat.} \left(-CH_2-CH=\underset{\underset{Cl}{|}}{C}-CH_2-\right)_n$$

Neoprene

2-Chlorobuta-1,3-diene is made from ethyne:

$$2CH\equiv CH \xrightarrow{CuCl\ as\ cat.} CH_2=CH-C\equiv CH \xrightarrow{HCl} CH_2=CH-\underset{\underset{Cl}{|}}{C}=CH_2$$

Polymerisation of 2-methylbuta-1,3-diene with a Ziegler catalyst gives a polymer which has very similar properties to natural rubber. The catalyst (triethylaluminium and titanium(IV) chloride) produces a regular molecular pattern similar to that of natural rubber, the distance being 8·1 Å between each recurring unit (p. 322).

Competition between natural rubber and man-made poly(2-methylbuta-1,3-diene) (polyisoprene) is one of economics. In recent years, rubber trees have been bred which give a very high yield of latex, while 2-methylbuta-1,3-diene is still expensive to make. 2-Methylbuta-1,3-diene at present is generally manufactured by the dehydrogenation of 2-methylbutane, which is present in naphtha (p. 302), or 2-methylbut-1-ene and 2-methylbut-2-ene, which are formed during the cracking of naphtha (p. 305). For example:

$$CH_3-CH_2-\underset{\underset{CH_3}{|}}{C}=CH_2 \xrightarrow[600°C]{Al_2O_3+Cr_2O_3\ as\ cat.} CH_2=CH-\underset{\underset{CH_3}{|}}{C}=CH_2 + H_2$$

21.5 Silicones

Silicones are polymers with alternating silicon and oxygen atoms and alkyl or aryl groups attached to the silicon atoms:

$$-\underset{\underset{R}{|}}{\overset{\overset{R}{|}}{Si}}-O-\underset{\underset{R}{|}}{\overset{\overset{R}{|}}{Si}}-O-$$

The —Si—O— framework gives the polymers stability towards heat, as in silica, while the nature of the organic groups determines other properties such as solubility in organic solvents, water-repellency and flexibility. For example, phenyl groups give more flexible polymers than alkyl groups.

Silicones are manufactured from chlorosilanes, e.g. R_2SiCl_2 and $RSiCl_3$, where R is an alkyl or aryl group. The chlorosilanes are made in a variety of ways. For example, if chloromethane is passed through heated silicon, with copper as catalyst, a volatile mixture of chlorosilanes distils over, which can be purified by fractionation. For example:

$$Si + 2CH_3Cl \xrightarrow[300°C]{Cu\ as\ cat.} (CH_3)_2SiCl_2$$

The dichlorosilanes are hydrolysed to form dialkylsilanediols, $R_2Si(OH)_2$, which polymerise spontaneously to form long linear chains. Cross-linking can be effected at a later stage by addition of an organic peroxide, to yield silicone rubbers.

Silicones can be sub-divided into three classes: (a) silicone fluids, (b) silicone rubbers, (c) silicone resins. Their physical form depends on the structure of the polymer.

(a) **Silicone fluids** are long-chain polymers of the alkylsilanediol, which are very stable. They have the structure:

$$HO-\underset{R}{\overset{R}{Si}}-O-\left[\underset{R}{\overset{R}{Si}}-O\right]_n-\underset{R}{\overset{R}{Si}}-OH$$

or

$$R-\underset{R}{\overset{R}{Si}}-O-\left[\underset{R}{\overset{R}{Si}}-O\right]_n-\underset{R}{\overset{R}{Si}}-R$$

Those with short chains are oils, which have a more or less constant viscosity over a wide temperature range (−50 to +200°C). They also have very low vapour pressures. Thus, fluids with phenyl groups attached to the silicon atom are used as oils for vacuum pumps. Fillers are added to silicone fluids to form heavy greases used when changes in viscosity or a high vapour pressure are undesirable.

The fluids are also used in polishes (a mixture of wax and a silicone fluid dissolved in an organic solvent), in paints and for water-proofing fabrics, paper and leather. They also have anti-foaming properties and have been used sometimes to suppress the foaming of detergents in sewage disposal plants, although the fluids are very expensive.

(b) **Silicone rubbers** are made by introducing some cross-linking into long chain linear polymers. Thus the structure is somewhat similar to natural rubber. The cross-linking is effected by addition of catalysts (e.g. di(benzoyl) peroxide).

Although their strength at normal temperatures is inferior to that of natural rubber, silicone rubbers are more stable at low temperatures (−70°C) and at high temperatures (200–300°C) and are generally more resistant to chemical attack. They are thus used for specialised purposes when ordinary rubber would be useless.

(c) **Silicone resins** have a three-dimensional structure similar to that of silica:

$$\begin{array}{cc} | & | \\ O & O \\ | & | \\ -Si-O-Si-O- \\ | & | \\ O & O \\ | & | \\ -Si-O-Si-O- \\ | & | \\ O & O \\ | & | \end{array} \qquad \begin{array}{cc} R & R \\ | & | \\ -O-Si-O-Si-O- \\ | & | \\ R & O \\ R & | \\ | & | \\ -O-Si-O-Si-O- \\ | & | \\ O & O \\ | & | \end{array}$$

Silica　　　　　　　　Silicone resin

Note that the structure is three-dimensional and that the atoms are tetrahedrally arranged about the silicon atoms. The resins are usually applied as a solution in an organic solvent, and are used as an electrical insulating varnish or for application to surfaces where water repellency is desired. They are also used to give an 'anti-stick' surface to materials coming into contact with 'sticky' materials such as dough and other foodstuffs.

Thus the word 'silicones' is a general term used for long-chain polymers (fluids), two-dimensional structures (rubbers) and three-dimensional macromolecules (resins). Silicones have several distinctive and valuable properties: (i) constancy of physical properties over a wide temperature range, (ii) water-repellency, (iii) electrical insulation, (iv) 'anti-stick', (v) 'anti-foam'.

21.6 Practical work

1. Preparation of poly(phenylethene)

Mix thoroughly 5 cm³ of phenylethene and 0·3 g of di(dodecanoyl) peroxide in a test-tube. Place the test-tube in a beaker of boiling water for about 15 minutes, and then cool.

The solid formed, poly(phenylethene), can be dissolved in methylbenzene and then precipitated from this solution by adding ethanol.

2. Preparation of a polyacrylic ester

Fill a test-tube about 1/3 full of small Perspex chips and heat them, collecting the distillate, which is methyl 2-methylpropenoate, in a test-tube surrounded by cold water (Fig. 21.2).

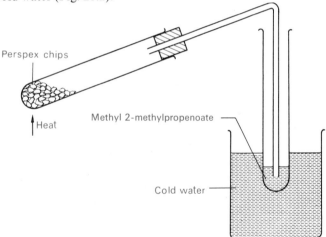

FIG. 21.2. *Preparation of methyl 2-methylpropenoate from Perspex*

To 5 cm³ of the monomer in a test-tube, add about 0·1 g of di(dodecanoyl) peroxide and shake the mixture until the catalyst has dissolved. Stopper the tube and place it in a beaker of water. Bring the water to the boil (take care, as the stopper may be expelled with considerable force). The liquid will polymerise after about 10 minutes.

The monomer may be discoloured (owing to oxidation during its preparation).

3. Preparation of nylon-6.10

In a 250 cm³ beaker, dissolve 1·5 cm³ of decanedioyl chloride in 50 cm³ of tetrachloromethane. In a second beaker make up a solution of 2·2 g of

1,6-diaminohexane and 4 g of sodium carbonate in 50 cm³ of water.

Add the aqueous solution of 1,6-diaminohexane from a pipette, *with a pipette bulb*, making sure that there is a minimum of mixing of the two layers.

Then, using a wire loop, draw a thread of nylon from the liquid interface. A glass rod can also be used, with the advantage that the nylon fibre can be wrapped around it as the fibre is drawn from the interface.

4. Preparation of nylon-6.6

Nylon-6.6 can be prepared with hexanedioyl chloride in place of decanedioyl chloride.

5. Properties of nylon-6

Warm, *very gently* over a small bunsen flame, about 1 g of nylon-6 pellets in a test-tube. Draw some fibres from the melt, using a thin wire.
  (a) Test the fibres for strength and elasticity by pulling them.
  (b) Examine a fibre using a polaroid, before and after stretching.

6. Preparation of a phenol-methanal resin

To 10 cm³ of formalin in an 'old' boiling-tube (which can be thrown away after being used), add about 4 g of phenol. Then add about 1 cm³ of *concentrated* sulphuric acid dropwise, with stirring. Keep the mixture for two or three days, when the condensation polymerisation should have taken place and the solid resin formed.

7. Preparation of a carbamide-methanal resin

Place about 0·5 g of carbamide in a test-tube, and add about 1 cm³ of *concentrated* hydrochloric acid and 5 cm³ of water. Add 1 cm³ of formalin and shake the mixture. Allow the mixture to stand and a white powder is deposited. This is a thermosetting plastic.

8. Preparation of an epoxy resin

To 10 cm³ of Epikote 815 in a small tin, add, with stirring, about 1·5 cm³ of bis-(2-aminoethyl)amine. After about 30 minutes, a hard mass of epoxy resin is formed. Observe whether heat is evolved during the reaction.

Bis-(2-aminoethyl)amine has a harmful vapour and safety glasses should be worn.

9. Preparation of a polyurethane foam

To about 5 cm³ of a triol (Caradol C2) in a small tin or old beaker, add 10 cm³ of a di-isocyanate (Caradate 30). Mix the two liquids together with a spatula for about 15 seconds. Clean the spatula at once. The foam is expanded by the carbon dioxide produced during the reaction, because some water, already added to the triol by the manufacturer, reacts with some of the isocyanate groups.

10. Preparation and properties of a silicone

Put a few drops of dichlorodimethylsilane (**CARE**) in a shallow dish.

POLYMERS

Allow the vapour of the chlorosilane (**CARE**) to come into contact with a filter paper, by placing the paper on top of the dish for a few minutes.

Although the paper may appear to be dry, there is a considerable number of layers of water molecules adsorbed on to the surface. These react with the dichlorodimethylsilane vapour to form diols, which then polymerise to yield a silicone.

Then pour water gently on to the treated paper, and contrast its water-repelling properties with the properties of an untreated filter paper.

21.7 Further reading

'Technology of Plastics'. *Penguin Technology Survey 1967*. N. Denton, ed. A. Garratt. Penguin Books Ltd.

21.8 Films

The Polyolefins. Shell Film Library.*
Catalysis. I.C.I. Film Library, Imperial Chemical Industries Ltd.*
* On free loan.

21.9 Questions

1 Give an account of the preparation, structure, properties and uses of high polymers. (O(S))

2 Explain using suitable examples the principles of addition and condensation polymerisation.
   What is the effect of chain-length and cross-linking on the physical properties of polymers? (L(X,S))

3 The structure drawn below is that of one end of a molecule of a compound, the total molecular weight of which is of the order of 10^4.

$$H_2N-CH-C-N-CH-C-N-CH-C-N-CH_2-C----$$

with side groups $(H_2C)_4$, CH, CH_2; H_2N, $CH_3\,CH_3$, C_6H_5; and carbonyl O's.

(a) Name two classes of compound to which the large molecule could belong.
(b) If the compound were hydrolysed by boiling with hydrochloric acid, four smaller molecules would be formed from that part of the structure shown above. Draw the structural formula of each of these molecules.
(c) To what general class of compound do these smaller molecules belong? (L(Nuffield))

4 Give (a) **two** general methods for increasing the length of a carbon chain and (b) **two** general methods for diminishing the length of a carbon chain. Illustrate by means of specific examples, with accompanying formulae, names and reagents.
   How is poly(ethene) made? Give **two** reasons for its widespread use as a container material. Show by reference to the respective chemical structures how poly(ethene) differs from the polyester Terylene. (JMB)

5 Name (a) **one** commercial plastic formed by addition polymerisation, (b) **one** commercial plastic formed by condensation polymerisation, (c) **one** naturally occurring polymer. Give the monomers from which these three polymers are formed and give an account of the new chemical bonds formed in polymerisation and of the conditions necessary for the formation of the polymers named in (a) and (b).
   What features of molecular structure should be present in polymers which are required to be (i) elastic, (ii) rigid? (C(T))

POLYMERS

6 Explain the meaning of (i) addition polymerisation, (ii) condensation polymerisation. Give **one** large-scale use of each of these processes. What is the effect of 'cross-linking' on the physical properties of a polymer?

Write the structural formulae for two different polymers you might expect to be formed from glycine, ($H_2N.CH_2.CO.OH$). (C(T))

7 What is meant by a 'condensation reaction'? Illustrate your answer with examples drawn from the chemistry of aldehydes and ketones. Explain why the condensation reaction between $HO.OC(CH_2)_4CO.OH$ and $H_2N.(CH_2)_6.NH_2$ is of commercial importance. Comment on the relative merits of various methods which you could employ to determine the molecular weight of the product of this reaction. (W(S))

Appendix I

Questions

1 Suggest the identity of compounds A, B and C from the following data, and explain the reactions which are described.

A is a crystalline solid. When a little water is added to A two liquid layers are formed; addition of more water gives a homogeneous aqueous solution. This liquid

   (i) has no reaction with sodium carbonate solution but indicators show that it has a pH value of less than 7;

   (ii) gives a purple coloration on the addition of a few drops of iron(III) chloride solution;

   (iii) gives a white precipitate on the addition of bromine water.

B is a white crystalline solid which, on heating with an excess of soda-lime, evolves a gas producing an alkaline solution in water. An aqueous solution of B

   (i) deposits a white crystalline precipitate on the addition of concentrated nitric acid;

   (ii) evolves carbon dioxide and nitrogen on the addition of sodium nitrite solution followed by dilute hydrochloric acid;

   (iii) evolves nitrogen on the addition of alkaline sodium hypochlorite solution.

C is a gas which dissolves in water to give a strongly alkaline solution. This solution, when neutralised by hydrochloric acid, gives a white crystalline compound on evaporation.

When C is burnt it produces twice its own volume of carbon dioxide, the gas volumes being measured at the same temperature and pressure. (C(T))

2 Identify the following compounds, explaining your reasoning and writing equations where possible for the reactions which are described.

   (a) A colourless solid, A, dissolves in water to give a neutral solution. When heated with aqueous sodium hydroxide an alkaline gas is expelled and the residual solution effervesces when it is treated with an excess of a dilute acid. A white crystalline precipitate is formed when an aqueous solution of A is treated with concentrated nitric acid.

   (b) A liquid B is miscible with water in all proportions to give solutions with a pH of less than 7. When B is warmed with concentrated sulphuric acid a gas is given off which burns with a blue flame. B reduces hot alkaline permanganate and ammoniacal silver nitrate.

   (c) A colourless liquid C, boiling at 184°C, is sparingly soluble in warm water to which it gives a feebly alkaline reaction. When it is treated with sodium nitrite, in the presence of an excess of dilute hydrochloric acid at 5°C, it yields a solution which reacts with an alkaline solution of phenol to give an orange yellow precipitate.

   (d) D is a blue solid which gives off an inflammable vapour and leaves a bright metallic residue when it is heated in a test tube. It dissolves in water to give a light blue solution which (i) turns to a deep blue colour with an excess of ammonia, and (ii) gives a reddish brown colour with a few drops of iron(III) chloride solution. When D is heated with concentrated sulphuric acid a sharp smell is produced, but if alcohol is also present a sweet fruity odour results.

   (e) E is a white solid which is almost insoluble in cold water but dissolves quite readily in hot water to give an acidic solution. When it is heated with soda-lime it gives off a vapour which burns with a luminous and smoky flame and is 39 times denser than hydrogen under the same conditions. (O)

QUESTIONS

3 Identify the organic compounds described below and where possible write equations to represent the reactions which occur.

   (a) A colourless solid, A, has a characteristic smell. Its aqueous solution is very feebly acidic and gives a white precipitate with bromine water and a violet coloration with aqueous iron(III) chloride. It reacts with phosphorus pentachloride to give a derivative with a vapour density of 56·3.

   (b) A white solid, B, has a high melting point and dissolves in water to give a neutral solution. It forms salts with acids and bases and yields a gas when heated with soda-lime which is alkaline to litmus paper but, unlike ammonia, burns in air.

   (c) A colourless oil, C, has a very characteristic smell. On exposure to the air it slowly forms a colourless solid which dissolves very sparingly in cold water to give an acidic solution. C reduces ammoniacal silver nitrate but not Fehling's solution, and reacts with phenylhydrazine to give a crystalline derivative.

   (d) A colourless fuming liquid, D, reacts violently with water to give two acids in equal molar proportions. Its vapour density is 39.

   (e) A white solid, E, dissolves in water to give a solution which is acidic and which decolorises a hot acidified solution of potassium permanganate. When heated with concentrated sulphuric acid it is completely decomposed without blackening and yields a mixture of gases, one of which turns lime water milky and one of which burns with a blue flame. (O)

4 Suggest, with reasons, the identity of the organic compounds A, B and C.

   (i) A colourless liquid hydrocarbon A has no reaction with bromine in the dark until iron powder is added, when a colourless gas, fuming in moist air, is evolved.

     When A is heated under reflux with alkaline potassium permanganate, and then an excess of sulphur dioxide is bubbled through the mixture, colourless crystals separate on cooling.

   (ii) A colourless liquid B reacts violently with water to give a strongly acidic solution. A portion of this solution on treatment with dilute nitric acid and silver nitrate solution gives a white precipitate. A further portion, after neutralisation with ammonia, reacts with 'neutral' iron(III) chloride (ferric chloride) solution to give a deep red brown solution. This solution gives a brown precipitate when boiled.

     1·00 g of B was allowed to react with an excess of water. The resulting solution required 21·6 cm^3 of 1M (1N) sodium hydroxide for an end point with phenolphthalein indicator.

   (iii) A colourless crystalline compound C gives an acidic solution in water. This solution gives a white precipitate on treatment with calcium chloride solution.

     When treated with hot concentrated sulphuric acid, C reacts to evolve a gas which burns with a blue flame together with a gas which gives a white precipitate with lime water. (C(T))

5 An optically inactive acid A, $C_5H_8O_5$, on being heated lost CO_2 to give an acid B, $C_4H_8O_3$, capable of being resolved.

   On action of sulphuric acid, B gave an acid C whose ethyl ester gave D on the action of hydrogen and platinum.

   D, with conc. ammonia gave E, C_4H_9ON, which with bromine and potassium hydroxide solution gave F, C_3H_9N. F with nitrous acid gave G.

   G on mild oxidation gave H. Both G and H gave the iodoform reaction. Elucidate the reaction scheme and suggest a synthesis of C.

6 Explain the following observations, and identify **all** the compounds mentioned.

   (a) Three isomeric compounds of formula (C_3H_9N) react differently with nitrous acid.

QUESTIONS

(b) Compound J ($C_4H_{10}O$) gives a mixture of three compounds K, L, M (C_4H_8) when passed over alumina at 300°C. Compound N ($C_4H_{10}O$) gives only one product P (C_4H_8) under the same conditions. Comment on the types of isomerism shown by compounds K, L, M and P. (C(N, S))

7 A compound A contains C, 66·4 per cent; H, 5·5 per cent; and Cl, 28·1 per cent. Show how these figures are used to derive the empirical formula C_7H_7Cl.

When A is treated with aqueous potassium hydroxide it is converted into a hydroxy compound B. Mild oxidation of B gives a compound which yields a white precipitate with hydroxylamine ($H_2N.OH$). Further oxidation of B gives a white crystalline solid C which liberates carbon dioxide from aqueous sodium carbonate. Benzene is obtained on heating C with soda-lime.

From these data deduce the structure of A and by means of equations trace the course of the above reactions.

How may A be synthesised from the parent hydrocarbon? (JMB)

8 Suggest a possible structural formula for each of the compounds A to D inclusive and explain the reactions involved, giving equations where possible:

A, molecular formula $C_3H_6O_2$, gives an effervescence with sodium hydrogen carbonate solution and reacts with phosphorus pentachloride to give a compound which contains 38·4 per cent by weight of chlorine.

B, *empirical* formula CH_2Br, on refluxing with aqueous sodium hydroxide gives a compound which reacts with sodium to give hydrogen, one mole of the compound giving one mole of hydrogen.

C, molecular formula $C_4H_{10}O$, is readily oxidised to give a compound C_4H_8O which can be further oxidised to give a compound $C_4H_8O_2$.

D, molecular formula C_7H_7Cl, on refluxing with aqueous sodium hydroxide and subsequent mild oxidation gives a compound C_7H_6O, which readily undergoes further oxidation to give a compound $C_7H_6O_2$. (L(X))

9 Account for the following observations, and identify the compounds G–S.

(a) The chlorine in compound G (C_7H_7Cl) is readily displaced by treatment with dilute aqueous sodium hydroxide. The chlorine in the isomeric compound H is unaffected by this treatment.

(b) Compound J ($C_4H_{10}O$) gives a mixture of three compounds K, L, M (C_4H_8) when passed over alumina at 300°C. Compound N ($C_4H_{10}O$) gives only one product P (C_4H_8) under the same conditions. Comment on the types of isomerism shown by the compounds K, L, M and P.

(c) If compound Q is treated first with bromine and then with ethanoyl chloride, compound R ($C_8H_6Br_3NO$) is obtained. If Q is treated first with ethanoyl chloride and then with bromine, compound S (C_8H_8BrNO) is the product. (C(T, S))

10 Give the structures of A, B and C in the following:

(i) Compound A, C_7H_8O, which is soluble in aqueous sodium hydroxide, is no more soluble in aqueous sodium carbonate than it is in water. If forms $C_7H_5OBr_3$ with bromine.

(ii) Compound B, $C_6H_4Br_2$, which is unaffected by boiling with aqueous sodium hydroxide, gives two isomers of formula $C_6H_3Br_2NO_2$ with a mixture of concentrated nitric and sulphuric acids.

(iii) Compound C, $C_8H_6O_4$, dissolves in aqueous sodium carbonate but not in cold water. When heated, the compound is converted into $C_8H_4O_3$. (O Schol.)

11 A compound A, $C_{13}H_{10}BrNO$, which is sparingly soluble in cold water, dissolved on boiling with concentrated hydrochloric acid. When cooled the resulting solution deposited a solid B, $C_7H_6O_2$, which displaced carbon dioxide from sodium carbonate solution. Basification of the solution which remained yielded a solid C, which contained C, 41·9 per cent; H, 3·5 per cent; Br, 46·5 per cent

and N, 8·1 per cent. Treatment of an ice-cold solution of C in hydrobromic acid with sodium nitrite followed by copper(I) bromide gives compound D. The reaction of D with fuming nitric acid and concentrated sulphuric acid gives only one compound E.

Suggest possible structures for the compounds A, B, C, D and E. (O Schol.)

12 Treatment of 2-bromobutane ($CH_3.CHBr.CH_2.CH_3$) with hot, alcoholic potassium hydroxide gives a mixture of three isomeric butenes, C, D, and E. Reaction with trioxygen and then water of the minor product, C, gives methanal and another aldehyde in equimolar amounts. Both D and E give the same single product, F, with trioxygen and then water.

Write down the structures of the compounds C, D, and E. Describe the type of isomerism shown by D and E, and explain how it arises.

What reactions would you carry out to identify F?

Draw clear, structural diagrams to represent the two molecular species obtained when either D or E is treated with hydrogen chloride. How would these species differ in properties? What difficulties would be encountered in their separation? (C(T,S))

13 How, by a chemical test, would you distinguish

(i) between $CH_3.C_6H_4.NH_2$ and $C_6H_5.CH_2.NH_2$;

(ii) between $CH_3.C_6H_4.OH$ and $C_6H_5.CH_2.OH$;

(iii) between $CH_3.C_6H_4.Cl$ and $C_6H_5.CH_2.Cl$;

(iv) between C_6H_6 (benzene) and C_6H_{12} (hexene)?

In each case, indicate which member of the pair is identified and give the reaction, if any, of the other compound with the reagent used. (C(T))

14 Describe **two** chemical tests which you would carry out in each case to distinguish between the following pairs of substances:

(a) ethene and ethyne;

(b) methylamine and ammonia;

(c) ethanal and propanone;

(d) ammonium ethanoate and ethanamide;

(e) methanoic acid and ethanoic acid.

Wherever possible give equations for the reactions. (O)

15 Describe, with equations, a chemical test you would employ to distinguish between members of each of the following pairs of compounds:

(a) ethyl ethanoate and ethyl benzoate;

(b) chlorobenzene and (chloromethyl)benzene;

(c) ammonium ethanoate and ethanamide;

(d) ethanoyl chloride and chloroethanoic acid;

(e) propanone and pentan-3-one. (W)

16 (a) Give **one** *chemical* test in **each** case to distinguish

(i) between benzoic acid and phenol;

(ii) between benzaldehyde and phenylethanone;

(iii) between nitrobenzene and phenylamine.

(b) Outline the essential practical steps that you would take to obtain pure samples of both substances from a mixture of phenol and phenylamine. (C(T))

17 For each of the following pairs of compounds describe one *simple* chemical test that would distinguish between its members. State exactly what you would do and what you would expect to see and write equations for all the reactions you describe.

QUESTIONS

(a) Propanone and methanol;
(b) phenol and benzoic acid;
(c) carbamide and ammonium ethanoate;
(d) hexane ($CH_3.CH_2.CH_2.CH_2.CH_2.CH_3$) and benzene;
(e) propene ($CH_3.CH=CH_2$) and propyne ($CH_3C\equiv CH$);
(f) stearin (a fat) and casein (a protein). (JMB)

18 Describe **one** chemical test in **each** case to distinguish:
(a) between methanoic acid and ethanedioic acid;
(b) between phenol and benzenesulphonic acid;
(c) between propanone and ethanal;
(d) between methylbenzene and benzene;
(e) between trichloromethane and 1,2-dichloroethane. (C(N))

19 Describe how you would distinguish between the following pairs of isomers by **two** simple chemical tests in **each** case. Give equations for the reactions involved.
(a) $CH_2=CH.CH=CH_2$ and $CH_3.CH_2.C\equiv CH$;
(b) $CH_3.CH_2.CH_2.OH$ and $(CH_3)_2CH.OH$;
(c) $o\text{-}CH_3.C_6H_4.NH_2$ and $C_6H_5.CH_2.NH_2$;
(d) $HO.CH_2.CO.NH_2$ and $NH_2.CH_2.COOH$. (C(S))

20 With the aid of a **named** example in each case, outline, giving conditions and reagents, how you would bring about the following change of functional group in *aliphatic* compounds:
(a) —CH_2OH to —$CO.OH$;
(b) —$CO.Cl$ to —CHO;
(c) —NH_2 to —OH;
(d) —CN to —$CO.OH$;
(e) —$CO.OH$ to —CN.
Name the product formed in each case. (W)

21 Describe briefly how you would carry out the following conversions, giving essential conditions for the reactions but no details of the apparatus:
(a) $CH_3I \rightarrow C_2H_5NH_2$;
(b) $C_6H_5CH_3 \rightarrow C_6H_5CHO$;
(c) $C_6H_6 \rightarrow C_6H_5NH_2$;
(d) $CH_3COOH \rightarrow CH_3NH_2$. (O)

22 Give conditions and equations to show how the following conversions may be made:
(a) ethylamine to ethanol;
(b) ethanoyl chloride to ethanal;
(c) ethanamide to methylamine;
(d) ethanonitrile to ethanoic acid;
(e) ethanoic acid to aminoethanoic acid. (W)

23 Indicate the steps by which you would carry out the following conversions:
(a) C to CH_3COOH;
(b) **either** D_2O to CD_4
    or CH_3COOH to CH_2NH_2COOH;
(c) $(COOH)_2$ to $HCOOH$;
(d) $C_6H_5NO_2$ to C_6H_6. (O(S))

QUESTIONS

24 Indicate the steps by which you would bring about **four** of the following conversions:
  (a) CH_3COOH to $C_2H_5NH_2$;
  (b) C_6H_6 to $C_6H_5NH.NH_2$;
  (c) CH_3OH to $CH_3.CO.CH_3$;
  (d) C_2H_5OH to $CH_3.CHOH.COOH$;
  (e) C_2H_5OH to $H_2N.CH_2.CH_2.CH_2.CH_2.NH_2$.
  Give equations for the reactions which take place. (O(S))

25 Write down **one** reaction scheme for each of the following conversions, indicating reagents and conditions for each step:
  (a) ethyne → ethanoyl chloride;
  (b) phenylamine → phenyl benzoate;
  (c) ethanol → chloroethanoic acid;
  (d) benzene → phenylmethanol;
  (e) propan-2-ol → 2-aminopropane. (O and C(S))

26 Write down **one** reaction scheme for each of the following conversions:
  (a) ethyne → ethanoic acid;
  (b) benzene → 2-nitrophenol;
  (c) ethanal → ethylamine;
  (d) diethyl propanedioate → propanonitrile;
  (e) nitrobenzene → benzoic acid. (O and C(S))

27 Write formulae to show the structures of the products and comment briefly on the reactions between the substances mentioned for **five** of the following cases:
  (a) ethanol and sulphuric acid;
  (b) ethanol and sodium hypochlorite;
  (c) benzene and fuming sulphuric acid;
  (d) propanone and sodium hydrogen sulphite;
  (e) phenylamine, hydrochloric acid and sodium nitrite at 0°C;
  (f) phenol and ethanoic anhydride. (O Schol.)

28 How, and under what conditions, does sodium hydroxide react with
  (i) 1,2-dibromoethane;
  (ii) ethanal;
  (iii) benzaldehyde;
  (iv) carbamide;
  (v) ethyl ethanoate?
  What chemical test will distinguish between ethanamide and carbamide? (C(T))

29 How, and under what conditions, does sodium hydroxide react with the following compounds:
  (a) chlorobenzene;
  (b) chloroethanoic acid;
  (c) carbamide;
  (d) tristearin (a fat);
  (e) sodium 2-hydroxypropane-2-sulphonate

$$\begin{pmatrix} CH_3 & OH \\ & C & \\ CH_3 & SO_3^- \; Na^+ \end{pmatrix}$$

(O and C)

QUESTIONS

30 Describe, with the aid of an illustrative example in each case together with any necessary conditions, **one** use of each of the following reagents in organic chemistry:
  (a) phosphorus pentachloride;
  (b) phosphorus pentoxide;
  (c) trioxygen;
  (d) lithium aluminium hydride;
  (e) hydrogen. (W)

31 Give an account of the uses in organic chemistry of:
  (a) sodium;
  (b) silver nitrate;
  (c) benzoyl chloride;
  (d) copper(II) sulphate;
  (e) phosphorus pentachloride;
  (f) nitric acid. (C Entrance)

32 Discuss the use of the following in the synthesis of organic compounds:
  (a) sulphuric acid;
  (b) nitric acid;
  (c) sodium;
  (d) aluminium chloride. (O(S))

33 How and under what conditions does sulphuric acid react with the following compounds:
  (a) propanone;
  (b) propene;
  (c) phenylamine;
  (d) 2-hydroxypropanoic acid;
  (e) *N*-phenylethanamide? (O and C(S))

34 How, and under what conditions, does sulphuric acid react with:
  (i) ethanamide;
  (ii) phenol;
  (iii) ethanedioic acid;
  (iv) ethanol;
  (v) phenylamine? (C(T))

35 Describe the reactions which can take place between sulphuric acid and each of the following substances: ethyne, ethanol, ethanal and benzene.
Give the conditions under which the reactions take place. (O)

36 What is the action of reducing agents on the following compounds:
  (a) ethyne;
  (b) ethanoyl chloride;
  (c) benzenediazonium chloride;
  (d) phenol;
  (e) bromoethane?
State the reducing agents and the conditions necessary in each case. (O and C(S))

37 By means of specific examples, illustrate the uses of the following as oxidising agents in organic chemistry: potassium permanganate, potassium dichromate, trioxygen, ammoniacal silver nitrate.

QUESTIONS

How would you convert compound (a), below, into compounds (b) and (c)?

(a) $CH_3.CHOH.CH_2.CH_2.CH=CH_2$
(b) $CH_3.CO.CH_2.CH_2.CO_2H$
(c) $HO.CO.CH_2.CH_2.CH=CH_2.$ (O Schol.)

38 By means of equations and brief notes on conditions of temperature and concentration, indicate the various ways in which sulphuric acid can react with the following compounds:

(a) ethanol;
(b) ethene;
(c) benzene;
(d) glycine;
(e) carbamide.

How and under what conditions does sodium hydroxide react with (i) the products from ethene and (ii) the products from benzene? (JMB)

39 Explain concisely the meaning of each of the following in organic chemistry, illustrating your answer with one example in each case:

(a) ethanoylation;
(b) unsaturation;
(c) nitration;
(d) polymerisation;
(e) sulphonation. (AEB)

40 Explain, with the aid of appropriate reactions and necessary conditions in each case, the significance of the following terms: *dehydration, decarboxylation, alkylation, acylation, condensation.* (W)

41 Explain and illustrate with an appropriate example, each of the following terms:

(i) alkane;
(ii) alkene;
(iii) ethanoylation;
(iv) saponification;
(v) polymerisation;
(vi) Cannizzaro reaction. (AEB)

42 Explain carefully the following terms illustrating each by one example of your own choice:

(a) photochemical chlorination;
(b) reaction with trioxygen;
(c) hydrogenation;
(d) 'cracking' of alkanes;
(e) saponification.

Describe very briefly the commercial application of any **three** of these processes. (JMB)

43 Describe, with the necessary conditions, the hydrolysis of **five** named compounds each selected from a different homologous series. (W)

44 Explain, and illustrate with **one** example in each case, the meaning of **five** of the following: unsaturation, polymerisation, homologous series, saponification, nitration, ethanoylation. (O)

45 Write down the structure of the functional group in each of the following organic compounds:

(a) an acid chloride;

(b) an acid anhydride;
(c) an aldehyde;
(d) an amide;
(e) a nitrile.

Describe (i) a method of preparation, and (ii) a characteristic chemical reaction of a *named* example in each case. (AEB)

46 'The carbonyl group, $>C=O$, modifies the properties of the group to which it is attached.' Comment on this statement by comparing:

(a) ethanoic acid with ethanol in (i) the degree to which ionisation occurs in water, (ii) the reaction with phosphorus pentachloride;

(b) the action of cold dilute acids on ethanamide and ethylamine;

(c) the action of cold water on ethanoyl chloride and chloroethane.

Show how the properties of the carbonyl group are modified by comparing the behaviour of propanone and ethanoic acid towards 2,4-dinitrophenylhydrazine, and interpret the result in terms of electronic theory. (S)

47 Carbon compounds are often classified according to the functional groups they contain. Examples of such groups are given below:

(i) $>C=C<$ (v) $_H^>C=O$

(ii) $-C\equiv C-$ (vi) $-\overset{|}{\underset{|}{C}}-O-\overset{|}{\underset{|}{C}}-$

(iii) $-CH_2-OH$ (vii) $-\overset{|}{\underset{|}{C}}-NH_2$

(iv) $>C=O$

(a) Name the *class* of compound you associate with each functional group.

(b) Give *one* named example from each class and write its structural formula.

(c) Give one typical reaction for each of the seven compounds you have chosen in part (b).

Write an equation for each reaction. (L(X))

48 In what types of organic compounds do the following functional groups occur:

(ii) $-COCl$, (ii) $-NH_2$, (iii) $-COOH$, (iv) $-O-$, (v) $-CONH_2$?

Describe **two** reactions characteristic of each group. (AEB)

49 For each of the following pairs of compounds give:

(a) **one** reaction in which the specified group behaves similarly in both;

(b) **two** reactions in which the specified group behaves differently in each.

(i) the $>C=O$ group in $CH_3.CH_2.CHO$ and $CH_3.CH_2.CO.CH_3$;

(ii) the $-NH_2$ group in $CH_3.CH_2.CH_2.NH_2$ and $C_6H_5.CO.NH_2$;

(ii) the $-Br$ atom in $CH_3.CH_2.Br$ and $C_6H_5.Br$.

Give brief *practical* details of chemical tests for distinguishing between any **two pairs** of the components in (i), (ii), and (iii).

(The results of negative tests must be made clear.) (S)

QUESTIONS

50 'A benzene ring influences the chemical behaviour of functional groups attached to it.' Illustrate and discuss with suitable examples. (O Schol.)

51 Name and give the structures of four compounds which contain respectively one of the following groups:

(i) HC≡C, (ii) COCl, (iii) $CONH_2$, (iv) OH in ROH.

Give **two** reactions of each compound which involve the above groupings, and mention any essential conditions that are needed for each reaction to take place. (O)

52 For each of the following, give the name and structural formula of one organic compound which contains the type of linkage mentioned:

(a) a triple bond between carbon and carbon;
(b) a triple bond between carbon and nitrogen;
(c) a triple bond between nitrogen and nitrogen;
(d) a double bond between carbon and oxygen;
(e) a double bond between carbon and nitrogen;
(f) a double bond between nitrogen and oxygen.

Specify the reagents and conditions necessary to reduce the substances you mention giving the appropriate equations and naming the products. (*Note.* The reagents should be different in each case.)

Describe one other characteristic reaction of each of the compounds with triple bonds. (JMB(S))

53 A compound X is shown to have the following structure

$$OHC.CH_2.CH=CH.CH_2.CONH_2$$

How would you expect X to behave towards the following reagents under varying conditions:

(a) sodium hydroxide;
(b) phosphorus pentachloride;
(c) potassium permanganate;
(d) bromine;
(e) sodium and ethanol? (L)

54 From your knowledge of the reactions of particular groups predict the reactions of the compound

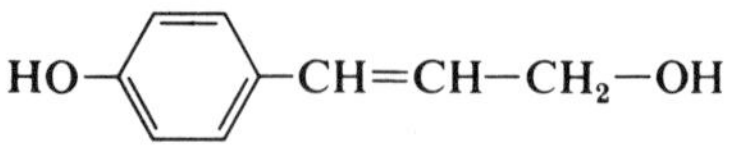

$$HO-\underset{}{\underset{}{\bigcirc}}-CH=CH-CH_2-OH$$

with:

(a) bromine water;
(b) hydrogen and finely divided nickel;
(c) sodium hydroxide;
(d) ethanoic acid;
(e) dilute, alkaline potassium permanganate. (C(T))

55 Name the organic compounds produced, write their structural formulae, and state the conditions under which they are formed when:

(a) phenol, ethanamide, and ethyne are each treated with bromine;
(b) phenylamine, ethanol and propene are each treated with hydrogen bromide. (W)

QUESTIONS

56 (a) Monochlorination of benzene in the presence of iron(III) chloride (ferric chloride), and treatment of the product with a mixture of concentrated nitric and sulphuric acids gives two isomers of formula $C_6H_4ClNO_2$. Write structures for these isomers, and suggest a method by which a third isomer could be prepared from benzene.

(b) A compound is believed to have structure X.

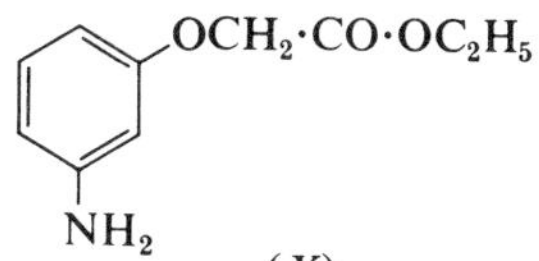

[Structure X: benzene ring with $OCH_2 \cdot CO \cdot OC_2H_5$ group and NH_2 group]

By what chemical reactions would you show the presence of:
(i) the benzene ring;
(ii) the amino group;
(iii) the ester group?

By what reactions would you distinguish between isomers X, Y and Z?

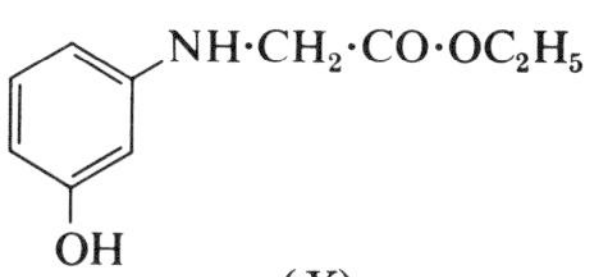

[Structure Y: benzene ring with $NH \cdot CH_2 \cdot CO \cdot OC_2H_5$ group and OH group]

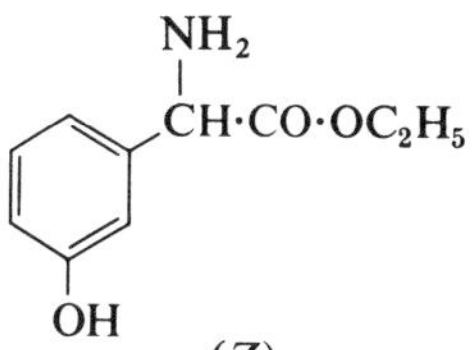

[Structure Z: benzene ring with $CH \cdot CO \cdot OC_2H_5$ (with NH_2 on the CH) group and OH group]

(C(T, S))

57 This question involves practical procedures in the chemistry of carbon compounds. Where you have direct experience describe the practical details as fully as you can, otherwise give an account of the reagents and conditions which you think would be appropriate.

(a) Suppose you were given a substance reputed to have the following formula:

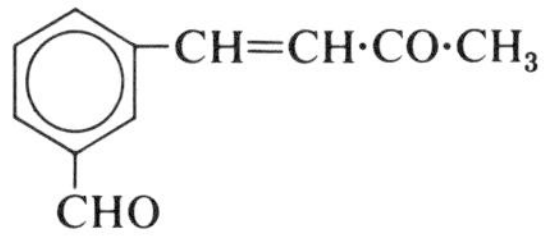

[Structure: benzene ring with $CH=CH \cdot CO \cdot CH_3$ group and CHO group]

Describe how you would attempt to confirm its formula. You should assume that the constituent elements and the molecular weight of the compound are known.

(b) Describe how you would attempt to prepare a substance of the following formula:

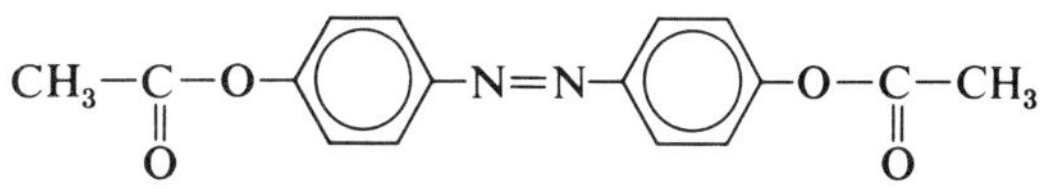

$$CH_3-\underset{\underset{O}{\|}}{C}-O--N=N--O-\underset{\underset{O}{\|}}{C}-CH_3$$

Assume that the only starting material containing the benzene ring which you are allowed is phenol. You are allowed to use any other material which you would normally expect to find in a school laboratory. (L(Nuffield,S))

QUESTIONS

58 Using your knowledge of simpler substances, set out the differences which you could expect to find in the properties and reactions of the compounds in the following pairs of substances:

(a) Ph-C(OH)(H)-CH₃ and HO-C₆H₄-CH(H)-CH₃

$$\text{(a)} \quad \text{C}_6\text{H}_5-\overset{\overset{\displaystyle OH}{|}}{\underset{\underset{\displaystyle H}{|}}{C}}-CH_3 \quad \text{and} \quad HO-C_6H_4-\overset{\overset{\displaystyle H}{|}}{\underset{\underset{\displaystyle H}{|}}{C}}-CH_3$$

(b) (i) 2-hydroxybenzamide (C₆H₄(OH)(CO·NH₂)) and (ii) 2-aminobenzoic acid (C₆H₄(NH₂)(CO·OH))

(c) C₆H₅-CH₂·CHO and CH₃-C₆H₄-CHO

Starting from methyl-2-nitrobenzene, outline a possible synthesis of amine, (b) (ii), and show how this, in turn, could be converted into the hydroxy-derivative (b) (i). (JMB(S))

59 An understanding of the 'nature of the chemical bond' allows one to explain some of the properties of known compounds and make predictions about the properties of unknown ones.

Discuss the chemical bonding in the following compounds and explain how chemical properties of each depend on the types of bonds they contain:

CH_4, $CH_2=CH-CH=CH_2$, $(CH_3)_3N$, CH_3CHO.

What predictions would you make about the properties of cyclopropane,

$$\begin{array}{c} CH_2 \\ \diagup \diagdown \\ H_2C-CH_2 \end{array}$$

(C Schol.)

60 How, and under what conditions, do the following pairs of substances react:

(a) propene and hydrogen iodide;
(b) ethanol and sulphuric acid;
(c) nitrobenzene and nitric acid;
(d) chlorine and ethanoic acid?

Explain the underlying chemical principles in (a) and (c).
What light does reaction (d) throw on the structure of ethanoic acid? (AEB)

61 Give names and structural formulae for the isomers of compounds having molecular formulae (a) $C_2H_4Br_2$, (b) C_4H_9I.

State what happens when each of the isomers you name reacts with potassium hydroxide, mentioning any essential conditions of reaction. In the cases of any hydroxy compounds formed, what happens to these on oxidation?
Outline how the isomers of $C_2H_4Br_2$ may be prepared. (AEB(S))

62 Suggest reasons why

(i) ethers do not react with sodium, but dissolve in concentrated acids and possess lower boiling points than the isomeric alcohols;
(ii) methanoic acid shows reducing and acidic properties, but ethanoic acid shows only acidic properties;

(iii) the compounds represented by CH$_3$.CHOH.COOH may or may not be optically active;

(iv) nitrous acid gives an alcohol on reaction with a primary aliphatic amine but not with a primary aromatic amine. (AEB(S))

63 Discuss the following statements:

(a) Benzoic acid is soluble in aqueous sodium hydrogen carbonate whereas phenol is not.

(b) The carbon–oxygen bond lengths in dimethyl ether are 1·43 Å; in propanone the carbon–oxygen bond length is 1·24 Å, and in carbon monoxide it is 1·13 Å.

(c) The addition of bromine to ethene is a very fast reaction, whereas the addition of bromine to propenonitrile (CH$_2$=CH—CN) proceeds slowly.

(d) Ethyne has a pK_a of 26 and forms salts with sodium, copper and Grignard reagents. Ethene does not form salts with these reagents. (C Schol).

64 Explain:

(a) The carbon–carbon bond lengths in the benzene nucleus are all identical.

(b) Nitric acid alone has no reaction with benzene but a reaction occurs if concentrated sulphuric acid is present.

(c) Propanone, but not methyl ethanoate, forms a phenylhydrazone.

(d) Chloroethanoic acid is stronger than bromoethanoic acid and both are stronger than ethanoic acid itself.

(e) Hydrogen bromide reacts with propene to give two isomeric derivatives with one isomer in much the greater proportion. (S(S))

65 Give a classified account of the addition of simple molecules to unsaturated linkages in organic compounds. Your account should include, if possible, any suggested mechanisms to explain the additions. (L(S))

66 Suggest mechanisms for the following changes and where possible give evidence in support of your suggested mechanisms:

(i) the addition of HCN to aldehydes and ketones;

(ii) the nitration of benzene;

(iii) the reactions which take place when ethene is passed into bromine water containing sodium chloride;

(iv) the conversion of ethanol into diethyl ether.

67 Comment on **four** of the following observations:

(a) Treatment of *trans*-butenedioic acid with dilute potassium permanganate solution gives ($\pm$)2,3-dihydroxybutanedioic acid.

(b) Bromobenzene is stable to dilute aqueous sodium hydroxide solution whilst 1-bromobutane is hydrolysed.

(c) Reaction of 1-chloropropane with benzene and aluminium chloride gives (1-methylethyl)benzene.

(d) When a limited amount of hydrogen is passed into benzene containing a nickel catalyst a mixture of cyclohexane and benzene is obtained, but no intermediate products can be isolated.

(e) When propanone is dissolved in heavy water containing sodium carbonate in solution hexadeuteropropanone is formed. (O Schol.)

68 Explain **three** of the following observations:

(a) On treatment with hot chromic acid ethanal and propanone both yield the same compound which is soluble in sodium carbonate solution with the evolution of carbon dioxide; but with cold chromic acid only ethanal yields such a compound.

(b) Both ethanol and propanone yield trichloromethane when they are heated with aqueous sodium hypochlorite.

(c) A compound, BrCH=CHBr, on being heated yields another compound of the same molecular formula.

(d) Optically inactive 2-hydroxypropanoic acid reacts with a laevorotatory base to give a mixture of two salts which can be separated by fractional crystallisation. (O Schol.)

69 Comment on **three** of the following:

(a) The pK_a values for propanoic, ethanoic, and fluoroethanoic acids are 4·88, 4·80 and 2·66, respectively.

(b) When treated with deuterium oxide containing a trace of sodium hydroxide, pentan-3-one incorporates four atoms of deuterium per molecule.

(c) Optically active butan-2-ol ($CH_3.CH_2.CHOH.CH_3$) retains its optical activity indefinitely in aqueous solution, but racemises rapidly in aqueous sulphuric acid. The alcohol undergoes no chemical change in aqueous sodium bromide solution, but is rapidly converted into 2-bromobutane upon addition of sulphuric acid to the solution.

(d) In addition of bromine to alkenes, the following order of reactivity is observed:
$$CH_2=CH_2 < CH_3.CH=CH_2 < (CH_3)_2C=CH_2.$$

(e) The heats of combustion of cyclopropane, cyclobutane, cyclopentane, and cyclohexane are 500, 656, 793·5, and 944 kcal/mole, respectively.

(O Schol.)

70 Comment on **four** of the following observations:

(a) Phenylamine dissolves in dilute hydrochloric acid solution to a much greater extent than it dissolves in water.

(b) While benzaldehyde reacts smoothly with boiling sodium hydroxide solution to give phenylmethanol, $C_6H_5CH_2OH$, and sodium benzoate, ethanal is polymerised by cold sodium hydroxide solution.

(c) Phenol dissolves in sodium hydroxide solution but not in sodium carbonate solution.

(d) Benzene tends to undergo substitution reactions (with bromine for example) rather than addition reactions.

(e) Phenol is brominated much more rapidly than benzene. (C Schol.)

71 Write balanced equations for the reactions which occur, name the organic products formed, and state what would be observed when:

(a) benzoyl chloride is added slowly to cooled concentrated aqueous ammonia, and the product is isolated and distilled with phosphorus pentoxide;

(b) benzene is added slowly to a mixture of fuming nitric acid and concentrated sulphuric acid, and the resulting mixture is poured into water;

(c) propan-2-ol is mixed with a little potassium iodide solution and a solution of sodium hypochlorite is added;

(d) ethanamide is treated with bromine, followed by dilute potassium hydroxide solution, and the resulting mixture is run into hot concentrated potassium hydroxide solution;

(e) benzaldehyde is shaken with concentrated potassium hydroxide solution and, after standing overnight, water is added. The mixture is then extracted with ether and the aqueous solution is acidified. (JMB)

72 You are given unlabelled bottles of the following substances: propan-2-ol (isopropanol), methanoic acid, benzoyl chloride, methanal, phenylamine, ethanonitrile.

In each case give a *single* positive chemical test (six texts in all) which would enable the bottle to be labelled correctly. (O and C)

QUESTIONS

73 For each of the following cases give the structural formula of a compound which fulfils the stated conditions:

   (a) an aldehyde, $C_5H_{10}O$, which shows optical activity;
   (b) a compound, $C_6H_5NO_3$, which is volatile in steam;
   (c) a hydrocarbon, C_4H_6, which gives a silver derivative;
   (d) an amine, C_3H_9N, which cannot be ethanoylated;
   (e) a dibasic acid which on heating gives a compound, $C_5H_{10}O_2$, which could show optical activity;
   (f) an aliphatic aldehyde, $C_5H_{10}O$, which is not polymerised by alkali;
   (g) an aldehyde, C_8H_8O, which polymerises with alkali.   (O and C(S))

74 Explain what impurities are likely to be present in crude samples of iodoethane, benzoic acid, ethyl ethanoate, and phenylamine prepared in the laboratory. Describe, in each case, how such impurities may be removed and what criterion of purity you would employ for the final product.

A yield of 13·04 g of phenylamine was obtained from 21·30 g of nitrobenzene. Calculate the percentage of the theoretical yield for the reaction.   (W)

75 Crude *N*-phenylethanamide is purified by washing successively with water, aqueous sodium hydrogen carbonate, dilute hydrochloric acid and water. Explain the purpose of each wash.

Why is it incorrect to attempt to dry (a) ethanol with calcium chloride, (b) ethanoic acid with potassium carbonate, (c) iodoethane with sodium? What alternative procedure would you adopt to dry ethanol, ethanoic acid and iodoethane?

What principle underlies the use of a mixed melting point as a criterion of purity?   (W)

76 (a) Mention, with brief comment, **one** example of each of the following:

   (i) A pure compound which cannot be adequately represented by a single structural formula.
   (ii) A reversible isomeric change.
   (iii) A reaction which is catalysed by acids and by bases.

(b) Predict the outcome of the following experiments:

   (i) 2-Methylpropene ($Me_2C=CH_2$) is treated with concentrated sulphuric acid and then with water.
   (ii) A mixture of 1 mole of ethyl benzoate and 1 mole of ethyl 2,4,6-trimethylbenzoate is heated with water containing 1 mole of potassium hydroxide.
   (iii) Phenyl ethyl ether is treated with concentrated hydriodic acid.   (C Schol.)

77 Explain **five** of the following observations:

   (a) Glycine, $C_2H_5O_2N$, is soluble in water and insoluble in ether.
   (b) Ethanal and methanal react differently with aqueous sodium hydroxide.
   (c) From a mixture of ethene and ethyne, ethene may easily be separated.
   (d) A ketone, $C_5H_{10}O$, is known which does not give the iodoform reaction.
   (e) The reaction between iodoethane and silver cyanide yields two *isomeric* compounds.
   (f) A compound, C_2H_3ClO, reacts violently with ethanol, evolving an acidic gas.   (O and C)

78 Explain:

   (a) The function of sulphuric acid in the nitration of benzene.
   (b) The ease with which phenol is nitrated compared with benzene.
   (c) The increase in acid strength as the hydrogen atoms of the methyl group in ethanoic acid are successively replaced by chlorine atoms.

(d) The low yield of ethanal when a mixture of calcium methanoate and ethanoate is heated.

(e) The reluctance of bromine to add on to ethene if the walls of the vessel containing the gases are covered with inert wax. (S(S))

79 Suggest explanations for the following observations:

(a) The apparent molecular weight of ethanoic acid calculated from physical measurements was found to be 55 using a solution of the acid in water and 120 using a solution of the same concentration in benzene.

(b) A solution of 2-hydroxypropanoic acid isolated from a natural source rotated polarised light to the right, whereas a solution of 2-hydroxypropanoic acid synthesised from ethanal did not rotate polarised light.

(c) Phenylethene, $C_6H_5.CH:CH_2$, changed into a hard, transparent solid on standing in air. (C(S))

80 Describe briefly how you would obtain **one** constituent in a pure condition from **each** of the following mixtures (a chemical method is required in **each** case):

(a) phenol and benzenesulphonic acid;
(b) ethanol and propanone;
(c) ethanoic acid and methanoic acid;
(d) phenylamine and chlorobenzene;
(e) benzene and methylbenzene. (C(N))

81 By using *simple laboratory procedures* indicate how you would obtain a pure specimen of the compound with the first structure from each of the following mixtures:

(a) $CH_3.CH_2.CH_2.CH_3$ $CH_3.CH_2.CH\!=\!CH_2$ $CH_3.C\!\equiv\!C.CH_3$

(b) $C_6H_5\text{—OH}$ $C_6H_5\text{—COOH}$

(c) $CH_3\cdot CO\cdot CH_3$ $CH_3\cdot CH_2\cdot OH$

(d) $C_6H_5\text{—NH}_2$ $C_6H_5\text{—NO}_2$

Physical methods alone will not be accepted and all chemical reactions must be fully explained. (S(S))

82 Describe and explain how you would obtain a sample of the first named substance from **each** of the following mixtures:

(a) ethanol and water;
(b) phenol and ethanoic acid;
(c) ethane and ethene;
(d) chloroethane and ethylamine. (C(N))

83 Give one example in each case of a reaction involving

(a) homolytic fission of a C—H bond;
(b) heterolytic fission of a C—halogen bond;
(c) formation of a C—C bond.

Suggest a possible mechanism for each reaction that you mention.

(O Schol.)

84 Write short notes on the types of reactions which establish new carbon–carbon bonds. (C Schol.)

85 (i) In **two** of the following cases, write formulae to illustrate reactions in which
  (a) a new carbon–carbon bond is made;
  (b) a new carbon–nitrogen bond is made;
  (c) an existing carbon–oxygen bond is broken.
[Give two or three examples in each case, with the names of the substances, the reagents and the essential conditions. *No* further account is required.]
  (ii) Show which bonds in the following molecule could be broken (a) by oxidation, and (b) by hydrolysis. Write the formulae of the products.

$$Cl-C_6H_4-CH=CH-CO-NH-CH_3$$

(O Schol.)

86 Give a well-ordered account of methods of forming carbon–carbon bonds in organic chemistry.
What would you expect the yield of propane to be in the Wurtz reaction between sodium and an equimolar mixture of iodomethane and iodoethane? (O Schol.)

87 Suggest syntheses for **five** of the following compounds, starting from readily available chemicals containing not more than 3 carbon atoms for aliphatic compounds, or 6 carbon atoms for aromatic compounds:

$CH_3CH_2CH(OH)CH_2CH_3$ $CH_3CH_2CH_2C(=O)OCH_3$ $CH_3CH=CHCH_3$

cyclopropane ($H_2C-CH_2-CH_2$)   phenol   3-methylaniline (m-toluidine)

$CH_3(CH_2)_2CH_2OCH_2(CH_2)_2CH_3$ (C Schol.)

88 Answer **three** of the following. On an examination paper, a student offered the suggestions given below for carrying out a series of conversions of organic compounds. Point out the errors made by this student, and where the method was not correct, offer suggestions by which the desired conversion *could* be carried out:

(a) $CH_3CH_2OH \xrightarrow{NaCl} CH_3CH_2Cl \xrightarrow[\text{(ii) HCHO}]{\text{(i) Mg, ether}} CH_3CH_2CHO$;

(b) $CH_3CH_2C(O)CH_3 \xrightarrow{NaOH} CH_3CH_2C(O)OH \xrightarrow[NaOH]{CH_3OH} CH_3CH_2C(O)OCH_3$;

(c) $CH_3CH=CH_2 \xrightarrow{H_2O} CH_3CH(OH)CH_3 \xrightarrow{NaOH} CH_3CH(ONa)CH_3 \xrightarrow{CH_3I} CH_2CH(OCH_3)CH_3$;

(d) $C_6H_6 \xrightarrow{CH_3OH} C_6H_5CH_2OH \xrightarrow[\text{or Pt, H}_2]{LiAlH_4} C_6H_5C(O)OH$

(C Schol.)

QUESTIONS

89 Write feasible schemes (formulae and reagents, and conditions where relevant) for **four** of the following transformations:

(a) $CH_3.CH_2.OH \rightarrow HC{\equiv}CH$;

(b) $C_6H_6 \rightarrow C_6H_5CN$;

(c) $CH_3.CO.CH_3 \rightarrow CH_3.CH(CO_2H).CH_3$;

(d) $C_6H_5{-}CH_3 \rightarrow C_6H_5{-}CH_2{\cdot}OCH_2{\cdot}CH_3$

(e) $CH_2{=}CH_2 \rightarrow CH_3.CO.OCH_2.CH_2.OCO.CH_3$;

(f) $NH_3 \rightarrow H_2N.CO.NH.CO.NH_2$. (O Schol.)

90 In each of these questions a statement is followed by five alternative responses. Only one of these alternatives is correct. Indicate your choice of A, B, C, D or E.

(i) Which one of the following is a polyester?

A 1,2-diaminohexane;
B poly(chloroethene);
C Terylene;
D Nylon 6.6;
E poly(phenylethene).

(ii) Which one of the following statements is necessarily an inaccurate description of the homolytic fission of a chemical bond?

A One of the atoms leaves with both of the electrons forming the bond.
B The fission is promoted by ultraviolet light.
C The fission gives atoms or radicals.
D The fission is assisted by heat.
E The fission can occur in either the gaseous or the liquid phase.

(iii) When equal volumes of ethanoyl chloride and ethanol are mixed and the mixture is then poured into excess cold dilute alkali, which one of the following statements best describes the change which takes place?

A A white precipitate is thrown down.
B A vinegar-like smell is discernible.
C A pleasant fruity smell is discernible.
D A pungent smell of hydrogen chloride is discernible.
E A vigorous effervescence ensues.

(iv) Which one of the following groups of products is formed when methane and chlorine are mixed in the dark?

A chloromethane and hydrogen chloride;
B chloromethane, dichloromethane and hydrogen chloride;
C trichloromethane, tetrachloromethane and hydrogen chloride;
D no products are formed;
E hydrogen chloride and a deposit of carbon.

(v) Which one of the compounds having the following structural formulae can be resolved into optical isomers?

A $NH_2.CH_2COOH$;
B $NH_2.CH(CH_3).COOH$;
C $CH_3.CH{=}CH.C_2H_5$;
D $HOOC.CH_2.CH(CH_3).CH_2.COOH$;
E $HO.CH_2.CH_2.C(CH_3)_2.CH_2.NH_2$.

(vi) Which of the following reagents does not react with ethanal?

A NH_3;
B CN^-;
C $CH_3.CO_2C_2H_5$;
D BH_4^-;
E acidified MnO_4^-.

(vii) The compound CH$_3$—CH(CH$_3$)—CO$_2$H was produced by oxidation. Which one of the following compounds was the most likely starting material?

A butan-1-ol;
B 2-methylpropan-1-ol;
C butan-2-ol;
D 2-methylpropan-2-ol;
E propan-2-ol.

(JMB (Specimen paper))

91 For each of the following reactions choose from the list A–E the most appropriate reaction type.

A nucleophilic addition;   B electrophilic addition;
C nucleophilic substitution;   D electrophilic substitution;
E redox

$$C_6H_6 + Br_2 \xrightarrow{AlBr_3} C_6H_5Br$$

$$C_6H_5.CHO \xrightarrow{\text{Fehling's solution}} C_6H_5.COOH$$

$$CH_2{=}CH_2 + HBr \longrightarrow CH_3{-}CH_2Br$$

$$C_6H_5.CHO + NaHSO_3 \longrightarrow C_6H_5.CH(OH).SO_3Na$$

The following formulae represent the structure of five organic compounds.

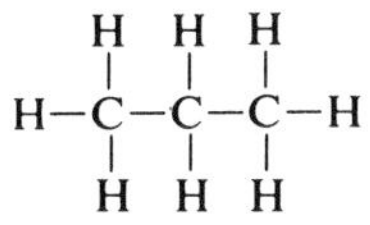

A: CH$_3$—CH$_2$—CH$_3$

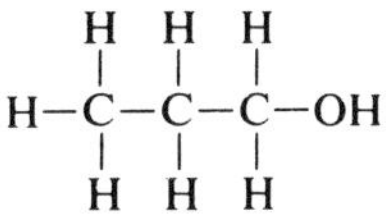

B: CH$_3$—CH$_2$—CH$_2$—OH

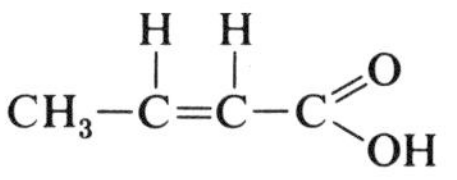

C: CH$_3$—CH=CH—C(=O)—OH

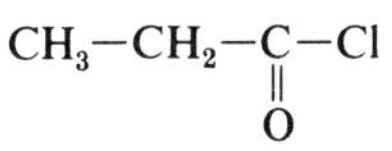

D: C$_6$H$_5$NH$_2$

E: CH$_3$—CH$_2$—C(=O)—Cl

By using the appropriate letter for the compounds answer the questions which follow and give the equation where indicated.

(a) Which compound will form a salt with mineral acids?
(b) Which compound will react with an organic acid in the presence of a mineral acid to form an ester?
(c) Which compound will decolorise both bromine water and dilute acidified potassium permanganate solution?
(d) Which compound would form an aldehyde on oxidation?
(e) Which compound would you expect to react vigorously with water?

(JMB (Specimen paper))

92 (a) Give the structural formula for each of the following compounds. (The formulae should be abbreviated to the type: CH$_3$.CHCl.CH$_2$.CH$_3$).

A sodium hexanoate
B 3-methylpenta-2,4-diene
C 4-bromo-2-methylpentan-1-ol
D 2-methylpentane-1,4-diol

QUESTIONS

(b) Which one of the compounds would be the most likely starting point for the development of a new polymer?

(c) Which one of the compounds would be most likely to leave a solid residue when it is strongly heated in air?

(d) State briefly how *C* could be converted into *D* (the isolation of the product is not required).

(e) (i) Which one of the compounds is most likely to be a solid at room temperature?
  (ii) State briefly the reasons for your answer.

(f) If you were given unlabelled samples of *B*, *C* and *D*, state briefly how you would quickly distinguish between them using the least number of tests and observations. (L(Nuffield))

Appendix II

Apparatus and chemicals

Apparatus with ground-glass joints is described throughout this book. In this way, preparations and reactions can be studied more rapidly than with older apparatus.

Sensible precautions ensure a long life for the apparatus. The joints must be clean and the flasks must be heated over a gauze, or in a water- or oil-bath, by means of a small flame from a microburner or bunsen burner.

The same apparatus as described in *Organic Chemistry through Experiment* (D. J. Waddington and H. S. Finlay, Mills and Boon Limited) is used, namely that produced in conjunction with Philip Harris Ltd., Birmingham. The number of pieces has been reduced to a minimum and the set consists of:

1. Stillhead;
2. Adaptor for thermometer, steam lead or air leak;
3. Stopper;

4 and 5. Adaptors for collection of gases;

6. Steam lead which can also be used as an air leak;
7. Dropping funnel, which can also be used as a separating funnel;
8. Water condenser, which can also be used as an air condenser;
9. Flask head;
10. 50 cm^3 pear-shaped flask. A 100 cm^3 round-bottomed flask may be used instead;
11. Receiver with side arm.

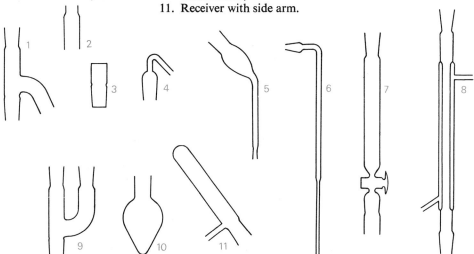

An air condenser, fractionating column and an air leak (for vacuum distillation) can also be obtained.

Test-tube preparations of gases and liquids are used throughout the book for they can be done quickly, and many of the properties of compounds are studied by experiments using test-tubes and other simple equipment. Standard sized Pyrex test-tubes (150 × 16 mm) and delivery tubes made from soda-glass tubing (4 mm i.d.) are convenient to use. Dropping pipettes may be made from soda-glass tubing (6 mm i.d.) and using cylindrical rubber teats. These should be calibrated in 0·5 cm^3 portions with a small dab of paint, the calibration saving much time during experiments.

A list of organic and inorganic reagents is given in this Appendix and a list of suppliers in Appendix III.

APPARATUS AND CHEMICALS

Organic reagents

Alkylbenzene (Appendix III)

Benzaldehyde
Benzene
Benzoic acid
Benzoyl chloride
Bis-(2-aminoethyl)amine (diethylene triamine) (Appendix III)
Bromobenzene
1-Bromobutane
2-Bromobutane
2-Bromo-2-methylpropane
Butan-1-ol
Butenedioic anhydride (maleic anhydride)

Caradate 30 (Appendix III)
Caradol C2 (Appendix III)
Carbamide (urea)
Chlorobenzene
2-Chlorobenzoic acid
1-Chlorobutane
2-Chlorobutane
(Chloromethyl)benzene (benzyl chloride)
Cotton wool
Cyclohexane
Cyclohexanol
Cyclohexene

Decanedioyl chloride (sebacoyl chloride) (Appendix III)
1,6-Diaminohexane
Dichlorodimethylsilane (Appendix III)
1,1-Dichloroethane
1,2-Dichloroethane
Di(dodecanoyl) peroxide (lauroyl peroxide) (Appendix III)
Diethyl ether
2,4-Dinitrophenylhydrazine. 1 g in 50 cm^3 of methanol to which 2 cm^3 of *concentrated* sulphuric acid is added. Filter if necessary
Dodecanol (Appendix III)

Epikote 815 (Appendix III)
Ethanal (acetaldehyde)
Ethanamide (acetamide)
Ethanedioic acid (oxalic acid)
Ethane-1,2-diol (ethylene glycol)
Ethanoic acid (acetic acid)
Ethanoic anhydride (acetic anhydride)
Ethanol
Ethanonitrile (acetonitrile)

Ethanoyl chloride
Ethyl benzoate

Formalin, 40 per cent solution of methanal in water
Fructose
Fuchsin

Glucose
Glycine

Hexane (or pentane)
Hexanedioyl chloride (adipyl chloride) (Appendix III)

1-Iodobutane

Lard (or olive oil)
Light petroleum, b.p. 60–80°C

Malachite green
Methanoic acid (formic acid)
Methanol
Methylamine, 33 per cent solution in water
Methylammonium chloride
Methylbenzene (toluene)
2-Methylbutan-2-ol
Methylene blue
2-Methylpropan-2-ol

Naphthalen-2-ol (β-naphthol)
Ninhydrin
Nitrobenzene
Nitroethane (Appendix III)
2-Nitrophenol
4-Nitrophenol
Nylon-6, pellets (Appendix III)

Oil, crude
Olive oil (or lard)

Paraffin oil
Pentane (or hexane)
Pentan-3-one (diethyl ketone)
Pent-1-ene (Appendix III)
Perspex, chips
Phenol
Phenylamine (aniline)
Phenylethanone (acetophenone)
Phenylethene (styrene) (Appendix III)
Phenylethyne (phenylacetylene) (Appendix III)
Phenylhydrazine
Phenylmethanol (benzyl alcohol)

(Phenylmethyl)amine (benzylamine)
Proline
Propane-1,2,3-triol (glycerol)
Propan-2-ol
Propanone (acetone)

Schiff's reagent. 0·1 per cent solution of fuchsin in water through which sulphur dioxide is passed until the solution is colourless
Sodium ethane-dioate (sodium oxalate)
Sodium ethanoate (sodium acetate), anhydrous

Sodium methanoate (sodium formate)
Sodium potassium 2,3-dihydroxybutanedioate (sodium potassium tartrate)
Starch, soluble
Sucrose

Tetrachloromethane
Trichloromethane (chloroform)
Triphenylchloromethane (Appendix III)
Tris-(2-hydroxyethyl)amine (triethanolamine)

Urease (Appendix III)

Inorganic reagents

Alumina, for chromatography
Aluminium, powder
Aluminium trichloride, anhydrous
Ammonia solution, (a) concentrated, (b) 2M solution

Bleaching powder
Bromine, (a) liquid, (b) saturated aqueous solution

Calcium dicarbide
Calcium chloride, anhydrous
Calcium hydroxide, saturated solution (lime-water)
Chlorosulphonic acid
Copper(II) carbonate
Copper(I) chloride
Copper(II) oxide
Copper(II) sulphate, anhydrous

Disodium pentacyanonitrosylferrate-(III) (sodium nitroprusside)

Fehling's solution I. 7 g of copper(II) sulphate pentahydrate in 100 cm³ of water

Fehling's solution II. 12 g of sodium hydroxide and 5 g of sodium potassium 2,3-dihydroxybutane-dioate (tartrate) in 100 cm³ of water

Hydrochloric acid, (a) concentrated, (b) 2M solution

Iodine, (a) solid, (b) 1 per cent solution in 20 per cent solution of potassium iodide
Iron, filings
Iron(III) chloride
Iron(II) sulphate

Lithium aluminium hydride

Magnesium, turnings for Grignard reactions
Magnesium sulphate, anhydrous
Mercury(II) chloride

Nitric acid, (a) concentrated, (b) 2M solution

Phosphoric acid, concentrated
Phosphorus pentachloride
Platinum, wire (about 28 s.w.g.)
Potassium bromide
Potassium carbonate, anhydrous
Potassium dichromate
Potassium hydroxide, pellets
Potassium permanganate

Rocksil

Silica gel (Appendix III)
Silver nitrate
Soda-lime
Sodium
Sodium carbonate, anhydrous
Sodium chloride
Sodium dichromate
Sodium hydrogen carbonate
Sodium hydroxide, (a) pellets, (b) 2M solution
Sodium metabisulphite
Sodium nitrate
Sodium nitrite
Sodium sulphate, anhydrous
Sulphur dioxide. Syphon
Sulphuric acid, (a) oleum, (b) concentrated, (c) M solution

Tin, granulated

Zinc, dust

Appendix III

Suppliers of apparatus and chemicals

Most of the chemicals and apparatus can be obtained from all laboratory suppliers. If the following chemicals provide some difficulty, they may be obtained from Philip Harris Ltd., Birmingham B3 1DJ, or from:

Alkylbenzene	Unilever Education Section, Unilever Limited, Unilever House, Blackfriars, London EC4
Bis-(2-aminoethyl)amine (diethylene triamine)	B.D.H. Ltd.
Caradate 30 and Caradol C2	Alfred Jeffery and Co., Marshgate Lane, Stratford, London E15
Decanedioyl chloride (sebacoyl chloride)	B.D.H. Ltd.
Dichlorodimethylsilane	B.D.H. Ltd.
Di(dodecanoyl) peroxide (lauroyl peroxide)	B.D.H. Ltd.
Dodecanol	B.D.H. Ltd.
Epikote 815	Shell Chemical Co. Ltd.
Hexanedioyl chloride (adipyl chloride)	B.D.H. Ltd.
Nitroethane	B.D.H. Ltd.
Nylon-6 pellets	Griffin and George Ltd.
Pent-1-ene	Koch-Light Laboratories Ltd.
Phenylethene (styrene)	B.D.H. Ltd.
Phenylethyne (phenylacetylene)	B.D.H. Ltd.
Silica gel (Kieselgel G-Nach-Stahl)	Anderman and Co. Ltd. London SE1
Triphenylchloromethane	B.D.H. Ltd.
Urease	Koch-Light Laboratories Ltd.

Appendix IV # Film libraries

British Petroleum: applications to borrow BP films should be made to the BP Film Library, 15 Beaconsfield Road, London NW10 2LE. Enquiries concerning the purchase of films should be addressed to Films/TV Branch, Information Dept, The British Petroleum Company Limited, Britannic House, Moor Lane, London EC2Y 9BU.

Educational Foundation for Visual Aids, National Audio-Visual Aids Library, Paxton Place, Gipsy Road, London SE27 9SS

Gas Council Film Library, 59 Bryanston Street, London W1A 2AZ

Guild Sound and Vision Ltd., Woodston House, Oundle Road, Woodston, Peterborough PE2 9PZ

I.C.I. Film Library, Thames House North, Millbank, London SW1P 4QG

Shell Film Library, 25 The Burroughs, Hendon, London NW4 4AT

Unilever films are handled by the National Audio-Visual Aids Library (see above) but special enquiries may be addressed to The Film Librarian, Unilever Films, P.O. Box 68, Unilever House, Blackfriars, London EC4P 4BQ. In Scotland, Unilever films are obtainable from the Scottish Central Film Library, Glasgow G3 7XN.

Appendix V

Physical constants

International relative atomic weights

($^{12}C = 12 \cdot 000000$)

Element	Atomic weight	Element	Atomic weight
Aluminium	26·9815	Nitrogen	14·0067
Bromine	79·909	Oxygen	15·9994
Calcium	40·08	Phosphorus	30·9738
Carbon	12·01115	Platinum	195·09
Chlorine	35·453	Potassium	39·102
Copper	63·54	Silicon	28·086
Fluorine	18·9984	Silver	107·870
Hydrogen	1·00797	Sodium	22·9898
Iodine	126·9044	Sulphur	32·064
Iron	55·847	Tin	118·69
Lead	207·19	Zinc	65·37
Magnesium	24·312		

Masses of isotopes relative to ^{12}C

^1H	1·0078246	^{35}Cl	34·9688531
^{14}N	14·0030738	^{37}Cl	36·9659034
^{16}O	15·9949141	^{79}Br	78·91839
^{31}P	30·973764	^{81}Br	80·91642
^{32}S	31·9720727	^{127}I	126·90466

Some aldehydes and ketones and the 2,4-dinitrophenylhydrazone derivatives

For Experiment 12, Section 12.7, p. 178

	B.p. (°C) of the aldehyde or ketone	M.p. (°C) of the 2,4-Dinitro-phenylhydrazone
Methanal	−21	166
Ethanal	21	168
Propanal	49	155
Butanal	75	123
2-Methylpropanal	64	187
Pentanal	104	98
Benzaldehyde	179	237
2-Hydroxybenzaldehyde	197	252
4-Methoxybenzaldehyde	248	254
4-Methylbenzaldehyde	204	233

PHYSICAL CONSTANTS

Propanone	56	128
Butan-2-one	80	115
Pentan-2-one	102	144
Pentan-3-one	102	156
Heptan-4-one	144	75
Phenylethanone	202	250
4-Methylphenylethanone	224	258
Diphenylmethanone	306 (m.p. 49°C)	238

Index

Page numbers in **bold** type refer to experimental work

Absolute ethanol, 141
ABS rubber, 322
Acetaldehyde—see Ethanal
Acetamide—see Ethanamide
Acetanilide—see N-Phenylethanamide
Acetic acid—see Ethanoic acid
Acetic anhydride—see Ethanoic anhydride
Acetoacetic ester—see Ethyl 3-oxobutanoate
Acetone—see Propanone
Acetonitrile—see Ethanonitrile
Acetophenone—see Phenylethanone
Acetylation—see Ethanoylation
Acetyl bromide—see Ethanoyl bromide
Acetyl chloride—see Ethanoyl chloride
Acetylene—see Ethyne
Acetyl iodide—see Ethanoyl iodide
Acid amides,
  physical properties, 214
  preparations, 207, 210–11, 213, 214
  reactions, 214–16, **221**
  structure, 214–15
Acid anhydrides,
  physical properties, 212
  preparations, 193, 211, 213
  reactions, 213, **220–1**, **257**, **258**, **289**
  uses, 205, 207, 213, 286, 315
Acid bromides, 209–212
Acid chlorides,
  physical properties, 209
  preparations, 209
  reactions, 209–212, **220**
  uses, 212
Acid halides,
  physical properties, 209
  preparations, 209
  reactions, 209–212, **220**
  uses, 212
Acid isonitriles, 219
Acid nitriles,
  physical properties, 218
  preparations, 117, 216, 218, 253
  reactions, 218, **221**
Acids, amino—see Amino-acids
  dicarboxylic,
    nomenclature, 188
    preparations, 188–9, 266
    reactions, 147, 189–190, 321
  monocarboxylic,
    dimerisation, 183
    dissociation constants, 59–62, 184–5
    distinguish from phenols, 185
    hydrogen bonding, 183
    manufacture, 184
    nomenclature, 182
    physical properties, 183
    preparations, 129, 183, **196–7**
    reactions, 184–7, **199–200**

  structure, 186–7
  uses, 184, 187, 194, **197**
  salts of—see Salts
  sulphonic, 61, 100–1, 150
Acrilan, 321
Acrolein—see Propenal
Acrylonitrile—see Propenonitrile
Activation energy, 115
Acylation, 209
Acyl chlorides – see Acid chlorides
Acyl group, 204
Acyl halides—see Acid halides
Addition polymerisation, 311
Addition reactions,
  of aldehydes, 167–9
  of alkenes, 76–9, **85–6**
  of alkynes, 90, **92**
  of aromatic hydrocarbons, 101, 103
  of ketones, 167–9
Adenine, 275–7
Adipic acid—see Hexanedioic acid
Adipyl chloride—see Hexanedioyl chloride
Adrenalin, 234
α-Alanine, 249, 272
β-Alanine, 249
Alcohol, absolute, 141
Alcohols, dihydric—see Diols
  monohydric,
    classes, 137–8
    distinguish between primary, secondary, tertiary, 145
    hydrogen bonding, 138–9
    manufacture, 140–1, 303, 308
    nomenclature, 137–8
    physical properties, 138–9
    preparations, 117, 128–9, 139–140
    reactions, 141–5, **154–6**
    uses, 146
  polyhydric, 146–9
Aldehydes,
  identification, 179
  manufacture, 166–7
  nomenclature, 164
  physical properties, 164–5
  preparations, 165–6, 212
  reactions, 167–175, **178–9**
  uses, 175–6
Aldol—see 3-Hydroxybutanal
Aldoses, 279–282, **288**
Alicyclic compounds, 7, 94–109
Aliphatic compounds, 8, 94–109
Alkanes,
  nomenclature, 65
  physical properties, 65–7
  reactions, 68–72, 301–2
  uses, 72, 301–2
Alkenes,
  manufacture of long chain, 303

  nomenclature, 75
  physical properties, 75
  preparations, **84–5**
  reactions, 76–81, 303–5
  uses, 81, 84, 303–5
Alkylation, 100, 295
Alkyl bromides,
  nomenclature, 111–12
  physical properties, 111–12
  preparations, 112–13, **129–130**
  reactions, 113–120, **132**
Alkyl chlorides,
  manufacture, 113, 301
  nomenclature, 111–12
  physical properties, 111–12
  preparations, 112, **130–1**
  reactions, 113–120, **132**
  uses, 120, 323
Alkyl cyanides—see Acid nitriles
Alkyl fluorides, 126–7
Alkyl halides,
  manufacture, 113, 301
  nomenclature, 111–12
  physical properties, 111–12
  preparations, 112–13, **129–131**
  reactions, 113–120
  uses, 120, 323
Alkyl iodides,
  nomenclature, 111–12
  physical properties, 111–12
  preparations, 113
  reactions, 113–120, **132**
Alkyl isocyanides, 219
Alkyl nitrites, 118
Alkyl radical (group), 5, 113, 193
  structure, 113
Alkynes,
  manufacture, 89
  physical properties, 89
  preparations, 91
  reactions, 90, **92–3**
  uses, 91
Allenes—see Propadienes
Allyl chloride—see 3-Chloropropene
Alumina,
  for chromatography, 18, 20, **26–7**
Aluminium lithium hydride—see Lithium aluminium hydride
Aluminium oxide,
  for chromatography, 18, 20, **26–7**
Amides—see Acid amides
Amines,
  basic strength, 62
  classification into primary, secondary, tertiary, 244
  manufacture, 245, 307
  nomenclature, 243–4
  physical properties, 62, 243–4

preparations, 117, 216, 218, 244–5, **254–7**
reactions, 245–9, **257–8**, **325–6**
uses, 249, 309, 317, 320
Amino-acids,
essential, 272
nomenclature, 249, 271–2
physical properties, 250
preparations, 250–1
proteins, 271–2
reactions, 251–2, **259**
2-Aminobutane, 243
Aminoethane—see Ethylamine
Aminoethanoic acid—see Glycine
Aminomethane—see Methylamine
2-Aminopropanoic acid—see α-Alanine
3-Aminopropanoic acid—see β-Alanine
Ammonia,
manufacture, 309
uses, 301, 304
Ammonium carbamate, 217, 309
Ammonium cyanate, 1
Amylase, 285
α-Amylose, 285
β-Amylose, 285
Analysis,
qualitative, **43–4**
quantitative, 30–2
Aniline—see Phenylamine
Anisole—see Methyl phenyl ether
Anti-freeze, 148
Anti-knock, 296
Araldite, 319
Aromatic halides,
manufacture, 125
physical properties, 125
preparations, 124–5, 252–3
reactions, 125–6
uses, 126
Aromatic hydrocarbons, 8, 94–109
Aryl compounds – see under Aromatic compounds (e.g. Aromatic halides) or under the title of the homologous series (e.g. Aldehydes)
Aspartic acid, 249
Asymmetric carbon atom, 233–4
Atomic orbitals, 47–8
Auxochromes, 254
Azines, 170
Azo compounds, 253–4
Azo dyes, 253–4, **258**

Bakelite, 316–17, **326**
Barbiturates, 217
Barbituric acid, 217
Base peak, 35
Beckmann rearrangement, 308
Benzal chloride—see (Dichloromethyl) benzene
Benzaldehyde,
physical properties, 164
preparations, 104, 165–6
reactions, 167–175, **178–9**
Benzaldehyde oxime,
isomerism of, 230
Benzaldehyde phenylhydrazone, 171

Benzanilide—see N-Phenylbenzamide
Benzene,
bonding, 56–8, 94–6
heat of hydrogenation, 94–5
manufacture, 96, 295–6, 296–8, 305
physical properties, 97
reactions, 97–101
stabilisation energy, 95–6
structure, 56–8, 94–6
uses, 150, 305
Benzenediazonium chloride,
preparation, 247, 252, **258**
reactions, 252–4, **258**
Benzenediazonium tetrafluoroborate, 253
Benzene-1,2-dicarboximide, 250
Benzene-1,2-dicarboxylic acid, 188
Benzene-1,4-dicarboxylic acid, 188, 306
Benzene-1,2-dicarboxylic anhydride, 306, 319
Benzene hexachloride—see Hexachlorocyclohexane
Benzenesulphonic acid,
preparation, 100–1
reactions, 150
sodium salt of, 150
uses, 150
Benzoic acid,
dissociation constant, 62, 185
manufacture, 184
physical properties, 181–3
preparations, 104, 183, **196**
reactions, 184–6, **200**
Benzoin—see 2-Hydroxy-1,2-diphenylethanone
Benzonitrile, 253
Benzophenone—see Diphenylmethanone
Benzotrichloride—see (Trichloromethyl) benzene
Benzoylation, 210, **220**
Benzoyl chloride,
physical properties, 209
preparation, 209
reactions, **155**, 210–12, **220**, **257**
Benzoyl group, 210
Benzyl alcohol—see Phenylmethanol
Benzylamine—see (Phenylmethyl)amine
Benzyl bromide—see (Bromomethyl) benzene
Benzyl chloride—see (Chloromethyl) benzene
Berzelius, 1
Bifunctional catalysts, 296
Bimolecular reaction, 115
Biphenyl-2,2′-disulphonic acid,
isomerism of, 236
Bis-(2-hydroxyethyl) ether, 305
Bitumen,
manufacture, 293, 298
uses, 299
Biuret, 217
Biuret test, 217, **221–2**, **287**
Blasting gelatin, 149
Bond energy, 4, 50–1
Bonding, 1–4, 47–64, 96
covalent, 1–4, 50
delocalised, 56–8, 96

electrovalent, 48–50
hydrogen—see Hydrogen bonding
ionic, 48–50
localised, 56, 96
pi (π), 55
sigma (σ), 55
tetrahedral, 1
Bond length, 50
Bromination,
of benzene, 99, **131**
of ethene, 76–8, **85**
of pent-1-ene, **86**
of phenol, 153, **155**
of phenylamine, 248, **258**
Bromine,
qualitative analysis of, **43–4**
quantitative analysis of, 32
Bromobenzene,
physical properties, 125
preparations, 99, 124, **131**, **253**
reactions, 125–6, **132**, **196**
1-Bromobutane,
physical properties, 112
preparation, **130**
reactions, **132**
2-Bromobutane, **132**
1-Bromo-1-chloro-2,2,2-trifluoroethane, 127
Bromoethane,
physical properties, 112
preparations, 78, 112–13, **129–130**
reactions, 116–120, **132**
Bromoethanoic acid,
dissociation constant, 61
2-Bromoethanol, 78
Bromoform—see Tribromomethane
Bromomethane, 112
2-Bromo-2-methylpropane, **132**
Bromophenylamines, 248
N-(Bromophenyl)ethanamides, 248
1-Bromopropane, 82–3
2-Bromopropane, 82
Buchner funnel, 17
Buta-1,3-diene, 302, 322
Butane,
manufacture, 291
physical properties, 65
preparation, 120
reactions, 72, 295
uses, 295, 302
Butanedioic acid, 188–9
Butanedioic anhydride, 189
Butanoic acid, 182, 183
Butan-1-ol, 28, 138
Butan-2-ol, 138
Butan-2-one, 164
But-2-enal, 172
But-1-ene, 75
cis-But-2-ene,
isomerism of, 230
physical properties, 75
trans-But-2-ene,
isomerism of, 230
physical properties, 75
cis-Butenedioic acid, 230, **239–240**
trans-Butenedioic acid, 230, **239–240**

Butenedioic anhydride, 230
Butyl alcohol—see Butan-1-ol
t-Butyl alcohol—see 2-Methylpropan-2-ol
Butyl bromide—see 1-Bromobutane
t-Butyl bromide—see 2-Bromo-2-methylpropane
Butyl chloride—see 1-Chlorobutane
t-Butyl chloride—see 2-Chloro-2-methylpropane
Butyl iodide—see 1-Iodobutane
But-1-yne, 89
But-2-yne, 89, 90

Cannizzaro reaction, 172, 174–5, **179**
Caprolactam, 307
Carbamic acid, 216
Carbamide,
  manufacture, 216, 309
  physical properties, 216
  preparations, 1, 216
  reactions, 217, **221**, **287**, **326**
  synthesis, 1
  uses, 217, 317, **326**
Carbamide-methanal plastics, 317, **326**
Carbanions, 122–3
Carbenes, 122–3
Carbohydrates, 278–287, **288–9**
Carbon,
  qualitative analysis, **43**
  quantitative analysis, 30–2
Carbon black, 301
Carbonium ions, 62–4
Carbon monoxide,
  manufacture, 308
  uses, 184, 308
Carbon suboxide—see Tricarbon dioxide
Carbon tetrachloride—see Tetrachloromethane
Carbonyl group, 164
Carboxylic acid group, 182
Carboxylic acids—see Acids
Carbylamine reaction, 122–3, 219, 247
Carius's method, 32
Carotenes, 19
Carothers, 320
Carrier gas, 22
Catalase, **287**
Catalysis, bifunctional, 296
Catalytic cracking, 96, 298, **299–300**
Catenation, 4
Celanese silk, 286
Cellophane, 287
Celluloid, 287
Cellulose, 286
Cellulose ethanoate, 286, **289**
Cellulose trinitrate, 149
Cellulose xanthate, 286
Chain isomerism—see Isomerism
Chain reaction, 71
Chloral—see Trichloroethanal
Chlorination,
  of benzene, 99
  of carboxylic acids, 186
  of ethane, 72
  of ethene, 76, 81
  of ethyne, 90
  of methane, 69–72
  of phenol, 153
Chlorine,
  qualitative analysis of, **43–4**
  quantitative analysis of, 32
Chlorobenzene,
  manufacture, 125
  physical properties, 125
  preparations, 253, **259**
  reactions, 105, 125–6, **269**
  uses, 126
2-Chlorobenzoic acid,
  reduction, **197–8**
2-Chlorobuta-1,3-diene, 323
1-Chlorobutane, 112, **132**
2-Chlorobutane, 112
Chloroethane,
  manufacture, 113, 120, 304
  physical properties, 112
  preparations, 72, 78, 112
  reactions, 113–120
  uses, 120, 304
Chloroethanoic acid,
  dissociation constant, 61
  preparation, 186
  reactions, 190
2-Chloroethanol, 78, 162
Chloroethene,
  manufacture, 81, 90, 304
  physical properties, 112
  polymerisation, 314–15
  reactions, 123–4
  structure, 123–4
Chloroethylbenzenes, 104–5
Chloroform—see Trichloromethane
Chloromethane,
  manufacture, 69, 112
  uses, 323
Chloromethylbenzenes, 103, 125
(Chloromethyl)benzene, 104, 125
2-Chloro-2-methylpropane,
  physical properties, 112
  preparations, **130–1**
Chlorophylls, 19, **27**
Chloroprene—see 2-Chlorobuta-1,3-diene
1-Chloropropane, 112
2-Chloropropane, 112, 113
3-Chloropropanoic acid, 182
3-Chloropropene, 148, 319
Chlorosilanes, 323
CHN analyser, 30–2
Chromatography,
  of amino-acids, 21, **28**, 273
  column, 18–19, **26–7**
  gas, 22–4, **28**, 237–8
  paper, 21–2, **28**
  thin-layer, 20–1, **27–8**
Cis-trans isomerism—see Isomerism
Co-enzyme A, 281
Co-enzymes, 275, 281
Collodion, 287
Condensation polymerisation, 311
Condenser,
  air, 13
  reflux, 12
  water, 13
Contact process, 309
Co-polymerisation, 321, 322
Copper(I) dicarbide, 90, **92**
Copper(II)-glycine, 252, **259**
Cordite, 149
Cotton, 286
Cracking,
  catalytic, 96, 298, **299–300**
  of gas oil, 296–8
  of kerosine, 296–8
  of naphtha, 303
  of paraffin oil, **299–300**
  thermal, 302, 303
Crick, 275
Crystallisation, 17
Crystal picking, 236–7
Cumene, 100
Cumene hydroperoxide, 150
Cumene process, 150–1
Cyanohydrins, 167–9, 191, 280, 315
Cycloalkanes, 7, 72–3, **74**, 101, 306–8
Cyclobutane, 73
Cyclohexane,
  manufacture, 101, 306–7
  reactions, 73, **74**, 307
  structure, 73
  uses, 307
Cyclohexanol,
  manufacture, 307
  uses, 307
Cyclohexanone,
  manufacture, 307
  uses, 307–8
Cyclohexanone oxime, 307
Cyclohexene,
  physical properties, 75
  preparation, **84**
  reactions, **85**
Cyclopentane, 7, 73
Cyclopropane, 72–3
Cysteine, 272, 273
Cytosine, 275–7

Dacron, 321
D.D.T., 126
Decane, 65
Decanedioic acid, 188
Decanedioyl chloride, **325**
Decarboxylation,
  of benzoic acid, **200**
  of disodium ethanedioate, **200**
  of ethanedioic acid, 189
  of glycine, 251
  of propanedioic acid, 189
  of sodium ethanoate, 193, **199**
Dehydration, 145
Delocalisation, 56–8
Delocalisation energy, 96
Denaturation, 278
Deoxyribonucleic acid, 275–8
DERV, 298
Detergents,
  alkylbenzene sulphonates, 195, **199**
  alkyl sulphates, 195, **198**
  ethoxylates, 195

preparations, **198–9**
Dettol, 154
Dextrose—see Glucose
Dialkylsilanediols, 324
1,6-Diaminohexane, 307, 320
Diastase, 285–6
Diastereoisomers, 237
Diazonium compounds,
  preparation, 252, **258**
  reactions, 252–4, **258**
Diazotisation, 252
Di(benzoyl) peroxide, 313, 314, 315, 324
1,2-Dibromoethane,
  physical properties, 112
  preparations, 76, 120
  reactions, 120–1
1,2-Dibromoethene, 90
1,3-Dibromopropane, 73
Dichlorobenzenes, 124, 228
Dichlorocarbene, 122–3
2,4-Dichloro-3, 5-dimethylphenol, 154
1,1-Dichloroethane,
  manufacture, 72, 90
  physical properties, 112
  preparations, 121, 170
  reactions, 121, **132–3**
1,2-Dichloroethane,
  manufacture, 72, 76, 120, 304
  physical properties, 112
  preparations, 72, 76, 120
  reactions, 81, 120–1, **132–3**, 304
  uses, 81, 304
Dichloroethanoic acid, 61
1,2-Dichloroethene, 90
Dichloromethane,
  physical properties, 112
  preparation, 69
  structure, 2–4
(Dichloromethyl)benzene, 104, 125
2,4-Dichlorophenol, 154
2,4-Dichlorophenoxyethanoic acid, 154
2,2-Dichloropropane, 170
Di(dodecanoyl) peroxide, **325**
Dienes,
  isomerism, 236
  manufacture, 302, 323
Diesel oil, 298
Diethylamine,
  physical properties, 243
  preparation, 244
  reactions, 247
Diethyl ether,
  manufacture, 160
  physical properties, 159
  preparations, 160
  reactions, 160–1
  uses, 161
Diethyl ketone—see Pentan-3-one
Diethyl malonate—see Diethyl propanedioate
Diethyl propanedioate, 207–8
gem-Dihalides, 121, **132–3**, 170
vic-Dihalides, 121, **132–3**
2,3-Dihydroxybutanedioic acid,
  isomerism, 234–6, 237
  meso-form, 235

physical properties, 235
Di-isocyanates, 317–19
Diketopiperazine, 252
Dimethylacetylene—see But-2-yne
Dimethylamine, 243
4-Dimethylaminoazobenzene, 254
Dimethyl benzene-1,4-dicarboxylate, 321
Dimethylbenzenes, 305, 306
1,2-Dimethylcyclobutane,
  isomerism, 230
Dimethyl ether, 159
2,2-Dimethylpropane, 66
1,3-Dinitrobenzene,
  preparation, 105, 267, **267–8**
2,4-Dinitrophenylhydrazine, 171, **178–9**, 350
2,4-Dinitrophenylhydrazones,
  physical properties, **354–5**
  preparations, 171, **178**
  uses, **179**
Diols,
  manufacture, 146
  nomenclature, 146
  physical properties, 146
  reactions, 147–8, **156**
  uses, 148, 321
Dipeptides, 271
Diphenylamine, 243
Diphenyl ether, 159
Diphenylmethanone, 164, 166
Diphenyls,
  isomerism of, 236
Dipolar ions, 250
Dipole moment, 58
Disaccharides, 278, 283–5, **288**
Disodium ethanedioate,
  preparation, 189
  reactions, 193, **200**
Distillation, 13–16
  fractional, 15–16
  steam, 16
  vacuum, 14–15
DNA, 275–8
Dodecane, 65
Dodecyl sulphate, **198**
Dyes, 253–4, **258**
Dynamite, 149

*E*1 reactions, 119
*E*2 reactions, 118–19
Electrophilic reactions, 78
Elimination reactions, 118–120
Empirical formula, 32
  determination of, 32–3
Enantiomers, 234
Enantiomorphs, 234
Energy,
  bond, 4, 50–1
  delocalisation, 96
  hydrogen-bond, 139
  promotion, 51
  stabilisation, 96
Energy level, 47
Enol group, 152, **156**
Enzymes, 275
Epoxides—see Epoxyethane and Epoxy-

propane
Epoxyethane,
  manufacture, 79
  physical properties, 161
  preparation, 79
  reactions, 161–2
  uses, 162, 195
Epoxypropane, 304, 318
Epoxy resins, 319–20, **326**
Essential amino-acids, 272
Esterification, 204–5
Esters,
  physical properties, 204
  preparations, 117, 193, 204–5
  reactions, 205
  saponification, 206
  uses, 207
Ethanal,
  manufacture, 90, 166
  physical properties, 164–5
  preparations, 165–6, **176**
  reactions, 167–176, **178–9**
  uses, 175–6, 303
Ethanal cyanohydrin, 167–9, 191
Ethanal 2,4-dinitrophenylhydrazone, **178**
Ethanal hydrogen sulphite, **178**
Ethanal oxime, 170
Ethanal resin, 172, **178**
Ethanal tetramer, 175, **179**
Ethanal trimer, 175
Ethanamide,
  basic strength, 214–15
  physical properties, 214
  preparations, 207, 211, 213, 214
  reactions, 214–16, **221**
  structure, 215
Ethane,
  manufacture, 72, 302
  physical properties, 65
  preparations, 76, 90
  reactions, 72
  uses, 72, 302
Ethanedioic acid,
  manufacture, 189
  physical properties, 188
  preparations, 188
  reactions, 189–190, **200**
Ethane-1,2-diol,
  manufacture, 146, 161
  physical properties, 146
  preparation, 79
  reactions, 147, **156**
  uses, 148
Ethanedioyl chloride, 189
Ethanoic acid,
  dimerisation, 183
  manufacture, 184, 302
  physical properties, 61, 62, 182–3
  preparations, 183
  reactions, 184–6, **199–200**
  uses, 315
Ethanoic anhydride,
  manufacture, 213
  physical properties, 212
  preparations, 211, 213
  reactions, 213, **220–1**, 257, **258**, 289

uses, 205, 207, 213, 286, 315
Ethanol,
  absolute, 141
  manufacture, 79, 140–1, 286
  physical properties, 138
  preparations, **86**, 139–140
  reactions, 141–5, **154–6**
  uses, 146, 303
Ethanonitrile, 218, **221**
Ethanoylation, 209–210
Ethanoyl bromide, 209
Ethanoyl chloride,
  physical properties, 209
  preparations, 209
  reactions, 209–12, **220**
  uses, 212
Ethanoyl iodide, 209
Ethene,
  manufacture, 72, 76, 298, **299–300**, 302, 303
  physical properties, 75
  preparations, **85**, 90
  reactions, 76–81, **85**
  structure, 6, 53–5
  uses, 81, 303–4
Ethene ozonide, 79–80
Ethenone, 213
Ethenyl ethanoate,
  manufacture, 315
  polymerisation, 315
  uses, 315, 321
Ether—see Diethyl ether
Ethers,
  nomenclature, 159
  physical properties, 159
  preparations, 117, 160
  reactions, 160–1
  uses, 161
Ethyl acetoacetate—see Ethyl 3-oxobutanoate
Ethyl alcohol—see Ethanol
Ethylamine,
  physical properties, 243
  preparation, 244–5
  reactions, 245–7
Ethylbenzene,
  manufacture, 305, 313
  preparation, 100
  reactions, 104–5, 313
  uses, 313
Ethyl benzoate,
  saponification of, **196**
Etyhl bromide—see Bromoethane
Ethyl chloride—see Chloroethane
Ethylene—see Ethene
Ethylene bromohydrin—see 2-Bromoethanol
Ethylene chlorohydrin—see 2-Chloroethanol
Ethylene dibromide—see 1,2-Dibromoethane
Ethylene dichloride—see 1,2-Dichloroethane
Ethyl ethanoate,
  physical properties, 204
  preparation, 204–5, **219–220**

  reactions, 205–7
  uses, 207
Ethylene glycol—see Ethane-1,2-diol
Ethylene oxide—see Epoxyethane
Ethyl ether—see Diethyl ether
Ethylidene dibromide—see 1,1-Dibromoethane
Ethylidene dichloride—see 1,1-Dichloroethane
Ethyl iodide—see Iodoethane
Ethyl methyl ether, 159
Ethyl nitrite, 118
Ethyl 3-oxobutanoate, 208, 229
Ethyl propanoate, 204
Ethyne,
  manufacture, 69, 89
  physical properties, 89
  preparations, **91**
  reactions, 90, **92–3**
  structure, 6–7, 55
  uses, 91, 323
Extraction, 16–17

Fats, 194, **220**
Fehling's solution, 174, **351**
Fehling's test, 174, **179**, 186–7, 280, 282, 284, **288**
Fermentation, 286
Fibroin, 274
Fillers, 311
Flame-ionisation detector, 22
Fluorobenzene, 125, 253
Fluorocarbons, 126–7
Fluoroethanoic acid,
  dissociation constant, 61
Fluoromethane, 127
Fluorotrichloromethane, 127
Fluothane, 127
Formaldehyde—see Methanal
Formalin, 164, 175
Formamide—see Methanamide
Formic acid—see Methanoic acid
Formula weight,
  determination of, 33–4
Fragmentation pattern, 35
Free radicals, 62–3, 71
Freons, 127
Friedel-Crafts reaction, 100, 166, **177–8**, 313
Fructose,
  physical properties, 282
  preparations, 280
  reactions, 282, **288**
Fructose-1,6-diphosphate, 281
Fructose-6-phosphate, 281
Fumaric acid—see *trans*-Butenedioic acid
Functional group, 5

Galactose, 284
Gammexane—see Hexachlorocyclohexane
Gas oil,
  cracking, 298
  distillation, 296–8
  manufacture, 293
Gasoline, 293, 294–5
Gas-turbine fuels, 296

Genes, 276
Geometrical isomerism—see Isomerism
Gluconic acid, 280
Glucosazone, 280–1, 282, **288**
Glucose,
  physical properties, 280
  preparations, 279
  reactions, 280–2, **288**
  uses, 281–2
Glucose-6-phosphate, 281
Glycerides, 194
Glycerine—see Propane-1,2,3-triol
Glycerol—see Propane-1,2,3-triol
Glycine,
  physical properties, 250, **259**
  preparations, 250–1
  reactions, 251–2, **259**
  uses, 271–2
Glycogen, 282
Glycol—see Ethane-1,2-diol
Glycollic acid—see Hydroxyethanoic acid
Glyoxylic acid—see Oxoethanoic acid
Glyptal resins, 319
Greases, 298
Griess, 252
Grignard, 128
Grignard reagents,
  preparations, 120, 128, **196**
  reactions, 128–9, **196**
Guanine, 275–7
Gun cotton, 149

Halides—see Alkyl halides, Aromatic halides, Dihalides and individual compounds
Halogen carrier, 99
Halogens,
  qualitative analysis of, **43–4**
  quantitative analysis of, 32
Heat of hydrogenation, 95
Heisenberg's Uncertainty Principle, 47
Helium, 291
α-Helix, 273, 274
Heptane, 65, 294
Heterolysis, 72
Hexachlorocyclohexane, 101, **108–9**
Hexachloroethane, 72
Hexamethylenediamine—see 1,6-Diaminohexane
Hexane,
  physical properties, 65
  reactions, **74**
Hexanedinitrile, 307
Hexanedioic acid,
  manufacture, 307
  physical properties, 188
  reactions, 189, 320
  uses, 320
Hexanedioyl chloride, **326**
Hexan-1-ol, 138
Hex-1-ene, 75
Hexoses, 278, 279–282
Hofmann reaction, 216, **254–5**
Homologous series, 5, 65
Homolysis, 72
Hormones, 275

Hund's rule, 48
Hybridisation, 53
Hydrocarbons—see Alkanes, Alkenes, Alkynes, Aromatic hydrocarbons, Cycloalkanes and individual compounds
Hydrocracking, 298
Hydrodealkylation, 96, 305
Hydroforming, 296
Hydrogen,
　manufacture, 308
　qualitative analysis of, **43**
　quantitative analysis of, 30–2
　uses, 309
Hydrogenation,
　of aldehydes, 170
　of alkenes, 76
　of alkynes, 90
　of benzene, 95, 101
　of ketones, 170
　of phenol, 153
Hydrogen bonding,
　in acid amides, 214
　in alcohols, 138–9
　in carboxylic acids, 183
　in diols, 146
　in ethanoic acid, 183
　in ethanol, 138–9
　in nitrophenols, 153–4
Hydrogen cyanide,
　manufacture, 301
　uses, 168, 191, 315
Hydrogen peroxide,
　decomposition, **287–8**
　manufacture, 304
Hydrogen sulphide, 291
4-Hydroxyazobenzene, 253, **258**
3-Hydroxybutanal, 171–2
2-Hydroxybutanoic acid, 182
2-Hydroxy-1,2-diphenylethanone, 175
Hydroxyethanoic acid, 190–1
Hydroxylamine,
　manufacture, 266
　reactions, 171, 280, 282
4-Hydroxy-4-methylpentan-2-one, 172
2-Hydroxypropanoic acid.
　biological importance, 282
　isomerism, 191–2, 230, 234
　physical properties, 191
　preparation of (-)enantiomer, 238
　reactions, 192

Imines, 171
Inductive effect, 58–9
Infrared spectroscopy, 36–8, 139
Insulin, 274–5
Internal compensation, 235
Inversion of sugar, 283
Invertase, 283
Invert sugar, 283
Iodine,
　qualitative analysis of, **43–4**
　quantitative analysis of, 32
Iodobenzene,
　physical properties, 125
　preparations, 99, 253, **258**
　reactions, 125–6
1-Iodobutane, **132**
Iodoethane,
　physical properties, 112
　preparations, 98, 113
　reactions, 113–120
Iodoethanoic acid,
　dissociation constant, 61
Iodoform—see Tri-iodomethane
Iodoform reaction, 144, **156**, 173, **178**
Iodomethane,
　physical properties, 112
　preparations, 113
　reactions, 113–120
Iron(III) chloride test,
　for acids, **199**
　for enols, 152, **156**
Isobutane—see 2-Methylpropane
Isobutene—see 2-Methylpropene
Isobutyl alcohol—see 2-Methylpropan-1-ol
Isobutylene—see 2-Methylpropene
Isobutyric acid—see 2-Methylpropanoic acid
Isocyano-compounds, 219
Isomerisation,
　of alkanes, 295
Isomerism, 8–9, 228–240
　chain, 228
　cis-trans, 55, 229–230, **239–240**
　functional group, 228–9
　geometrical, 55, 229–230, **239–240**
　optical, 230–8, **239**
　position, 228
　stereo, 228, 229–238, **239–240**
　structural, 228–9, 238
Iso-octane—see 2,2,4-Trimethylpentane
Isoprene—see 2-Methylbuta-1,3-diene
Isopropyl alcohol—see Propan-2-ol
Isopropyl bromide—see 2-Bromopropane
Isopropyl chloride—see 2-Chloropropane

Jet fuels, 296

Katharometer, 22–3, 31–2
Keratin, 274
Kerosine,
　cracking, 296–8
　manufacture, 293
　uses, 296
Keten—see Ethenone
Ketones,
　identification, 179
　manufacture, 167
　nomenclature, 164
　physical properties, 164–5
　preparations, 165–6, **177–8**, 193–4
　reactions, 167–175, **178–9**
　uses, 176, 315
Ketoses, 279–280, 282, **288**
Kjeldhal's method, 32
Knock, 294
Kolbe, 1
Kolbe's reaction, 193

Lactic acid—see 2-Hydroxypropanoic acid

Lactides, 191
Lactose, 284–5
Lassaigne test, **43–4**
Lauroyl peroxide—see Di(dodecanoyl) peroxide
Le Bel, 234
Lithium aluminium hydride,
　reduction of acid halides, 212
　reduction of acids, 186, **197**
　reduction of aldehydes, 169–170
　reduction of 2-chlorobenzoic acid, **197–8**
　reduction of esters, 207
　reduction of ketones, 169–170
　reduction of nitro compounds, 266
Lubricating oil, 293, 298
Lucas test, 142
Lycra, 319
Lysine, 249, 272

Maleic acid—see *cis*-Butenedioic acid
Maleic anhydride—see Butenedioic anhydride
Malonic acid—see Propanedioic acid
Malt, 285–6
Maltase, 286
Maltose, 284–5
Markownikoff's rule, 81–2
Mass spectrometry, 33–6
Melamine, 317
Melaware, 317
Melting-point method, 24–6
Mesomerism, 58
Metaldehyde—see Ethanal tetramer
Methanal,
　manufacture, 167, 308
　physical properties, 164
　preparations, 165
　reactions, 167–175, **178–9**
　uses, 175, 316–17
Methanamide, 214
Methane,
　chlorination, 69–72
　occurrence, 68
　physical properties, 65
　reactions, 68–72
　structure, 1–2
　uses, 301
Methanoic acid,
　dissociation constant, 62, 185
　manufacture, 184, 302
　physical properties, 182–3, 185
　preparations, 183
　reactions, 186–7, 199
Methanol,
　manufacture, 69, 140, 308
　physical properties, 138–9
　preparations, 139–140
　reactions, 141–5, **155–6**
　uses, 146
Methylacetylene—see Propyne
Methyl alcohol—see Methanol
Methylamine,
　dissociation constant, 62, 246
　physical properties, 62, 243
　preparations, 244–5, **254–5**

reactions, 245–7, **257**
Methylammonium chloride, 246, **254–5**
Methylbenzene,
  manufacture, 101, 295, 305
  physical properties, 101–2
  preparations, 100
  reactions, 102–4, **108**, 267, **269**, 305
  uses, 267, 305, 306, 317, 318
Methyl benzoate, 204
Methyl bromide—see Bromomethane
2-Methylbuta-1,3-diene, 312
2-Methylbutane, 295
2-Methylbutanoic acid, 182
Methyl chloride—see Chloromethane
Methyl cyanide—see Ethanonitrile
Methylcyclohexane,
  manufacture, 103, 295
  uses, 295
Methylene chloride—see Dichloromethane
Methylene dichloride—see Dichloromethane
Methylene group, 5, 65
Methyl ethanoate,
  physical properties, 204
  preparations, 204–5
  reactions, 205–7
Methylethylamine, 243
(1-Methylethyl)benzene, 100
Methyl ethyl ketone—see Butan-2-one
Methyl group (radical), 5, 63
Methyl iodide—see Iodomethane
Methyl methanoate, 204, 210
Methyl 2-methylpropenoate,
  preparation, 315, **325**
  uses, 315–16, **325**
Methyl-3-nitrobenzene, 265
Methylnitrobenzenes, 102
Methylphenols, 149
Methyl phenyl ether, 159
2-Methylpropane,
  manufacture, 295
  physical properties, 66
  uses, 295
Methyl propanoate, 204
2-Methylpropanoic acid,
  dissociation constant, 182
2-Methylpropan-1-ol, 138
2-Methylpropan-2-ol, 138
2-Methylpropene,
  manufacture, 298
  physical properties, 75
  uses, 295, 298
Methyl-2,4,6-trinitrobenzene, 267
Molasses, 283
Molecular formula, 4
  determination of, 34
Molecular ion, 33–4
Molecular orbitals, 52–3
  acid amides, 214–15
  benzene, 56–8
  carbon-carbon double bond, 54–5
  carbon-carbon single bond, 54
  carbon-carbon triple bond, 55
  carboxylic acid, 186–7
  chloroethene, 123–4

ethene, 54–5
ethyne, 55
phenoxide ion, 151–2
phenylamine, 246
Molecular sieves, 302
Monochloroacetic acid—see Chloroethanoic acid
Monomer, 311
Monosaccharides, 278, 279–282, **288**
Mutarotation, 279

Naphtha,
  cracking, 303, 305
  distillation, 294–5
  manufacture, 293
  oxidation, 302
Naphthalen-2-ol, 253–4
Natta, 312–13
Natta process, 312–13
Natural gas, 291
Neoprene rubber, 323
Nicol prisms, 232
Nicotine, 234
Nitration,
  of benzene, 97–8, 266, **267**
  of chlorobenzene, 105–6, **269**
  of methylbenzene, 102–3, 267, **269**
  of nitrobenzene, 105, **267–8**
  of phenol, 106–7, 153–4, **269**
  of phenylamine, 106–7, 249
Nitrile rubber, 322
Nitriles—see Acid nitriles
Nitroalkanes,
  manufacture, 265
  nomenclature, 265
  physical properties, 265
  preparations, 118, 265
  reactions, 266–7, **269**
  uses, 269
Nitrobenzene,
  manufacture, 266
  physical properties, 265
  preparation, 97–8, 266, **267**
  reactions, 266–7, **268**
Nitro compounds—see Nitroalkanes and Nitro compounds, aromatic
Nitro compounds, aromatic,
  manufacture, 266
  nomenclature, 265
  physical properties, 265
  reactions, 266–7, **268**
  uses, 267
Nitroethane, 265, **269**
Nitrogen,
  qualitative analysis of, **43–4**
  quantitative analysis of, 30–2
Nitroglycerin, 148–9
Nitromethane,
  manufacture, 265, 302
  physical properties, 265
  preparations, 118, 265
  reactions, 266
  uses, 267
Nitronium ion, 97
Nitroparaffins—see Nitroalkanes
Nitrophenols,

hydrogen bonding, 153–4
physical properties, 149, 153–4
preparations, 153, **269**
separation, 27
2-Nitrophenylamine, 243, 249
N-(Nitrophenyl)ethanamides, 249
1-Nitropropane, 265, 266, 302
2-Nitropropane, 265, 266, 302
Nitrosamines, 247
NMR, 38–42
Nobel, 149
Nobel Prizes, 128, 275, 277, 313
Node, 48
Nomenclature,
  of acid amides, 214
  of acid anhydrides, 212
  of acid chlorides, 10
  of acid halides, 209
  of acid nitriles, 218
  of acids, dicarboxylic, 188
  of acids, monocarboxylic, 10, 182
  of alcohols, 10, 137–8
  of aldehydes, 10, 164
  of alkanes, 9, 10, 65
  of alkenes, 9, 10, 75
  of alkyl halides, 10, 111–12
  of alkynes, 10, 89
  of amines, 10, 243
  of aromatic amines, 244
  of aromatic halides, 124
  of carboxylic acids, 10, 182, 188
  of esters, 204
  of ethers, 159
  of ketones, 10, 164
  of nitroalkanes, 265
  of nitro compounds, 265
  of phenols, 149
Nonane, 65
Nuclear magnetic resonance spectroscopy, 38–42
Nucleic acids, 275–8
Nucleophiles, 115
Nucleophilic reagents, 115, 168–9
Nucleotides, 275
Nylon, 320–1, **325–6**

Octadecanoic acid, 194
Octane, 65
Octane number, 294
Oil,
  distillation, 293–4
  formation, 290–2
Olefins—see Alkenes
Optical isomerism—see Isomerism
Orange II, 254
Orbitals—see Atomic orbitals and Molecular orbitals
Orlon, 321
Osazones, 280–1, 282, **288**
Oxalic acid—see Ethanedioic acid
Oxalyl chloride—see Ethanedioyl chloride
Oximes,
  isomerism, 230
  preparation, 170, 307
Oxoethanoic acid, 191
OXO process, 308

2-Oxopropanoic acid, 192
Ozonolysis, 79–80

Paraffin hydrocarbons—see Alkanes
Paraffin oil,
  cracking of, **299–300**
Paraffin wax—see Wax
Paraformaldehyde—see Polymethanal
Paraldehyde—see Ethanal trimer
Pasteur, 236
Penicillium glaucum, 234, 238
Pentachloroethane, 72
Pentane,
  physical properties, 65
  reactions, **74**
Pentanoic acid, 182
Pentan-1-ol, 138
Pentan-3-one, 164, **178–9**
Pent-1-ene,
  physical properties, 75
  reactions, **86**
Pentoses, 278
Pent-1-yne, 89
Peptide link, 221, 271, **287**
Peptides, 271
Peroxide effect, 82–3
Peroxobenzoic acid, 79
Peroxoethanoic acid, 321
Perspex, 315–16, **325**
Petrol, 294–6, 303
Petroleum, 290–300
  distillation, 293–4
  formation, 290–2
  stabilisation, 293
  uses, 294–9, 301–9
Phenol,
  manufacture, 150–1
  physical properties, 149, 150
  preparation, 150, 252
  reactions, 151–4, **154–6**, **269**
  structure, 151–2
  uses, 154, 306, 316–17
Phenols, 149–156, 252
Phenoxide ion, 151–2
Phenylalanine, 272
Phenylamine,
  dissociation constant, 246
  manufacture, 245
  physical properties, 244, 246
  preparations, 245, **256–7**
  reactions, **220**, 245–9, **257–8**
Phenylammonium chloride, **257**
Phenyl benzoate,
  preparation, 210, **220**
Phenyl cyanide—see Benzonitrile
N-Phenylethanamide,
  physical properties, 258
  preparations, 247, **258**
Phenyl ethanoate,
  preparation, 151, 210
Phenylethanoic acid,
  dissociation constant, 62, 185
Phenylethanone,
  physical properties, 164–5
  preparations, 100, 165–6, **177**
  reactions, 167–174, **178–9**

Phenylethanone 2,4-dinitrophenylhydrazone, 171
Phenylethene,
  manufacture, 313
  physical properties, 75
  reactions, 313–14, **325**
  uses, 314, 322
Phenylethyne, 89, **92–3**
Phenylhydrazine,
  preparation, 254
  reactions, 171, 280–1, 282, **288**
Phenylhydrazones, 171
Phenyl isocyanide, 247
Phenylmagnesium bromide,
  preparation, **196**
  reactions, **196–7**
Phenylmethanol, 138, **154–6**
(Phenylmethyl)amine, 243, 246, **257**
Photochemical reactions, 69
Phthalic acid—see Benzene-1,2-dicarboxylic acid
Phthalic anhydride—see Benzene-1,2-dicarboxylic anhydride
Phthalimide—see Benzene-1,2-dicarboximide
Picric acid,
  acidity, 152
  preparation, 154
Plasticisers, 311
Plastics, 311–325, **325–7**
  thermosetting, 316–19, **326**
  thermosoftening, 311–16, **325–6**
Platforming, 296
Polarimeter, 232
Polarised light, 231–3
Polaroids, 232, **326**
Polyacrylic esters, 315–16, **325**
Poly(chloroethene), 314–15
Polyesters, 321
Poly(ethene), 312
Poly(ethenyl ethanoate), 315
Polymerisation, 311–327
  addition, 311
  co-, 321–2
  condensation, 311
Polymers,
  atactic, 313
  isotactic, 312–13
  natural, 271–8, 285–7, **287–9**
  synthetic, 311–325, **325–7**
Polymethanal, 174, **179**
Polyoxymethylene, 316
Polypeptides, 271
Poly(phenylethene), 313–14, 322, **325**
Poly(propene), 312–13
Polysaccharides, 278, 285–7, **288–9**
Poly(tetrafluoroethene), 316
Polythene, 312
Polyurethanes, 317–19, **326**
Position isomerism—see Isomerism
Potassium benzene-1,2-dicarboximide, 250
Promotion energy, 51
Propadienes, 236
Propanal, 164
Propanamide, 214

Propane,
  manufacture, 302
  physical properties, 65
  reactions, 72
  uses, 72, 302
Propanedioic acid,
  physical properties, 188
  preparation, 188–9
  reactions, 189–190
Propane-1,2-diol, 146
Propane-1,3-diol, 146
Propane-1,2,3-triol, 148–9, **156**, 194, 318, 319
Propane-1,2,3-triyl trinitrate, 148–9
Propanoic acid,
  dissociation constant, 62, 185
  manufacture, 302
  physical properties, 182
Propan-1-ol, 138
Propan-2-ol,
  manufacture, 141
  physical properties, 138–9
  preparations, 139–140
  reactions, 141–5, **156**
  uses, 146
Propanone,
  manufacture, 150–1, 167, 304
  physical properties, 164–5
  preparations, 165–6, **177**
  reactions, 167–174, **178–9**
  uses, 176, 315
Propanone cyanohydrin, 315
Propanone hydrogen sulphite, 169
Propanone oxime, 170
Propanoyl chloride, 51
Propenal, 148
Propene,
  manufacture, 72, 81, 302, 303
  physical properties, 75
  reactions, 81–3
  uses, 84, 148
Propenonitrile, 304, 321, 322
Propenonitrile fibres, 321
Propyl alcohol—see Propan-1-ol
Propyl bromide—see 1-Bromopropane
Propyl chloride—see 1-Chloropropane
Propylene—see Propene
Propylene oxide—see Epoxypropane
Propyl hydrogen sulphate, 73
Propyne, 89–90
Protection, 248
Protein hormones, 275
Proteins, 271–8
  denaturation, 278, **287**
  fibrous, 274–5
  α-helix, 273
  hormones, 275
  structural, 274–5
  structure, 272–4
PTFE, 316
Purines, 275–6
Purity,
  criteria of, 24–6
PVC, 314–15
Pyridine, 112
Pyrimidines, 275–6

Pyroxylin, 287
Pyruvic acid—see 2-Oxopropanoic acid

Quaternary ammonium salts, 244
Quinine, 237

Racemates, 234
Racemic form, 234
Racemic mixture, 235
Racemisation, 238
Raschig process, 125
Rayon, 286–7
Recrystallisation, 17
Rectified spirit, 141
Reduced crude, 293, 298
Refinery gas, 298
Reflux, 12
Reforming, 96, 295–6
  catalytic, 296
  of heptane, 295
  of hexane, 96
Refractive index, 231
Resolution, 236–8
Resonance, 58
Resonance hybrid, 58
Restricted rotation, 54–5, 229, **240**
Retention time, 23
Ribonucleic acid—see RNA
RNA, 275–7
  messenger, 276
  transfer, 276
Rosenmund reaction, 212
Rubber,
  natural, 322
  synthetic, 322–3

Saccharic acid, 280
Salts,
  of carboxylic acids, 192–3, **199–200**
Sandmeyer reaction, 253, **258**
Sanger, 274, 275
Saponification,
  of esters, **196**, 205–6, **220**
  of ethyl benzoate, **196**
  of a fat, **220**
Saturation, 6
Schiff's base, 171
Schiff's reagent, 174, **350**
Schiff's test, 174
Schotten-Baumann reaction,
  of phenol, **155**, 210
  of phenylamine, **220**, 257
Sebacic acid—see Decanedioic acid
Sebacoyl chloride—see Decanedioyl chloride
Silica gel, 20
Silicones, 323–5, **326–7**
Silk, 274
Silver dicarbide, 90
Silver mirror test, 174, **178–9**, 186, **199**, 280, 282, 284, **288**
S_N1 reaction, 115–16
S_N2 reaction, 115
Soaps, 194–6
Sodium alkylbenzenesulphonate,
  preparation, **198–9**

Sodium ammonium 2,3-dihydroxybutanedioate,
  isomerism, 236–7
Sodium benzenesulphonate, 150
Sodium benzoate,
  reactions, **200**
Sodium borohydride,
  reducing properties, 169
Sodium dodecyl sulphate,
  preparation, **198**
Sodium ethanedioate—see Disodium ethanedioate
Sodium ethanoate,
  reactions, 192–3, **199–200**
Sodium ethoxide,
  preparation, 142, **155**
  reactions, **155**, 207–8
Sodium hydrogen ethanedioate, 189
Sodium methanoate,
  reactions, 193, **199**
Sodium phenoxide, 150–2, **155**
Sodium potassium 2,3-dihydroxybutanedioate, 174, **351**
Specific rotation, 233
Spin-spin coupling, 41–2
Stabilisation energy, 96
Stanley, 277
Starch,
  fermentation, 286
  hydrolysis, 279–280, 285, **288**
  reactions, 285–6, **288**
  structure, 285
Stationary phase, 22
Stearic acid, 194
Stereoisomerism—see Isomerism
Stick diagram, 35
Strain energy, 73
Structural formula, 34
  determination of, 34–5
Structural isomerism—see Isomerism
Structure,
  of benzene, 56–8, 94–6
  determination of, 34–5
  of dichloromethane, 2–3
  of ethene, 6, 54–5
  of ethyne, 6, 55
  of methane, 1–2
Styrene—see Phenylethene
Substitution reactions, 69
  of aldehydes, 113–18
  of alkanes, 69, 72
  of aromatic compounds, 97–107
  of benzene, 97–101
  of carboxylic acids, 186
  of chlorobenzene, 105–6
  of ethylbenzene, 104–5
  of ketones, 113–18
  of methylbenzene, 102–4
  of nitrobenzene, 105, 267, **267–8**
  of phenol, 106–7, 153–4, **156**, 269
  of phenylamine, 106–7, 248–9
Succinic acid—see Butanedioic acid
Sucrose, 283, **288**
Sugar—see Sucrose
Sugar charcoal, 283
Sugars, 278, 279–285

Sulphonation, 101
Sulphur,
  manufacture, 291, 309
  qualitative determination of, **43–4**
Sulphuric acid,
  manufacture, 309
Sulphur tetrafluoride, 127
Synthesis gas,
  composition, 308
  manufacture, 308
  uses, 308–9

Tartaric acid—see 2,3-Dihydroxybutanedioic acid
Tautomerism, 229
Teflon, 127, 316
Terephthalic acid—see Benzene-1,4-dicarboxylic acid
Terylene, 321
Tetrabromoethane, 90
Tetrachloroethane, 72, 90
Tetrachloromethane, 112, 123
Tetraethyllead,
  manufacture, 120
  uses, 296
Tetrafluoroethene,
  manufacture, 127
  polymerisation, 316
Tetrahydrofuran, 161
Tetramethylsilane, 40
Thymine, 275–7
TMS, 40
TNT, 267
Tobacco mosaic virus, 277
Toluene—see Methylbenzene
Transition state, 115
Triacontane, 65
Trialkylaluminiums, 303, 312, 323
Tribromomethane, 173
2,4,6-Tribromophenol, 107, 153, **156**
2,4,6-Tribromophenylamine, 248, **258**
Tricarbon dioxide, 190
Trichloroethanal, 126
Trichloroethanoic acid, 61
Trichloroethene, 124
Trichloromethane,
  physical properties, 121
  preparations, 69, 121, 173
  reactions, 121–3
(Trichloromethyl)benzene, 104, 125
2,4,6-Trichlorophenol, 153
Triethanolamine—see Tris-(2-hydroxyethyl)amine
Triethylaluminium, 303, 312, 323
Triethylamine, 243
Trifluoroethanoic acid, 127
Tri-iodomethane,
  physical properties, 112
  preparation, 144, **156**, 173, **178**
Trimethylamine, 234
2,4,6-Trinitrophenol,
  acidity, 152
  preparation, 154
Trinitrotoluene—see Methyl-2,4,6-trinitrobenzene
Trioxan, 174

Tripeptide, 271
Triphenylamine, 243
Triplet coding, 276
Tris-(2-hydroxyethyl)amine, 198

Undecane, 65
Unimolecular reaction, 115–16
Unsaturated halides, 123–4
Unsaturation, 6–7, 53–6
Uracil, 275–7
Urea—see Carbamide
Urease, **287**

Valine, 272

van't Hoff, 234
Veronal, 217
Vinyl chloride—see Chloroethene
Virus, 277–8
Viscose rayon, 286
Vulcanisation, 322

Wacker process, 166–7
Watson, 275, 289
Wax,
  cracking, 298–9
  manufacture, 293, 298
  uses, 298
Wilkins, 275
Wöhler, 1

Wurtz reaction, 120

X-rays, 274
Xylenes—see Dimethylbenzenes

Yeast, 286

Zeolites, 302
Ziegler, 313
Ziegler catalyst, 312–13, 323
Ziegler process, 312
Zwitterions, 250
Zymase, 286

£55.65
ML

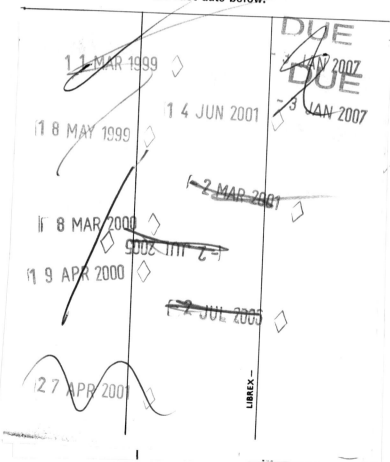

Harry's Card 27AP